Design and Forecasting Models for Disease Management

Scrivener Publishing
100 Cummings Center, Suite 541J
Beverly, MA 01915-6106

Publishers at Scrivener
Martin Scrivener (martin@scrivenerpublishing.com)
Phillip Carmical (pcarmical@scrivenerpublishing.com)

Design and Forecasting Models for Disease Management

Edited by

Pijush Dutta
Dept. of Electronics & Communication Engineering, Greater Kolkata College of Engineering and Management, Kolkata, India

Sudip Mandal
Dept. of Electronics & Communication Engineering, Jalpaiguri Government Engineering College, Jalpaiguri, India

Korhan Cengiz
Dept. of Computer Engineering, Istinye University, Istanbul, Turkey

Arindam Sadhu
Dept. of Electronics and Communication Engineering, Dr. Sudhir Chandra Sur Institute of Technology and Sports Complex, Kolkata, India

and

Gour Gopal Jana
Dept. of Electronics & Communication Engineering, Greater Kolkata College of Engineering and Management, Kolkata, India

Scrivener Publishing

WILEY

This edition first published 2025 by John Wiley & Sons, Inc., 111 River Street, Hoboken, NJ 07030, USA and Scrivener Publishing LLC, 100 Cummings Center, Suite 541J, Beverly, MA 01915, USA

For more information about Scrivener publications please visit www.scrivenerpublishing.com.

Wiley Global Headquarters
111 River Street, Hoboken, NJ 07030, USA

For details of our global editorial offices, customer services, and more information about Wiley products visit us at www.wiley.com.

Library of Congress Cataloging-in-Publication Data

ISBN 978-1-394-23404-2

Cover image courtesy of Pixabay.com
Cover design by Russell Richardson

Set in size of 11pt and Minion Pro by Manila Typesetting Company, Makati, Philippines

Printed in the USA

10 9 8 7 6 5 4 3 2 1

Dedication

We dedicate the book to our family members and the contributing authors. Without their patience, understanding, and support, the completion of this book would not have been possible.

Dr. Dutta is extremely grateful to his father Late Kajal Dutta and his mother, Smt. Tulu Rani Dutta for their affection and constant support. He is thankful to his friends for their love, understanding, prayers, and continuing support to complete this book project.

Dr. Mandal is grateful to parents and family members for their continuing support.

Dr. Cengiz would like to acknowledge and thank the most important people in his life, his parents and his partner, for their support.

Dr. Sadhu wishes to thank his parents, wife, and family members for their continuous support.

Prof. Jana is grateful to his parents, wife, and family members for their continuous support. This book has been a long-cherished dream, which would not have been turned into reality without the support and love of these amazing people.

Pijush Dutta
Sudip Mandal
Korhan Cengiz
Arindam Sadhu
Gour Gopal Jana
Editors

Contents

Preface

Overview

The novel application of artificial intelligence (AI) techniques, including machine learning, and deep learning analytics for design and forecasting models for disease management, is an emerging area in computer science, computation biology, information technology, bioinformatics, bioinformatics, and epidemiology. During the last 20 years, various AI approaches have been successfully applied in diverse fields. Medicine is one of the most active domains for AI techniques, since they facilitate model diagnostics information based on statistical data and reveal hidden dependencies between symptoms of major diseases.

The present capacity to develop, evaluate, manufacture, distribute, and administer effective medical countermeasures is inadequate to meet the burden of both recurrent and emerging outbreaks of infectious diseases. When such interventions are unavailable, public health measures and supportive clinical care remain the only feasible tools to slow an emerging outbreak. Decision-making under such circumstances can be greatly improved by the use of appropriate data and advanced analytics, such as infectious disease modeling. Furthermore, these analyses can guide decision-making when medical countermeasures become available, allowing them to be used more effectively.

Forecasting is an emerging analytical capability that has demonstrated value in recent outbreaks by informing policy and enabling real-time epidemic management decisions. Since healthcare is one of the key parameters in assessing the gross domestic product (GDP) of any country, it has become crucial to transition from traditional healthcare practices to a smart healthcare system. New healthcare technologies provide numerous opportunities to maximize disease recognition, analysis, and other natural variables that may affect it. Therefore, it is necessary to understand how computer-assisted technologies can best be utilized and adopted in the conversion to smart healthcare.

This book is an essential publication that widens the spectrum of computational methods that can aid in smart disease management. It contains several statistical and AI techniques that can be used to acquire data of many different diseases.

Objective

AI techniques for early disease detection and forecasting can reveal useful information to medical experts. As such, this volume intends to serve as a resource to elicit and elaborate on possible intelligent mechanisms for helping in the modeling, prediction, and diagnosis of the infected individuals, as well as providing means for detecting early signs of the disease. As such, this book examines numerous ML and DL techniques in the biomedical field that are used for detecting and forecasting disease management at the cellular level. It discusses applications for image segmentation, classification, neural networks, and a variety of computational intelligence approaches. It also discusses DL techniques for noise elimination and filtering of brain disease and hybrid ML techniques for diabetes prediction, mental illness, and stress level monitoring approaches.

We hope this book encourages an even wider adoption of these concepts and methods to assist physicians in problem solving and stimulates research that will lead to additional innovations in this area.

Organization

The volume comprises thirteen comprehensive chapters that encompass the modelling, prediction, and diagnosis of disease data. The arrangement of the chapters guides the reader from effective modelling of different disease data to detection and diagnosis. Chapter 1 presents an overview of how AI is transforming healthcare, as well as examples of consumer resistance to it, the economic implications of AI aversion, and ethical consideration and governance. Hand in hand with the emerging technology, healthcare systems are becoming progressively complex while constantly adapting to socioeconomic, epidemiological, and demographic changes, contributing to intricate global healthcare objectives.

Chapter 2 explores the use of integrated, and systematic moral cognitive therapy as a means of administering AI in the healthcare sector. The integration of AI in healthcare has the potential to improve prognosis with the help of an enormous database, but concerns have arisen regarding the vulnerability of a large section of end-users and the ethical implications of its use.

Social inequities, poverty lack of education, and risk of lifestyle diseases are mutually complementary to each other. An unhealthy lifestyle claims maximum responsibility among the factors of becoming a victim of lifestyle diseases. Zero preventive options other than lifestyle modification have enhanced the complication of management of such mass health issues.

Chapter 3 investigates a conceptual model that can bridge the gaps between healthcare, policy, and disease prevention. Such a model can assist with the recording and reporting of data to analyze the victim's pattern of lifestyle adoption in place of the current healthcare alternative. Color Filter Array (CFA) data can be efficiently compressed using existing image compression algorithms such as JPEG. In addition, it tested the overall performance of CFA compression methods and demosaicing algorithm selection. These CFA compression methods exploit how CFA images do not contain repetitive data introduced by demosaicing.

Chapter 4 establishes a relationship between the discovery of different CFA compression techniques and describes how the choice of demosaicing algorithm affects visual quality. Using a simple bilinear demosaicing method, CFA image compression produces better images compared to the traditional compression scheme.

Through image processing techniques, digital images can be compressed, monitored, and transmitted effectively in remote settings. Digital images have a significant effect on modern society, science, technology, and art. Some techniques utilized in image enhancement include filtering, segmentation, object recognition, and image fusion. Image processing techniques have played a crucial role in advancing healthcare, particularly in the field of smart healthcare. Chapter 5 reviews image processing for medical applications using secure smart healthcare techniques. Furthermore, the various challenges that scientists must resolve because of disease are also discussed. Image enhancement techniques engage an essential and fundamental task and interpretability of biomedical images, enabling accurate diagnosis and analysis. This chapter provides a comprehensive overview of image enhancement techniques that are specifically tailored for biomedical images. Chapter 6 presents various image enhancement methods designed specifically for biomedical imaging. The chapter evaluates and contrasts the usefulness of various approaches and the extent to which they are beneficial to the biomedical field.

Disease progression prediction is essential for patient care and therapeutic development. Clinical, genetic, and molecular data related to Parkinson's disease (PD) are collectively gathered by the Accelerating Medicines Partnership (AMP). Chapter 7 presents parameters linked to Parkinson's disease progression and evaluates the appropriateness of the AMP PD dataset for predictive modeling through an Exploratory Data

Analysis (EDA). The onset of the Covid-19 pandemic and the practice of social isolation have led to a rise in instances of mental health issues. Advances in machine learning in the present decade have created additional opportunities for detection and the identification of mental health issues. Chapter 8 investigates the automatic prediction and diagnosis of mental illness using machine learning algorithms.

One of the most common neurodegenerative illnesses that affect older adults and mostly impair memory in the brain is Alzheimer's disease. Chapter 9 presents three models, efficient net B2, efficient net B3, and efficient net B4 architecture-based deep learning solutions for detecting and categorizing Alzheimer's disease.

One of the biggest issues in today's world is mental stress. Now that age is not taken into account when calculating stress levels, it makes no difference. Depression, heart attacks, and even suicide can result from extreme stress. Stress may have an impact on many facets of our existence, such as our behavior, emotions, and capacity for thought. Chapter 10 presents a machine-learning approach for measuring a person's stress level using three crucial parameters: body temperature, step count, and humidity.

Diabetes Mellitus is commonly known as a metabolic issue in which the body cannot utilize insulin, store glucose for energy, or produce insulin. Diabetes patients can suffer various sicknesses that incorporate kidney disappointment, stroke, visual impairment, cardiovascular failures, and lower appendage removal. Chapter 11 presents a hybrid-based machine learning approach for identifying the potential chances of contracting diabetes diseases.

The increasing adoption of smart healthcare care and Internet of Things (IoT) devices has revolutionized the healthcare industries, offering enhanced patient monitoring, diagnosis, and treatment. However, the integration of these devices into the healthcare ecosystem raises serious concerns regarding the security and privacy of sensitive medical records. Chapter 12 contributes to strengthening the overall cyber security framework within the healthcare industry, ensuring patients' trust in smart healthcare IoT devices and fostering the continued growth of remote patient monitoring.

The proliferation of smart healthcare systems has introduced a new era of patient care, leveraging interconnected medical devices to improve monitoring and treatment. However, this enhanced connectivity also exposes healthcare environments to potential cyber security threats, including malicious devices. Chapter 13 shows how to employ advanced machine learning algorithms, anomaly detection techniques, and behavioural profiling to continuously monitor and analyze the behavior of interconnected devices. By establishing a baseline of normal device behavior, our system can identify deviations indicative of potential malicious intent.

We are deeply grateful to everyone who helped with this book and greatly appreciate the dedicated support and valuable assistance rendered by Martin Scrivener and the Scrivener Publishing team during its publication.

Pijush Dutta
Sudip Mandal
Korhan Cengiz
Arindam Sadhu
Gour Gopal Jana
Editors

Part 1

SAFETY AND REGULATORY ASPECTS FOR DISEASE PRE-SCREENING

1

A Study of Possible AI Aversion in Healthcare Consumers

Tanupriya Mukherjee[1] and Anusriya Mukherjee[2]*

[1]Department of MBA, Budge Budge Institute of Technology, Kolkata, India
[2]Department of Basic Science and Humanities, Budge Budge Institute of Technology, Kolkata, India

Abstract

Computer systems, that are capable of executing activities that require human intelligence, often referred to as Artificial Intelligence (AI), are making their mark in all industrial territory lately. The fact that enormous data sets can be used to teach these AI systems to recognize patterns and make predictions accordingly, is highly advantageous adding the agility of the present humanized systems at work. In recent times, it has been noted that artificial intelligence and machine learning technology have reshaped the healthcare sector as well. AI has the potential to transform the healthcare industry by increasing efficiency, lowering costs, and improving the prognosis for patients. Integrating AI in healthcare presents a few hurdles, of which the two most important are meeting compliance requirements and resolving issues of confidence with machine learning outcomes. Despite these obstacles, introducing machine learning, artificial intelligence and other technologies to the healthcare business has resulted in various benefits for both healthcare organizations and the patients they serve. Both machine learning and AI have shown an array of advantages in the healthcare industry by optimizing processes and assisting with routine chores, as well as assisting users in promptly finding solutions to critical concerns, enabling improved services for patients as well as consumers. Most healthcare providers are offering user-driven experiences and increasing operational efficiency in making the best possible use of collected data, assets, and resources by evaluating data trends, enhancing coherence and improving the accomplishments of clinical and operational procedures. Yet, even after exiting a major part of two decades in the medical industry, insufficient information exists about consumer perceptions towards AI in medical treatments

**Corresponding author*: anusriyam@gmail.com

Pijush Dutta, Sudip Mandal, Korhan Cengiz, Arindam Sadhu and Gour Gopal Jana (eds.) Design and Forecasting Models for Disease Management (3–46) © 2025 Scrivener Publishing LLC

and procedures. Our study is aimed at learning consumers' apprehensiveness to embrace AI-assisted healthcare in both tangible and hypothetical choices, be it independent or collaborative evaluations.

***Keywords*:** Artificial intelligence, machine learning, automated healthcare application, healthcare consumer, consumer behavior

1.1 Introduction to AI in Healthcare

The advent of artificial intelligence (AI) has grown exponentially to have practical influence in myriads of segments for over the last few decades and continues to permeate into various sectors of the global economy in recent years. The promise of automation, improved efficiencies, and reduced human error has created a rush of innovations integrating AI systems into their core offerings. Seeping deep into various sectors like healthcare, education, banking, and so on, the enormous spectrum of AI's offerings is ruling the roost by extending benefits ranging from mundane tasks, such as speech recognition and decision-making, to financial forecasting and advanced medical diagnosis. The world is gradually acknowledging the potential transformative power of AI-based diagnostic assistance, predictive tools, patient monitoring, and surgical robotics. The use of AI in healthcare is projected to register significant growth in the future. AI's essence in the health sector is underpinned by the ability of these systems to consume vast populations of information, emulate human cognition, and improve themselves (Obermeyer *et al.*, 2016) [1]. Consequently, these abilities have positioned AI as a powerful tool redefining access, affordability, accuracy, and even our understanding of healthcare. Deep learning algorithms are being trained to recognize patterns in radiology images and predict diseases even before any physical symptoms manifest. Google's DeepMind achieved over 94% accuracy in diagnosing eye diseases, a similar degree of precision to top human experts (De Fauw *et al.*, 2018) [2]. This improvement in early detection would significantly enhance treatment outcomes. Additionally, AI has shown compelling capabilities in surgery. Robotic-assisted surgeries have become more precise, stable, and minimally invasive as compared to conventional surgical techniques (Barr *et al.*, 2018) [3]. The Da Vinci Surgical System heralds a new era in precision medicine by utilizing AI functionalities to aid surgeons in intricate surgical procedures, thereby improving patient safety. AI has also proven instrumental in innovative drug discovery, reducing the length and cost of R&D cycles. An AI system developed by Insilico Medicine designed a novel drug in just 46 days, a process that would usually take a few years of human research (Zhavoronkov *et al.*, 2019) [4] to design medications and vaccines. However, like any other technology, AI is a double-edged sword-eliciting both apprehension and enthusiasm simultaneously and there is still potential AI aversion among consumers that may stifle this evolution. In particular, it has been observed in this study that the older generation exhibits aversion towards AI more prominently than the young demographic. Most of these systems rely on vast amounts of data for accuracy which raise critical concerns over patient privacy and data security. Breaching these might cause immense harm (Song *et al.*, 2019) [5]. While policies and procedures are being developed to mitigate these risks, they remain a persistent concern. Additionally, the question of oversight and accountability also arises. If an AI system makes a wrong diagnosis or surgical mistake, who is to be held accountable? These ethical and legal dilemmas further complicate AI's integration into healthcare (Sperry *et al.*, 2018) [6]. To demystify this conundrum, this book chapter aims to understand and examine the underlying factors behind possible AI aversion in healthcare consumers.

1.1.1 The Role of AI in Transforming Healthcare

Artificial Intelligence has become an integral transformative element of healthcare delivery by revolutionizing medical care, from improving diagnosis and prescribing treatments to streamlining hospital operations and enhancing patient experiences. A recent report by Accenture (2021) states that AI applications, when utilized fully in healthcare, could save up to $150 billion annually for the US alone by 2026 (Basole *et al.*, 2021) [7].

AI's role in advancing medical research - AI has demonstrated remarkable capabilities in advancing medical research. By analyzing vast quantities of data, AI can predict disease outbreaks, accelerate drug discovery, and provide new insights into disease mechanisms. For instance, AI technology assisted researchers in identifying potential treatments for COVID-19 significantly quicker than traditional research methods would have allowed (Bullock *et al.*, 2020) [8].
AI's assistance in diagnostics and treatment - AI's contribution extends to improved patient diagnostics and treatment plans. Machine learning models can analyze complex medical images with higher accuracy and in less time than human clinicians, leading to quicker diagnosis and treatment initiation. For instance, researchers at Stanford University developed an AI model that achieved a performance comparable to human radiologists in detecting pneumonia in chest X-rays (Rajpurkar *et al.*, 2018) [9].

Tackling healthcare inefficiency - Furthermore, AI is effective in tackling the inefficiency in healthcare. It reduces paperwork, expedites administrative duties, and optimizes the schedule of health professionals, resulting in time and cost savings. As (Satava *et al.*, 1995) [10] highlight, about 36% of healthcare tasks—mostly managerial and administrative — can be automated, thereby freeing up medical staff for patient-facing duties.

The Echo Dot case - Patient experiences are significantly enhanced through AI as well. A notable example is the use of Amazon's Echo Dot by NHS England to supply information to patients at the comfort of their homes. This initiative not only eased the burden on healthcare professionals but also boosted patients' morale and wellbeing (Hasanuzzaman *et al.*, 2019) [11].

The above examples cited demonstrate how AI integrated tools can work alongside traditional healthcare delivery methods to deliver optimum patient outcomes. To fully actualize AI's potential in healthcare, it requires continuous exploration and investment in AI capabilities, revised regulatory guidelines and proactive stakeholder engagement.

1.1.2 The Unfolding Paradigm: Potential Benefits and Challenges of AI Implementation in Healthcare

Healthcare in the modern age is increasingly becoming intertwined with burgeoning technology, the potential of which holds tremendous promise for improving patient outcomes on one hand and challenges that hinder its full realization on the other, whilst posing significant challenges.

Benefits of AI implementation in healthcare: The integration of AI in healthcare offers a plethora of benefits. AI can streamline diagnostic procedures, make predictions about health trends, and contribute to personalized medicine. Regular health monitoring can be performed more efficiently with AI algorithms, reducing the workload of healthcare professionals. Deep learning techniques, a subset of AI, can help uncover complex patterns in voluminous datasets, improving diagnosis accuracy. For instance, AI algorithms have been used to accurately diagnose skin cancer, matching or even surpassing dermatologists' success rates (Esteva *et al.*, 2017) [12]. AI has shown promise to enhance early detection and prevention of diseases, thereby improving patient outcomes. AI can also be pivotal in the analysis of healthcare trends. Various researchers have successfully used AI to predict disease outbreaks. For example, BlueDot, a Canadian AI firm, correctly predicted the COVID-19 outbreak a week earlier than the Centers for Disease Control and Prevention (CDC) (Noah *et al.*, 2000) [13]. Moreover, AI's ability to process and learn from a patient's medical history opens doors to personalized medicine, tailored to each patient's unique genetic make-up. This can substantially increase treatment effectiveness and reduce potential side effects (Krittanawong *et al.*, 2018) [14]. **Challenges of AI implementation in healthcare:** Despite the immense benefits, AI implementation in healthcare also comes with substantial challenges. The complex nature of medical data with numerous variables and the need for meticulous data labeling makes it a difficult task. Incorrect data labeling in AI-based systems can lead to inaccurate clinical decisions (Atkins *et al.*, 2021) [15]. Moreover, patient privacy concerns are also pertinent given that AI operates using vast amounts of patient data. Strict regulatory checks are imperative to maintain patient anonymity and avoid data misuse. Regulatory bodies must establish safeguards to maintain data security and ensure that AI applications adhere to privacy laws [16]. Furthermore, while AI systems' decision-making is based on algorithmic patterns, these algorithms are not infallible. They may lack the clinician's expertise in interpreting unusual clinical scenarios, emphasizing the importance of clinician oversight on AI-based decisions. Also, the initial financial cost of integrating AI technology into healthcare systems can be high, which may limit its adoption, particularly in resource-constrained settings.

1.1.3 Overview of Consumer Receptivity Towards AI in Medicine: A Comparative Analysis

Artificial intelligence (AI) in medicine is a revolutionary concept that has the propensity for wide-reaching implications. However, acceptance and adoption of this innovation remain variable, thus necessitating evaluation of consumer receptivity towards AI in medicine by compare consumer attitudes and receptivity, drawing insights from different demographic segments across the globe. Many consumers display a remarkable openness to AI in medicine, particularly in routine tasks and facilitative roles (Haynes *et al.*, 2020) [17]. Foremost, AI provides unmatched efficiency and accuracy when it comes to medical image analysis, medical records examination, diagnostics, and drug discoveries (Topol *et al.*, 2019) [16]. In particular, consumers view AI as a complementary tool for healthcare professionals, improving precision and allowing more time for direct patient care (Shorvon *et al.*, 2018) [18]. However, the receptivity becomes significantly lowered when AI is introduced into more involved, perceived high-stake roles such as surgical procedures and disease prognosis. Medical decisions significantly impact personal health and wellbeing, thus the reluctance to entrust such responsibilities to AI. This contrast suggests that the scope of acceptance for AI in medicine appears to hinge on its application magnitude. Medical practitioners' endorsement or resistance towards AI in medicine critically influences consumer receptivity. For example, a global survey revealed that those within healthcare were positive regarding AI's dynamic capabilities but remained reserved about its ethical implications and data security (Arnold *et al.*, 2017) [19]. Their main concern revolves around patient's confidentiality and the potential misuse of personal data, which require stringent regulations (Braunack-Mayer *et al.*, 2010) [20]. Significantly, demographic factors influence consumer receptivity towards AI in medicine. Younger consumers appear more accepting of AI's role in healthcare provision, while older individuals are skeptical, perhaps due to unfamiliarity or mistrust in technology. Geographic location is also crucial; countries with robust technological infrastructure and awareness show a higher acceptance rate of AI in healthcare (Williams *et al.*, 2020) [21]. Different cultural attitudes towards AI in the medical field also need consideration. Countries where technology is deeply enmeshed in societal daily functions, such as Japan and South Korea, show greater acceptance of AI in medicine than those where technology is seen as a potential threat to traditional ways of life. Thus consumer receptivity towards AI in medicine is a multifaceted issue that depends on AI's role, professionals' endorsement, demographic factors, and cultural perspectives. Although AI provides numerous benefits, its adoption in the medical field will remain variable unless it satisfies these different cohorts' particular concerns. Balancing the benefits and potential risks is key to fully unlocking AI's potential in healthcare. The robustness of regulations governing AI's use and data protection measures will be crucial moving forward. As we continue to explore AI's capabilities, it is paramount that digital innovation functions ethically, securely, and to the advantage and safeguarding of patients.

1.2 Consumer Reluctance to Utilize AI in Healthcare: Present Scenario

In the world increasingly inured to digital intervention in practically every facet of existence, it would seem natural for AI to permeate healthcare services which holds substantial promise, the challenges are equally compelling. As we stand on the brink of what could be a revolutionary change in healthcare delivery, it is paramount to confront these challenges proactively and constructively, to ensure that AI is a tool that augments human ability, rather than complicating it. Recent developments reveal significant consumer reluctance in harnessing AI's potential advantages in healthcare. Despite the apparent advantages, the healthcare industry's AI potential is met with considerable skepticism and resistance from the very beneficiaries it seeks to serve. Exploring the reasons behind this reluctance provides insights into various factors such as privacy concerns, trust issues, and potential job loss (Asmundson *et al.*, 2020) [22].

From an economic perspective, the high costs associated with acquiring, implementing, and maintaining AI systems can be a deterrent for many healthcare organizations. While AI promises significant long-term gains, the short-term financial strain may be too high for some, particularly in resource-limited settings. Consequently, measures such as public-private partnerships, shared-risk models, and cost-effectiveness studies could be beneficial (Cassar *et al.*, 2020) [23].

Healthcare AI is also not immune to algorithmic bias, which may result in unethical and harmful treatment recommendations. Panch *et al.* warn of biases arising due to the misrepresented of certain demographic groups in training data, leading to potentially discriminatory practices (Panch *et al.*, 2019) [24]. Steps to mitigate this bias might encompass using diverse datasets for training AI models, regular model audits, and comprehensive bias training for AI users in healthcare. Nonetheless, studies indicate an upturn in consumer acceptance of AI in healthcare is foreseeable, given specific conditions. Greater transparency, stronger regulations protecting privacy, and emphasizing the importance of these technologies to augment rather than replace human effort would lead to a paradigm shift in consumer behavior (Gerotziafas *et al.*, 2020) [25].

Another prominent impediment towards consumer acceptance of AI in healthcare is privacy concerns. The most formidable challenges are the issue of data security in operation of AI technologies due to processing of large amounts of personal data, causing significant privacy apprehensions for consumers. With AI, the use of EHRs (electronic health records), predictive modeling, and algorithms necessitates the collection and analysis of vast amounts of patient data (Majeed *et al.*, 2019) [26]. This can then prompt concerns about the breach of patient confidentiality, data mismanagement, and misuse, posing serious ethical dilemmas. Moreover, varying legal constructs around data privacy globally further intensify these challenges. As suggested by Blease *et al.*, efforts towards strengthening security protocols, and adhering to a standardized international regulatory framework could pave the way forward [27]. The Cambridge Analytica scandal and similar instances of data breaches further exacerbate these concerns Greenwald *et al.*, 2013) [28].

Closely related to privacy are the issues of trust and ethics. Trust in AI technologies tends to be low due to potential implicit biases in algorithmic training and decision-making. Ethical dilemmas surrounding the deployment of AI technologies further underline the need for transparent AI healthcare systems (Nemitz *et al.*, 2018) [29].

In addition to these, the possibility of job losses has cast a shadow over AI healthcare's prospect. With machines potentially usurping medical professionals' roles, workers fear a decline in employment opportunities (Chui *et al.*, 2016) [30]. Another difficulty in AI implementation stems from its integration into existing healthcare systems. Healthcare organizations often confront resistance to change from staff who are unaccustomed to or fearful of new technology. The issue is complex and multifaceted, involving factors such as lack of digital literacy and the fear of job redundancy (Akter *et al.*, 2017) [31]. Addressing this may require an investment in training and educating healthcare providers to acclimate them to the new environment.

Additionally, there are concerns about the transparency, as high-stakes decisions in healthcare require understanding the underlying reasoning, the 'black-box' nature of some AI models can be a considerable barrier (Scott *et al.*, 2019) [32]. Research on explainable AI models and regulatory mandates for transparency may help alleviate these concerns.

Automation in healthcare thus remains a terrain marked by both potential and apprehension. It is vital that the apprehensions be addressed earnestly, to establish a future where AI and humans coexist symbiotically, furthering our common goal of improving the quality of healthcare services.

1.2.1 Top Factors Influencing Consumer Resistance to Medical AI

Understanding, addressing and mitigating the factors of consumer resistance to medical will pave the way for the mainstreaming of AI in healthcare. With proper regulations, transparency, and communication, integrating the vast potentials of AI in healthcare will be a foreseeable future.

Lack of Understanding and Trust: The most frequently cited reason for resistance to medical AI is the lack of understanding and subsequent distrust due to the complex nature which often generates fear and suspicion amongst the lay population, creating resistance to its integration with healthcare (Jordan *et al.*, 2019) [33].

Privacy Concerns: According to N. Bostrom (2014), Concerns about personal privacy rank high in the list of consumer resistances to medical AI as most clients express apprehension about the potential misuse of personal medical data by unscrupulous entities (Bostrom *et al.*, 2014) [34].
Fear of Job Displacement: Both medical and non-medical staff fear that AI might supplant human roles in the healthcare sector creating a significant source of resistance against medical AI (Jungmann *et al.*, 2020) [35].
Transference Errors: Floridi *et al.* conducted a study (2018) found that possible data transference errors feed into the resistance against medical AI (Floridi *et al.*, 2018) [36].
Over-reliance on Technology: As clients, people value the human touch in healthcare, and the perceived loss of this aspect through over-reliance on technology might degrade the human aspect of medical can foster considerable resistance (Arya *et al.*, 2019) [37].
Costs: The potential costs associated with integrating AI into healthcare could also evoke resistance in patients with the concern that these costs might reflect in their medical bills, creating opposition to AI (Goodman *et al.*, 2014) [38].
Potential Cyberattacks: The susceptibility of AI platforms to threat of cyberattacks and hacking raises security concerns, fostering resistance amongst consumers leading to AI aversion (Topol *et al.*, 2019) [39].
Liability Issues: Ensuring accountability for medical errors made by AI systems is a significant issue in healthcare, and the nebulous nature of AI accountability can drive away potential users (Haugsten *et al.*, 2023) [40].
Potential Misdiagnosis: The gravity of a misdiagnosis, especially in crucial health cases, can foster significant opposition (Kong *et al.*, 2020) [41].
Limited Access: Lastly, the limited access to AI technology in some regions fosters resistance against medical AI. The perceived elitism of AI healthcare can deter the broader population from accepting AI (Brynjolfsson, 2014; Challen, 2019; Price, 2021) [42–44].

1.2.2 Uncovering the Psychological Barriers and Concerns Associated with AI Adoption in Healthcare

Fostering a culture that embraces AI and automation, rather than fearing their potential impacts, is critical. Navigating the psychological barriers and concerns associated with AI adoption in healthcare isn't an overnight achievement but a journey towards a more efficient, effective, and empathetic healthcare provision. (Hafner *et al.*, 1995) [45], only 39% of healthcare providers understood AI's capacities and potentials, signaling psychological barriers that impede the adoption [45]. The inexplicable capabilities of AI (Rao *et al.*, 2013) [46] and the lack of comprehensive knowledge about its inner workings, breeds anxiety among healthcare professionals (Castelvecchi *et al.*, 2016) [47]. This 'black box' nature of complex AI systems is perceptually intimidating, which can impede its acceptance in a field where life-altering decisions are the norm.

Another psychological concern is the fear of replacement and this perceived potential of job loss due to automation tends to foster resistance towards AI, (Mahdawi, *et al.*, 2017) [48]. However, research suggests that AI will augment human endeavors in healthcare by freeing up time for better patient interaction and care, fostering a patient-physician relationship that no machine can replace.

Trust and reliability form another significant barrier as healthcare providers themselves struggles to trust an algorithm over their experience and intuition. Building certainty and reliance concerning AI's accuracy, security and precision is crucial for its seamless integration into healthcare systems and resist aversion (Ross *et al.*, 2009) [49].

Also, ethical considerations loom around AI's utilization in healthcare. Professionals worry about AI systems making errors, data privacy breaches, or target patients with machine-driven biases. Effective AI planning and accountability measures in healthcare should address such apprehensions to foster confidence in its adoption.

Lastly, the concern of increased workload is palpable. The potential demand for additional skills to operate these sophisticated AI systems might spawn an initial additional workload for the healthcare workforce. The upskilling demand can be offset via creating training programs, fostering digital literacy, and establishing ongoing education pertaining to AI technologies.

Impressively, views about AI in healthcare are not uniformly gloomy, since automated healthcare can actually expedite diagnosis, treatment planning and enhance patient care. There exists an understanding for AI's potential in mitigating medical error and enhancing patient safety. Ironing out the psychological wrinkles requires strategies that alleviate fear, enhance trust, tackle ethical issues, and balance workload expectations. Emphasis should be placed on transparency, understanding, and clear communication about AI's role, scope, limitations, and benefits. Interestingly, a change-centric, proactive approach can successfully navigate these psychological barriers. Healthcare providers must be involved in AI policy-making, system design, and evaluation. More in-depth, accessible knowledge about AI can alleviate fears and misconceptions. Interactive AI training, comprehensive user support, and practical demonstrations can reinforce trust and confidence in its capabilities, thus accelerating healthcare's digital transformation.

1.2.3 Case Studies and Research Findings on Consumer Aversion to AI-Based Healthcare Services

AI-based healthcare services hold immense promise. However, without catering to consumers' concerns and fears, we risk alienating them, fuelling a continued aversion to such technology. Hence it is imperative to delve into the complexities surrounding this evolving discourse, basing on several case studies grappling with this issue.

The aversion towards AI-based healthcare services in certain sectors of the population continue to resist the increased onslaught of healthcare AI, leaning towards the familiar and traditional healthcare models (Rodriguez *et al.*, 2020) [50] which involves using complex algorithms and software to mimic the human cognitive function in analyzing complex medical data (Adadi *et al.*, 2018) [51]. A study by Ipsos MORI (2018) (Ipsos *et al.*, 2017) [52] found a lack of trust among consumers regarding the use and protection of their health data by AI systems. They worry about data breaches, misuse or theft of personal information, which may lead to identity theft (Bello-Orgaz, *et al.*, 2016) [53]. Consumers are understandably concerned about how their data is stored, used and secured, often leading to skepticism and resistance (Kostkova *et al.*, 2016) [54]. One noteworthy case reflecting this issue is the collaboration between Google and the National Health Service (NHS). Google's DeepMind Health was accused of improper access to patient records causing privacy activists and patients to question the integrity of handing over their healthcare data (Powles *et al.*, 2017) [55]. This example illustrates how a breach of trust can disrupt AI adoption in healthcare. Moreover, a level of mistrust is directed towards the AI algorithms itself, leading to misdiagnosis or incorrect treatment suggestions (Makin *et al.*, 2019) [56]. The study cites a case where an AI chatbot wrongly advised a patient, demonstrating that these concerns are not merely hypothetical.

Adding to that is the fear of the perception of AI as a 'faceless' entity has struck a discordant note with consumers who value the human touch in healthcare provision. A study by Accenture (2017) (Jeffery *et al.*, 2017) [57] found that consumers exhibited an increased level of unease with AI-based services. A case study of a Boston-based hospital shows how older patients found it difficult to adapt to an AI system, longing for personal interactions (Rodriguez *et al.*, 2020) [50]. This resistance stemmed from a fear of losing the vital human interaction like the compassion and understanding of a human clinician that forms an integral part of the consultation and healing process, that AI is yet to encapsulate (Luxton *et al.*, 2014) [58]. The real-world case studies of Babylon Health, an AI-based app designed to provide health consultations, reiterate the above issues. Despite the app's capabilities, it was met with considerable consumer resistance in the UK (Petkovic *et al.*, 2019) [59]. Users echoed concerns over the impersonal nature of consultations, indicating a preference for human doctors who can empathize and offer comfort, suggesting AI cannot replace the human healthcare provider.

Sadly, these fears have been somewhat validated by actual occurrences or nascent possibilities. It is crucial that healthcare providers, technology developers and respective legislation synchronize efforts to bridge the gap between AI advancements and consumer sentiments. Greater transparency, stronger data security measures, and a balanced model of care are pertinent in turning the tide towards a more accepting environment. The forward march of technology is inexorable but assessing and recognizing consumer sentiments and concerns will ensure a smoother and more beneficial integration of AI into the healthcare service landscape.

1.2.4 Impact on Consumer Decision-Making

Consumer decision-making forms the backbone of the modern competitive business world. While the advent of AI in healthcare presents promising prospects, its full potential would only be realized once the consumer's innate preference for human interaction and aversion to AI is understood and addressed. The key lies in bridging the gap between healthcare consumers and AI technology, creating a synergistic amalgamation of man and machine. There exists an increasing level of consumer engagement with AI in healthcare. A growing interest has been observed towards wearable technology, virtual assistants and patient portals integrated with AI, with consumers showcasing active involvement in tracking their health (Arnold *et al.*, 2017) [19]. Consumers, therefore, are seemingly amicable to adopt this new healthcare trajectory.

In parallel, there prevails a prominent sense of skepticism, shaping into AI aversion among a faction of healthcare consumers (Kumar *et al.*, 2017) [60]. This cognitive dissonance among healthcare consumers cements the argument that while these impressive technological advancements can be beneficial, they have also brewed concerns around privacy, preference for human interaction, and fear of the machine-made decisions.

User-oriented studies have continually demonstrated that despite the benefits of AI, people prefer human decision-making in sensitive areas, like healthcare (Dietvorst *et al.*, 2015) [61]. This aversion potentially impacts the domains of consumer decision-making. It is crucial to remember that healthcare decisions are innately personal and typically involve a significant emotional aspect. It can be argued that AI, despite its extraordinary capabilities, lacks the ability to understand, empathize, and respond to these emotional factors that contribute towards making healthcare decisions (Botrugno *et al.*, 2021) [62].

However, this does not mean that the future of AI in healthcare is bleak. Consumer aversion towards AI can be addressed through measures like transparency in AI operations, robust research and development in the area of AI integration and improving AI's ability to communicate empathetically. Leveraging technology to humanize AI, sculpting it to understand and respond empathetically to emotional factors, could potentially change the tide of AI aversion in healthcare consumers [63]. In the end, consumers hold the reins of their healthcare journey, and technology is but a tool in their hands to make this voyage smoother and more informed.

1.2.5 Effects of AI Aversion on Consumer Decision-Making Processes: An Analysis

AI aversion is a genuine concern that has significant implications on the decision-making process of healthcare consumers as it can provide customized experiences for consumers, significantly influencing their behavioral tendencies. For instance, AI can utilize customer data to generate personalized product recommendations, which potentially sways the consumer's purchasing decisions (Ngai *et al.*, 2009) [64]. Consequently, it may negatively impact these companies' sales and growth, illustrating the far-reaching implications of AI aversion (Brinker *et al.*, 2020) [65]. Companies that excessively depend on AI technology might alienate a segment of their customer base who deem AI as a replacement rather than a complement to human service (Van *et al.*, 2019) [66]. Furthermore, AI aversion can catalyze a resistance to change by consumers. Many individuals, comfortable in their routines, might refuse to incorporate AI into their decision-making process (Ransbotham *et al.*, 2017) [67]. Thus, businesses must be sensitive in implementing AI not to alienate consumers with heightened AI aversion. However, AI resistance potentially mitigates the beneficial applications of AI in shaping consumer behavior. Striking a balance between AI-driven services and the human touch is crucial for gaining customers' acceptance, trust, and loyalty.

1.2.6 Understanding How Consumer Perceptions Influence Their Choice Between Human and AI Healthcare Providers

Consumer perception is a fundamental facet in healthcare where the provider, whether human or AI, is tasked with addressing personal and sensitive health concerns (Joshi *et al.*, 2020) [68]. The importance of comfort, trust, and understanding cannot be understated. Interactions with healthcare providers are inherently personal since patients disclose sensitive information hoping for the provider to offer the requisite physical care, emotional support, guidance, and understanding. Hence consumer choice between human and AI healthcare providers is a multifaceted decision influenced by various factors, including perceived ability to empathize, provide personalized care, availability, and the ability to democratize healthcare.

AI healthcare providers can complement or substitute human expertise in several medical areas (Li *et al.*, 2022) [69]. Machine learning enables AI to adapt, learn, and improve over time. However, AI is often perceived as impersonal and lacking the human touch of empathy — a vital component of healing. This perception can inadvertently discourage patients from utilizing AI healthcare solutions, highlighting the essential role consumer perspectives play in AI adoption.

Human healthcare providers, on the other hand, offer irreplaceable personal interaction — empathy, emotional support, and in-depth consultation. Trust is a critical factor that enhances patient willingness to adhere to health advice and share personal information (Meyer *et al.*, 2018) [70]. In scenarios where a human touch is deemed integral, AI healthcare providers may not be the preferred choice due to their inherent inability to empathize or comprehend complex human emotional states.

Yet, it is essential to note that AI also offers unique benefits that could sway consumer perception in its favor. Firstly, AI can provide real-time health data analysis (Marr *et al.*, 2019) [71]. With the rise of wearable IoT devices, AI's ability to monitor health vitals continuously and alert patients of impending health issues is an invaluable asset. Secondly, AI providers are available around the clock and capable of delivering instantaneous medical advice without waiting times customary to human providers (Gupta *et al.*, 2021) [72]. Lastly, AI allows for democratization of healthcare — offering the same level of service regardless of geographic location or socioeconomic status. Therefore, to fully integrate AI into healthcare, stakeholders must not overlook the importance of consumer perception and work towards enhancing consumers' trust and understanding of AI applications in healthcare.

1.2.7 Exploring Role of Trust, Perceived Competence and Empathy in Consumer Preferences

It remains pertinent to remember that AI aversion in healthcare consumers is influenced by a complex interplay of the trust in healthcare providers, their perceived level of competence, and extent of empathy exhibited, which requires a comprehensive understanding of consumer preference.

Further in-depth research and analysis, might improve patient satisfaction and overall healthcare delivery standards.

Trust, a psychological state encompassing the intent to accept vulnerability based upon positive expectations or predictability, is foundational to any form of healthcare service provision (Möllering *et al.*, 2001) [73]. In the context of consumer behavior, trust can significantly influence the decision-making process (Hubble *et al.*, 1999) [74]. Patients' trust in physicians is imperative, as it enhances patient compliance, fosters patient satisfaction, and eventually elevates the quality of healthcare (Goudge *et al.*, 2005) [75]. Nonetheless, perceived competence signifies an equally crucial role within healthcare contexts. Consumers frequently appraise healthcare providers' competence based on their technical proficiency, knowledge, problem-solving capabilities, and the overall quality of provided healthcare services (Delgado *et al.*, 2019) [76]. The consumer's confidence in healthcare providers is pivotal as it can be dictated by the perceived competence of the provider, which further encourages adherence to medical advice and treatments (Delgado *et al.*, 2019) [76]. Notwithstanding the roles of trust and perceived competence, empathy's influence persists and intertwines in intriguing ways within healthcare contexts. Empathy, typically defined as the ability to understand and share the feelings of others, holds incredible sway in the realm of healthcare (Greenberg *et al.*, 2001) [77]. Empathy fosters therapeutic relationships, facilitates patient satisfaction, and leads to better clinical outcomes (Derksen *et al.*, 2013) [78]. As such, it has definite and profound impacts on consumer preferences when it comes to choosing a healthcare provider. Through comparative analysis, it is evident that trust can foster long-term patient-doctor relationships, fostering a positive environment for treatment. On the other hand, perceived competence can influence patients' confidence in their healthcare providers' abilities, thereby influencing treatment adherence. In contrast, empathy plays an essential role in understanding and validating patient issues, thereby increasing patient satisfaction.

1.3 Economic Implications of AI Aversion

Adopting AI technologies in healthcare requires significant financial commitment. Noteworthy expenses include the purchase of AI software and hardware, system integration, and maintenance. Providers also incur recruitment and training costs to ensure staff can use the AI systems effectively. However, these costs should be considered against the backdrop of potential contributions AI could bring to healthcare. AI technologies promise enhanced healthcare delivery with improved accuracy, efficiency, and cost-effectiveness. One financial benefit of adopting AI in healthcare is the potential for predictive analysis and improved patient outcomes (Delgado *et al.*, 2019) [76]. AI can be used to predict illness or disease progression, allowing for early intervention and reducing the costs associated with late diagnosis and treatment. This not only contributes to patient benefits, but it can also translate to significant cost savings for healthcare providers. Furthermore, AI has the potential to streamline administrative tasks, thereby reducing labor costs. An analysis by the Harvard Business Review found that AI could potentially automate approximately 80% of administrative work in healthcare, such as data entry, appointment scheduling, and billing (Greenberg *et al.*, 2001) [77]. The rising global healthcare costs and increasing pressures on healthcare systems necessitate innovative approaches, and AI may be part of the solution. Healthcare providers who ignore the AI revolution risk falling behind in the increasingly competitive healthcare market. Although incorporating AI may necessitate an initial outlay, the risk associated with ignoring AI may be even greater, with potential loss of competitive advantage and revenue. The adoption of AI has the potential to place the healthcare industry onto a more value-based trajectory. This transformation from volume-based to value-based care may be catalyzed by AI due to improved efficiency, reduced error rates, and personalized patient care (Derksen *et al.*, 2013)[78]. This not only has potential financial benefits for healthcare providers but also promises healthier communities.

While AI in healthcare offers promising cost-effective and efficient solutions, the aversion among consumers for AI remains a significant impediment. The possible version of AI among healthcare consumers poses considerable economic challenges to the healthcare sector, leading to underutilization of AI-enabled services, affecting healthcare providers, developers, and indirectly the economy. As per economist (Gordon *et al.*, 2017) [79], institutions might encounter a lower return on investment from AI projects due to decreased demand, possibly delaying the attainment of economies of scale. If unaddressed, this aversion might have profound economic implications by impeding the optimized utilization of AI that could otherwise enhance efficiency and reduce costs. Furthermore, according to Price Waterhouse Coopers, AI is predicted to contribute $15.7 trillion to the global economy by 2030; if consumer aversion continues (Belkacemi *et al.*, 2022) [80], this estimate may not be reached. From another perspective, AI aversion might indirectly affect the economy through its impacts on healthcare outcomes. A review by (Rajkomar, *et al.*, 2019) [81], suggested that AI could significantly improve diagnostic accuracy and efficiency, potentially reducing healthcare costs. Thus, tighter regulation, transparent communication, and education are needed to increase the acceptance of AI in healthcare. While the current economic implications of AI aversion are sobering, with proper measures in place, there is potential to mitigate negative outcomes and fully realize the benefits inherent to AI in healthcare.

1.3.1 Investigating Influence of AI Aversion on Consumer Willingness to Pay for Healthcare Services

Amidst the burgeoning influence of Artificial Intelligence in the delivery of healthcare services, (Dutta *et al.*, 2021) [82] articulates, not falsely, that AI has often been accused of harboring a series of ethical and moral dilemmas. The ethical quagmires are severe enough to engender AI aversion amongst consumers impacting their willingness to pay for AI healthcare services. A study by (Dutta *et al.*, 2021) [83] found a significant relationship between AI aversion and consumer willingness to pay. Despite appreciating the proficiency, efficiency, speed and analytical skills of AI, consumers preferred human involvement, fearing errors, inaccuracies, lack of transparency, and potential misuse of their personal data (Castelvecchi *et al.*, 2016) [47]. It seems that consumers perceive human agents are more likely than AI to hold patients' best interests at heart, fostering trust, security and acquiescence which AI currently falters. Furthermore, (Longoni *et al.*, 2019) [84] discovered that the scope of AI aversion stretches beyond trust and security issues. Interestingly, their study uncovers affective reactions, such as 'Creepiness', that distorts consumers' willingness to pay for AI services. The finding is evocative of the 'uncanny valley' effect wherein human's emotional responses slide from empathy to revulsion when anthropomorphic machines mimic realistically unsettling human traits. However, this is a nascent area of study that requires further investigation. More empirical research should be conducted to provide concrete evidence, and to create strategies on how to encourage consumers to accept AI technologies in the healthcare sector, thereby, potentially improving their willingness to pay for these services.

1.3.2 Influence of Patient Education on AI Aversion in Healthcare

AI aversion hinges on multiple factors as discussed. Interestingly, recent studies distinctly show the role of education in understanding and embracing AI technology in healthcare. Recent research suggests that the complex relationship between one's educational attainment and the degree of AI resistance in healthcare might have an underpinning role in this AI aversion, that education may equip individuals with 'Digital Health Literacy' (DHL), a term popularized by Norman and Skinner in 2006. This implies one's ability to seek, find, understand and appraise health information from electronic sources and apply this knowledge to health-related issues (Norman *et al.*, 2006) [85]. With the acceleration of technology into healthcare, the concept of DHL is becoming increasingly valid. In a prominent quantitative study conducted in mid-2020 by O'Sullivan and colleagues, a comparative metric showed a distinctively low level of AI aversion in the group of participants with higher education (Dutta *et al.*, 2021) [86]. Their samples suggested that knowledge about AI technology and a better understanding of its implications led to more acceptance. Similarly, a compelling research project undertaken demonstrated a direct positive correlation between the degree of education and acceptance of AI in healthcare (Hackl *et al.*, 2021) [87]. The finer points of this research affirmed that would-be healthcare consumers with doctoral degrees reported significantly lower anxiety regarding AI compared to those with a high school degree or less. Ethical apprehensions, often a determinant of AI aversion, often decrease with education level, as shown in a 2022 study by (Wyckof *et al.*, 2023) [88]. This research claimed that higher education consumers showed transparency and fairness in their apprehensions about AI, unlike lower-educated consumers who often had misconceptions. Global AI Attitude Study, a significant annual study conducted in 2023, mirroring similar findings to the above-mentioned studies (Neudert *et al.*, 2020) [89]. The data indicated that the higher one's educational attainment, the more likely they are to trust and accept AI. On the flip side, less educated or ill-informed patients typically exhibit a higher degree of AI aversion. Paradoxically, they are usually the very category of patients whose health management program would most profit from the seamless integration of AI. Their aversion originates mostly from a fear of the unknown, misconceptions about AI replacing human practitioners, and an over-reliance on sensationalist media reports painting AI in a sinister light. However, effective patient education coupled with a comprehensive understanding of AI's capabilities and limitations can help answer many misconceived notions. It is not only about building technology, but also about building its acceptance. AI is designed to supplement human doctors and not replace them, and this understanding is crucial to flip the perception of AI on its head, from an ominous threat to a life-saving tool.

1.3.3 Influence of Patient Awareness on AI Aversion in Healthcare

Public apprehension towards AI, particularly in healthcare, has been rearing its head, with patient awareness deemed to be a significant influencing factor. Despite the benefits, AI adoption in healthcare institutions has met resistance mostly from the patient's end, largely due to limited understanding and awareness about AI technology (Dutta *et al.*, 2023) [90]. Recent studies note that a significant percentage of patients express distrust in AI's ability to provide healthcare services (Ting *et al.*, 2019) [91]. Additionally, research indicates that most AI aversion in healthcare stems from poor patient awareness and knowledge of AI's full capabilities. Patient awareness plays a significant role in reducing AI aversion. High patient awareness contributes to trust in AI applications, stemming from an understanding of the benefits and safeguards in place (Dutta *et al.*, 2021) [92]. Patient education initiatives to improve AI understanding can effectively demystify misconceptions and ensure patients are aware of the benefits and limitations of AI (Ganie *et al.*, 2023) [93]. This illustrates the need for increased efforts towards patient education and clear communication about AI to foster acceptance and diminish aversion. As healthcare continues to evolve with the technological age, it is imperative that patient concerns are addressed to enhance trust and open more doors for innovative solutions like AI.

1.3.4 Influence of Age of Patient on AI Aversion in Healthcare

Despite the ameliorating potential, AI adoption within the healthcare sector seem to be backed by a constant specter of AI aversion, particularly amongst older cohorts of the population. As AI continually impacts healthcare services, it becomes critical that the user's age is taken into consideration. According to recent investigations, a negative correlation exists between the age of the patient and their aversion to AI (Chu *et al.*, 2023) [94]. This AI aversion amongst older patients could be attributed to multitude of reasons. One possible cause is the digital literacy gap. Another possible explanation is the comfort with traditional modes diagnosis and treatment. Many older adults are less flexible adapting to the changing healthcare landscape due to their long-standing reliance on human healthcare providers. For them, the advent of AI presents an impersonal approach to healthcare that departs alarmingly from their comfort zone. After all, older adults symbolize a key demographic segment in healthcare due to their unique and often wide-ranging healthcare needs. Therefore, it is crucial that this user agnostic technology is understood and accepted by the older population. Evidence from a study conducted by the Pew Research Centre (2019) strongly suggests that an individual's age significantly influences their level of comfort and trust in AI applications. In this study, 60% of adults aged 50 and above expressed concerns about the use of AI in healthcare, contrasting starkly with the popularity of AI within younger age brackets [95]. A systematic review suggests that older patients are less likely than their younger counterparts to possess the necessary technology proficiency needed to navigate AI, which may spike their aversion towards it [96]. The negative correlation between age and acceptance of AI also arises from the cognitive decline and fragility typically associated with old age. According to a meta-analysis, people at advanced age may struggle with cognitive tasks and decision-making processes required in AI-innovated medical processes, thereby propelling further resentment and mistrust towards AI implementation [97]. The anxiety associated with potential privacy breaches is substantiated by research conducted at the Oxford Internet Institute (2019), where a major hurdle observed in AI adoption was indeed the fear of privacy violations [98]. To surmise, the influence of a patient's age on AI aversion cannot be disregarded. Old age must not be a barrier to harnessing the immense potential of AI in healthcare. Therefore, acknowledging and addressing the underlying causes of this AI aversion is not merely advisable but pivotal in the realm of automated healthcare. By understanding the roots and impact of AI aversion within older patients, the healthcare industry can tailor AI applications more thoughtfully. Stakeholders can ensure that digital literacy programs are accessible to older adults, and stringent data protection measures are put in place. More importantly, while promoting AI within healthcare, the humanity of care must remain at the forefront. AI technologies unquestionably demand a significant initial financial investment; however, the long-term financial implications are potentially beneficial. The potential cost savings through improved patient outcomes, reduced labor costs, enhanced competitive advantage, and a value-based care model suggest adoption of AI could prove to be a strategically sound financial decision for healthcare providers.

1.4 Overcoming Resistance to Medical AI

AI, with its robust computational power and machine learning capabilities, may not fully comprehend human emotion and context. To respond to these concerns, the medical AI field could place greater emphasis on incorporating human-in-the-loop (HITL) systems, consisting of a trained AI model that continuously learns and adapts by utilizing a qualified professional's input, guaranteeing a synergistic and personal human-machine relationship. With HITL systems, patients no longer have to worry about impersonal interactions as it ensures consistent human oversight and feedback [99].

The second challenge arises from healthcare professionals' skepticism about AI's reliability and accountability. Practitioners fear a potential lack of precision in AI systems that might jeopardize patients' health and are concerned about who bears the responsibility in the case of misdiagnosis [100]. One suggested solution is the practice of Explainable AI (XAI). Unlike the traditional AI models that offer predictions without explanations leading to a black-box conundrum, XAI provides clear reasoning behind every prediction it makes. This allows healthcare practitioners to visually verify the decision-making process of AI, building trust in its reliability and precision [101].

It is true that technological innovations, like Medical AI, are not without glitches and errors. However, the consistent practice of refining and perfecting these AI systems by feedback-based learning can minimize the margin of error over time. Regular simulation tests, similar to a concept called *in silico* trials, may build healthcare professionals' confidence in the technology as it enhances the accuracy of the algorithms [102].

Concerns about privacy and confidentiality when using medical AI technologies is another stinging source of resistance. Patients and healthcare professionals worry about the potential misuse of sensitive data stored digitally [103]. Addressing these concerns requires a stringent data protection infrastructure and sophisticated encryption technology, backed by watertight cyber laws.

Advanced encryption technologies like Public Key Cryptosystems can ensure that only authorized professionals can access sensitive patient data, thus effectively safeguarding the patient's privacy and confidentiality [104]. Addressing these challenges requires consistent dialogue between AI developers, healthcare practitioners, and patients. Only through mutual understanding can we hope to bridge the divide between the limitless potential of medical AI and those who stand to benefit from it.

1.4.1 Strategies for Enhancing Consumer Trust and Acceptance of AI in Healthcare

The augmentation of trust in AI can be fostered through the creation of trustful relationships established amongst patients, healthcare providers, and AI developers. To attain this, healthcare providers need to relay open, frank, and accurate information about the AI models employed in their setting, including their benefits, risks, and any limitations (Lewis *et al.*, 2022) [105]. Central to this is transparency and the regular auditing of AI systems to ensure accuracy and minimize potential risks. Furthermore, there's a necessity for cautious and stringent monitoring of AI-based solutions to corroborate safety and effectiveness. The rigorous regulatory processes enacted by organizations such as the FDA and MHRA are paramount, such as the Pre-Market Approval (PMA) processes, which subject medical devices to safety and efficiency reviews. By encouraging these actions, consumer trust in the safety and reliability of AI systems is strengthened. Another critical avenue is increasing AI literacy and knowledge among consumers. When users comprehend the operations and advantages of AI solutions, they are much more likely to value and adopt them propose public engagement programs educating consumers about AI to promote more informed decisions about their healthcare and improve positive perception. Moreover, involving consumers in AI development phases propounds a key strategy. This co-design principle brings about services equipped with the capacity to handle specific needs more effectively (Sanders *et al.*, 2008) [106]. The co-creation tactics build trust as they reflect AI systems that are more aligned with user needs and desires.

1.4.2 Approaches to Alleviate Consumer Concerns and Misconceptions: Communication and Education

In an era characterized by a ceaseless stream of information, consumers are inevitably exposed to variegated perspectives and narratives, which often propel the development of concerns and misconceptions. To confront and alleviate such issues, the roles of communication and education have enhanced in significance. Effective communication is a fundamental framework that shapes attitudes, belief, and behavior. One predominant approach adopted in communication is the interactive communication model which emphasizes on active engagement between the sender and the receiver as pointed by McQuail & Windahl, 1993. Research conducted by Corner and Richardson (1993) reveals that the involvement of the audience in the communication process invokes better comprehension and retention of information. Active engagement, specifically in an online medium, involves using feedback mechanisms that could involve likes, comments, or shares, and aids in addressing concerns and rectifying misconceptions. Another contemporary communication approach is storytelling. According to [107] stories have a deep-rooted cultural framework that is understood and appreciated universally. By incorporating narratives into communication, companies are in a better position to convey the intended information while alleviating concerns and debunking misconceptions with undeniable impacts on consumers' perspectives. Education has a significant role to play in discouraging misinformation. Critical literacy, an educational approach, aims at equipping consumers with skills and knowledge to discerningly analyze and critically interpret the information they come across. This approach fosters skepticism and scrutiny, encouraging consumers to question the authenticity of information and subsequently debunk their preconceived misconceptions, thereby reinstating trust in companies. Narrative Simulation, another instructive approach, can be used extensively for businesses dealing with health-conscious consumers or technological products. It teaches consumers how to handle certain situations by creating a simulated environment, subsequently reducing apprehensions and misinterpretations arising in the minds of consumers. By evaluating the above-mentioned approaches, it becomes apparent that an efficacious blend of communication and education systems substantially contributes to alleviating consumer concerns and misconceptions. Businesses should strategically employ interactive communication and storytelling techniques to break down complex information into understandable narratives. Concurrently, an emphasized focus on critical literacy and narrative simulation would help educate consumers and encourage them to critically analyze the flood of information they deal with daily.

1.4.3 Cases of Successful Implementation of AI Technologies in Healthcare and Lessons Learned

C1: One of the notable applications of AI technologies in healthcare has been seen in Google's DeepMind Health project, an AI system developed to predict different diseases' onset, was used by the Moorfields Eye Hospital in London to predict eye disease before the manifestation of symptoms (Wakefield *et al.*, 2022) [108]. The findings revealed a 94% accuracy in the system predictions, serving as a significant breakthrough in early disease detection.

C2: IBM's Watson for Oncology is another case that demonstrates AI's revolutionary impact in healthcare. The AI-driven platform provides individualized treatment options for cancer patients, making recommendations based on the patient's medical records, a wealth of medical literature, and training by oncologists at the Memorial Sloan Kettering Cancer Center (MSKCC, 2015) (Ross *et al.*, 2017) [109].

C3: Woebot, a chatbot, uses AI to offer Cognitive Behavioural Therapy (CBT) to users dealing with mental healthcare conditions such as depression and anxiety. This not only increased access to mental health services but also reduced the stigma around seeking such support (Darcy *et al.*, 2022) [110].

C4: Outside of diagnostics and treatment, AI has also demonstrated great efficacy in administrative tasks. One such case is Zebra Medical Vision. With an AI model capable of reading and interpreting medical imaging, Zebra helps to identifies signs of diseases like cancers, cardiovascular conditions, liver diseases, and more for early intervention (Parush *et al.*, 2017) [111].

C5: AI Technologies in the Singapore-based healthcare tech company, ConnectedLife, have also managed to derive crucial insights that aid in managing and predicting Parkinson's disease, a neurodegenerative disorder (Gochoo *et al.*, 2021) [112].

Each of these case studies offers fundamentally valuable insights, underpinning some critical success factors for the implementation of AI technologies in healthcare. Firstly, collaboration between AI developers and healthcare professionals is crucial, enabling the creation of AI applications that align with practitioners' practical needs, as depicted in the DeepMind and IBM Watson case. Moreover, AI applications' proactivity in early disease detection, as seen in Zebra Medical Vision and DeepMind, not only enhances patient outcomes but also contributes significantly to cost efficiency across the healthcare system. This serves as a potent reminder that applications of AI should focus on tasks that directly improve patient care quality and the system's sustainability. Importantly, the use of AI should comply with rigorous privacy and ethical considerations, especially when handling sensitive patient data, as seen in the Woebot and ConnectedLife cases. As AI technologies continue making inroads into healthcare, legislation and policies will play a crucial role in ensuring these considerations are upheld.

1.5 Ethical Considerations and Governance

One of the chief ethical concerns entwining AI adoption is the potential compromise of individual privacy. As AI algorithms become more advanced and pervasive, they collect, synthesize, interpret and share a significant volume of user data, raising concerns about how data is being managed and utilized (Tom *et al.*, 2020) [113]. While AI can offer enhanced services through personalized experiences, an overt reliance on data might infringe on data privacy norms and rights (www.hrw.org/news/2019/04/24/morocco-arrest-over-twitter-posts). Practicing ethical data handling protocols becomes pivotal for maintaining trust in AI. Discrimination is another ethical dilemma associated with AI, caused by biased or unrepresentative data inputs into algorithms (Zou *et al.*, 2018) [114]. Notably, Amazon abandoned an AI recruitment tool when it was scrutinized for gender bias, fundamentally favoring male candidates. The ethical implications of such biases are serious, potentially fostering unfair systems and practices, underpinning the importance of fairness when implementing AI. Transparency in AI operations also poses serious ethical considerations. If AI systems cannot provide an understandable explanation of their decisions, it becomes difficult to hold them accountable (Dignum *et al.*, 2017) [115]. Without transparency, users may suffer a 'liability gap' where no party can be held accountable for AI-induced harm. Hence, developing explainable AI models that are intelligible to users whilst maintaining high performance remains a serious challenge. Interestingly, AI governance models can address these ethical quandaries. A robust governance model should ideally regulate the use, development and effects of AI in accord to societal norms and values (Jobin *et al.*, 2019) [116]. Establishing an AI ethics committee is one such model, where a diverse group investigates ethical issues, generates dialogue and creates guidelines for fair AI use. This encourages wider participation in the discussion about AI ethics, fostering collective decision-making. Additionally, external audits of AI systems can also be part of a governance model. Auditing makes sure AI systems are employed ethically and legally, and also helps in identifying and mitigating any unintentional biases in the system (Aprilicia *et al.*, 2021) [117]. Third-party audits make AI outcomes interpretable and understandable, reducing the opacity that often surrounds AI systems. The thrust of AI technology comes with ethical considerations that demand our attention and governance models to mitigate their effects. As we forge ahead into a future navigated by AI, it is imperative we craft this new reality to be beneficial for all, not just a privileged few. The failing on our parts to effectively counter these ethical and governance challenges would not be a testimony to AI's failure, but a reflection of our own short-sightedness in harnessing its potential.

1.5.1 Regulatory Frameworks for Ethical AI Operations to Fight Aversion in Healthcare Consumers

Regulatory frameworks and guidelines are the tools used to mold, inform, or impose ethical standards on AI's operations. However, there are inherent challenges. One issue lies in the global disparity in AI ethics regulation—the absence of globally harmonized AI standards generates a 'regulatory patchwork' further complicating efforts to safeguard ethics (Dutta *et al.*, 2022) [118]. Yet, the importance of setting and enforcing such frameworks cannot be overstated. Regulatory frameworks have a strong role in transcending the development of AI from mere machinations towards structures imbued with human values. Moreover, they serve to enhance trust in AI systems, which is crucial for their wider acceptance and integration into social structures. Evidence suggests that various global players recognize this importance. The European Commission, for example, is actively advocating for a human-centric approach towards AI's development. Fundamental rights and ethical aspects are integral to the commission's guidelines. Meanwhile, the UK government's Centre for Data Ethics and Innovation (CDEI) focuses on advising on the governance of data-driven technologies. Its strategy underscores the necessity for regulation to maintain public trust in AI (CDEI, 2020). Regulation can also alleviate specific ethical concerns associated with AI. For example, privacy infringements can result from the broad-scale data collection and analysis efforts inherent to AI. Therefore, regulatory frameworks like the European General Data Protection Regulation (GDPR) are key permissions for using individuals' data, and maintaining the ethical integrity of AI.

1.5.2 Addressing the Potential Cost-Effectiveness and Affordability Concerns Associated with AI-Based Healthcare Solutions

In the contemporary global healthcare scene, several innovations aim to provide optimal care and efficient services. One notable example is the incorporation of Artificial Intelligence (AI) in healthcare. Although AI possesses incredible potential, it abounds with concerns centered on cost-effectiveness and affordability that necessitates exhaustive exploration. AI has revolutionized countless sectors, including healthcare, by demonstrating its capability to assist with diagnosis, prognosis, and patient care. From using algorithms to detect melanoma better than human dermatologists to predicting patient readmission rates, AI's impact is profoundly apparent. Despite this, the healthcare providers, policy makers, and patients alike grapple with understanding the cost-effectiveness of AI-based healthcare solutions. Cost-effectiveness indicates whether a treatment or intervention provides sufficient health benefits relative to its cost. The quandary about AI's cost-effectiveness in healthcare emerges from the juxtaposition of the significant upfront implementation and maintenance costs against the potential long-term savings. It's often challenging to quantify the economic value of AI interventions meaningfully, as their benefits may transcend the straightforward parameters of monetary or health outcome gains. One study conducted on diabetic retinopathy screening using AI concluded that while deploying AI was cost-effective in the long term, the high upfront costs and ongoing maintenance expenses could present barriers for some institutions. Another investigation examining the use of AI in predicting patient readmission found that the cost savings through avoided readmissions far outweighed the input cost of AI technologies. These investigations illuminate the nuanced and context-specific nature of AI's cost-effectiveness in healthcare. Simultaneously, there are apprehensions regarding the affordability of AI-based healthcare treatments. The initial cost of acquiring the prevailing AI-based technology, coupled with the cost of training healthcare professionals to use these tools appropriately, makes AI affordability a legitimate concern. The expressed fear is that the wide-scale implementation of AI will lead to markedly high healthcare costs which may be unintelligible for the average individual or healthcare institution. However, this perspective overlooks the potential for AI to streamline healthcare services; by automating repetitive tasks, AI can save time for healthcare professionals to focus on more complex tasks, thereby creating potential cost savings in the long term. Furthermore, by facilitating early and accurate diagnosis, AI can contribute to reduced healthcare expenditure associated with late-stage disease treatment. In conclusion, while apprehensions regarding the cost-effectiveness and affordability of AI-based healthcare solutions are warranted, they should be addressed in a comprehensive and nuanced manner. By analyzing the specific context of implementation, and accounting for both the immediate and long-term implications, a more balanced perspective can be attained. As we continue to incorporate AI in healthcare, the commitment should not only be on leveraging technology for better health outcomes but also on ensuring these interventions are economical and accessible.

1.5.3 Balancing Privacy, Data Protection and Need for Transparency in AI Healthcare Applications

To establish a baseline understanding, it is vital to exemplify what constitutes AI's ethical boundaries. Essentially, these boundaries encompass principles like transparency, fairness, privacy, and security. The European Commission's 'Ethics Guidelines for Trustworthy AI' serves as a notable example, outlining requirements that should underpin the design and use of AI systems, including transparency, diversity, non-discrimination and fairness (Ala-Pietilä *et al.*, 2020) [119]. The importance of setting and enforcing such frameworks cannot be overstated. Regulatory frameworks have a strong role in transcending the development of AI from mere machinations towards structures imbued with human values. Moreover, they serve to enhance trust in AI systems, which is crucial for their wider acceptance and integration into social structures. Evidence suggests that various global players recognize this importance. The European Commission, for example, is actively advocating for a human-centric approach towards AI's development. Fundamental rights and ethical aspects are integral to the commission's guidelines. Meanwhile, the UK government's Centre for Data Ethics and Innovation (CDEI) focuses on advising on the governance of data-driven technologies. Its strategy underscores the necessity for regulation to maintain public trust in AI (CDEI, 2020). Regulation can also alleviate specific ethical concerns associated with AI. For example, privacy infringements can result from the broad-scale data collection and analysis efforts inherent to AI. Therefore, regulatory frameworks like the European General Data Protection Regulation (GDPR) are key permissions for using individuals' data, and maintaining the ethical integrity of AI.

The general importance of privacy in healthcare cannot be overstated. As Mordini rightly notes, "Privacy is a human dignity that touches upon the deepest aspects of our existence" (Camenisch *et al.*, 2009) [120]. The advent of AI technologies, given their inherent dependency on vast amounts of data for effective functioning, has significantly amplified this concern. The sensitive nature of health data poses real challenges as the risk of exposure can lead to stigmatization, discrimination, and other adverse effects, thereby necessitating robust privacy measures. In tandem with privacy concerns, the issue of data protection within AI healthcare applications emerges as paramount. The General Data Protection Regulation (GDPR), operative within the European Union, makes it clear that data protection is not just an optional extra, but a fundamental right. Its principles of data minimization, purpose limitation, and strict accountability are particularly noteworthy in the context of AI applications. For example, an algorithm trained to identify potential cancer signs should only access pertinent windowed data and not unrelated patient information, adhering to the data minimization principle. AI's characteristic "black box" problem amplifies the need for transparency in healthcare applications. This refers to the difficulty of understanding the mechanism by which an AI system makes decisions - adding to the complexity of managing privacy and data protection. Appropriate transparency measures make it possible to audit AI systems, discern potential bias, and ascertain compliance with data privacy and protection norms. The need for explainability in AI decisions in increasingly legislated, as seen in the GDPR's "right to explanation" – although it's often a challenge to practically implement in the healthcare context. Striking the right balance between patient privacy, data protection, and transparency in AI healthcare applications is no easy task. However, there is an emerging consensus that beyond regulations and technical solutions, a culture of privacy is essential. The healthcare industry must foster an environment where privacy and data protection are viewed as paramount, positioning transparency as their cornerstone.

1.6 Future Outlook and Opportunities

Currently, the healthcare industry is experiencing profound transformation propelled by AI and machine learning. These technologies automate administrative tasks, treat patients more efficiently and provide personalized treatment options. For instance, AI-based analytics provide real-time monitoring of patients, enable early disease detection, and predict health outcomes by analyzing patient patterns. With these capabilities, AI augments healthcare providers' operations and develops mechanisms for proactive response. Advancement in AI technologies are accelerating, pushing boundaries of what is possible in healthcare. Notably, predictive algorithms, machine learning and automated robotic surgeries are among trending AI implementations. These technologies enable precision medicine, advancing patient's diagnosis and reducing human error. For example, machine learning algorithms can consider millions of data points in seconds, surpassing human capability, thus improving diagnosis accuracy. Another notable innovation is AI-powered wearable technology. These devices offer promise in monitoring health, predicting illness, and analyzing patient symptoms. As these technologies become more sophisticated, the prospect for remote health management and patient-centered care becomes achievable. However, despite these evident benefits, several challenges face AI implementation. Primarily, data privacy and security concerns are marquee issues. As healthcare data is sensitive, the adoption of AI technologies necessitates robust protection measures. Effective legislation and correct ethical considerations will be crucial in upholding public trust. Additionally, as the propensity of AI increases, workforce disruption is likely. Concerns circulate around AI replacing physicians, clinicians and staff. However, views are gradually shifting, with an understanding that AI can complement rather than replace healthcare professionals. With continuous education and skill development, the transition could lead to hybrid roles, combining human expertise with AI capabilities. However, AI in healthcare is not an immediate panacea. The technical challenges need addressing, such as the adaptability to changing healthcare practices, integration with existing systems, and interpretability of AI output. Solving these challenges will necessitate concerted efforts from policymakers, regulators, healthcare providers, and AI developers.

1.6.1 The Future of AI in Healthcare and Its Impact on Consumer Aversion

As artificial intelligence (AI) continues to take a central role in modern healthcare, interested stakeholders have grown optimistic about its future implications. From complex surgical procedures to efficient patient data handling, AI has displayed considerable potential in health services. However, there exists a significant divergence between the optimistic realm of advancement and the realm of consumer aversion, prompted by confidentiality concerns, cyber security threats, and maladaptive system changes. Many industry predictors anticipate numerous transformative changes in the healthcare industry due to AI. They envisage advancements directed towards improving patient outcomes, streamlining healthcare processes, and general productivity increments. Despite these benevolent predictions, consumers remain apprehensive about the full integration of AI into healthcare systems. Evidence suggests an impending dominance of AI in the healthcare industry. One report prophesizes that by 2026, the global AI healthcare market would have swelled to $45.2 billion, marking a whopping 44.9% Compound Annual Growth Rate (CAGR) since 2018. Essential drivers for this growth include increased investments in AI and machine learning, and the demand for reducing healthcare costs while increasing the speed and accuracy of diagnoses. AI is predicted to contribute to better patient treatment plans, disease pattern identification, personalizing health services, and enhancing clinical decision-making. However, with these remarkable improvements come adverse consumer reactions – the most notable being aversion. Aversion towards AI in healthcare arises from several angles. From an emotional perspective, consumers often exhibit cynicism towards AI-driven healthcare due to perceived alienation.

The reduction of human interaction and empathy in health services causes concern amongst consumers, who fear they may be treated as 'just another dataset'. These transformations may result in a discouraging patient experience, exacerbating aversion and resistance. On a practical front, citizens fear the erosion of their data privacy as adoption of AI demands increased collection, storage, and analysis of personal health information. In an era where information is the new currency, consumers resist AI integration due to data breach concerns, with the increasing prevalence of cyber-attacks on healthcare data serving as a sober reminder of these vulnerabilities. These apprehensions may significantly hinder consumer acceptance, dampening the anticipated benefits of AI in the sector. As we brace ourselves for the future of AI in healthcare, stakeholders should foster a holistic approach. There must be a careful balance between capitalizing on the benefits of AI while simultaneously addressing consumers' privacy and emotional concerns. Moreover, a stronger and more resilient cyber security framework is required to mitigate cyber threats effectively. Only then can we maximize the true potential of AI in the healthcare industry, transforming aversion into acceptance.

1.6.2 Exploring Emerging Technologies and Trends That May Alleviate Consumer Concerns

Chatbots, AI's customer service adjunct, not only address queries in real-time, but also anticipate consumer needs based on behavioral analysis, making an enormous stride towards customer satisfaction. Blockchain, another rising technology, is progressing rapidly, particularly in financial sectors, where transparency, trust, and data security are integral. Through its decentralized, immutable system, Blockchain provides an effective solution for consumer's data privacy and security concerns. Moreover, with growing digitization and online transactions, consumers worry about safety. Blockchain's ability to provide secure and verifiable transactions offers a principal resolution for such concerns. Virtual and augmented reality (VR and AR) are novel trends that employ immersive experiences to mitigate consumer challenges. Especially in the era of COVID-19, where shopping behaviors drastically changed, AR and VR came to the rescue of many businesses, economically and strategically. Showing products in a simulated environment empowers consumers to experience products without physically visiting stores, substantially reducing their anxiety about in-person shopping during a pandemic. On the flip side, as with all revolutionary changes, these technologies also carry potential setbacks. One can argue that AI may lead to job losses due to automation, and unfamiliarity of Blockchain may foster hesitance among consumers. Furthermore, VR and AR, despite their interactive features, can never fully replicate the tangible experience that traditional shopping offers.

1.6.3 Opportunities for Collaboration Between AI Developers, Healthcare Providers, and Consumers

AI's capabilities in healthcare are broad, including aiding diagnostic procedures, predicting health deteriorations, enhancing personalized treatment, and improving drug discovery. To maximize its potential, collaboration is paramount. AI developers can benefit from healthcare professionals' knowledge, interpreting complex medical data patterns, unexpected findings, or difficulties encountered during system development. Simultaneously, consumers represent the user aspect, hence playing an influential role in AI tools' functionality and usability. This tripartite collaboration is key to realizing AI's unprecedented potential in healthcare. Healthcare providers possess the critical insight that AI developers can leverage for their algorithm development. First-hand information such as patients' needs, treatment protocols, and usage of existing tools could serve foundational for AI tools. Further, providers can help developers understand medical protocols, terminologies, disease pathologies, and healthcare dynamics in general, optimizing the AI tool's practicality. Consumers' collaboration guarantees that AI tools will be user-friendly, functional, and align with their healthcare needs. Consumers can share what they desire in an AI tool, hence facilitating the tool's customization. As well, consumers may provide feedback on existing AI tools, creating an essential loop of continuous improvement, thereby meeting the challenges of AI in healthcare head-on. Collaboration between these three parties promotes shared expertise, resource pooling, co-designing, and co-developing healthcare AI technology. It creates an environment that fosters not only technical developments but also the acceptability of AI technologies in the healthcare sector. This synergy would harness AI's potential in healthcare in a methodologically robust, ethically sound, and clinically relevant manner. AI has the potential to redefine healthcare, bringing efficiency and precision that is needed for futuristic healthcare. The simultaneous dialogue between AI developers, healthcare providers and consumers could shape the future of healthcare provision more robustly. Only through this collaboration can we ensure that the transformation of AI in healthcare is effective, efficient, and ethically sound.

1.6.4 Summary of Key Findings on Consumer Aversion to AI in Healthcare

Artificial Intelligence (AI) has increasingly penetrated various sectors, including healthcare, where it has gained immense promise for the possibilities it holds. Yet, despite the undeniable potential, a palpable level of consumer aversion exists towards its adoption in healthcare, calling for an exploration to understand the reasons for this apprehension and ways to mitigate the same. This chapter have underscored the pervasive fear among consumers about AI in healthcare, the crux of which are primarily issues related to trust, privacy, and the perceived impersonality of AI. Moreover, the data privacy concern also directly links to the usage of AI. As data is the fuel driving AI, there is a legitimate worry about personal health data being shared extensively and potentially falling into unscrupulous hands. The perceived impersonality of care delivered by AI is another significant concern. Although advanced, AI lacks the human touch, empathy, and personal understanding inherent in human doctors, which is integral to the patients' recovery process. As (Dutta *et al.*, 2023) [121] discerned, consumers were far more comfortable with AI being used to provide suggestions and advice to human doctors rather than replacing them entirely. However, these concerns should not spell doom for AI in healthcare; rather, they necessitate approaches for AI instigation that directly tackle these pressing fears. Consumer trust may be maximized through transparent and constant communication about the AI's success stories and its logical reasoning process. "OpenAI," an AI research lab, suggests regular and transparent communication to build trust with the users (https://openai.com). To assuage the privacy concerns, strict compliance to data ethics and privacy norms should be enforced. Moreover, a transparent data policy comprehensible to consumers about their data's usage, storage, and sharing will help build confidence. O'Neil (2020) underlines the importance of stringent privacy policy norms for building user trust in AI systems. On the front of AI's impersonality, it can primarily be employed as an au pair to doctors, providing them with comprehensive information to make informed decisions, thereby keeping the human touch alive. Moreover, ongoing advancements in AI technology oriented towards improving its emotional intelligence should be vigorously pursued. To summarize, the key revelations about consumer aversion to AI in healthcare hinge on trust, privacy, and the perceived impersonality of AI, which are not entirely unfounded. However, this does not consign AI's future in healthcare to oblivion. With strategic planning, thoughtful policy-making, and sensitive implementation, AI can very well surmount these roadblocks and unlock its full potential in healthcare, beneficial for providers and receivers alike. As with all powerful technologies, the key lies in the judicious and ethical use of AI, turning the tide of consumer aversion towards acceptance and embrace.

1.6.5 Implications for Healthcare Practitioners, Policymakers and Researchers

As societies become increasingly complex, often with significant shifts in demographics and disease patterns, the challenges facing healthcare systems multiply. These changes necessitate a timely and appropriate response from healthcare practitioners, policymakers, and researchers. Healthcare practitioners are significantly affected by these changes. An aging population brings with it the prevalence of chronic diseases, significantly altering the patient's profile that healthcare practitioners (HCPs) encounter (Condelius *et al.*, 2008) [122]. Not only does this necessitate a revision in the types of skills required by HCPs, but it also results in increased workload. The new norms necessitate a shift from acute, episodic care towards a more holistic and continual care strategy focused on patient centeredness (Berwick *et al.*, 2008) [123]. This shift represents one of the critical implications for practitioners. Their role is no longer limited to providing care. They must also act as navigators of a complex healthcare system, advocating for their patients' needs, and educating them on the management of their conditions. Simultaneously, rapid technological advancements provide both opportunities and challenges. These advancements have the potential to revolutionize healthcare delivery through telemedicine, precision medicine, and digital records, among other things. However, HCPs need to keep up to date with these changes, often requiring extensive and continuous education (Dutta *et al.*, 2023) [124]. Policymakers, on the other hand, face the challenge of orchestrating a complex array of variables to ensure the optimal performance of the health system. To do this effectively, they need to engage in continual health systems research, identifying areas of inefficiency and proposing solutions (Dutta *et al.*, 2023) [125]. Using evidence to inform policy, while a seemingly straightforward process, is often fraught with conflict. Political interests, resource constraints, and varying stakeholder priorities can significantly influence the process of policy-making. Thus, policymakers need to master the art of negotiation and consensus-building, recognizing that the perfect solution may not always be achievable. This focus on evidence-based health policy brings to light the researcher's role. Their ability to adapt and focus their investigations on pressing health issues contributes to informed decision-making. Effective communication of research findings is crucial to ensure they inform policy dialogue and decision-making processes (Greenhalgh *et al.*, 2005) [126].

1.7 Conclusion

In scrutinizing the results, we pay particular attention to correlations between respondents' demographic data and their attitudes towards AI. Artificial Intelligence (AI) is undeniably transforming numerous sectors globally, notably our healthcare, where AI applications promise significant transformations. Nonetheless, while advocating for crucial technological advancements, skeptics caution about unanticipated consequences. This essay explores the implications, both positive and negative, of integrating AI in healthcare. Global research firms, such as McKinsey, anticipate exponential growth in AI's use, suggesting that its advent in healthcare can revolutionize the delivery of quality patient care. However, the public's acceptance of AI interventions could be a potential bottleneck in the successful implementation of these initiatives. Although some patients have shown enthusiastic acceptance of AI-integrated treatments, others harbor concerns regarding matters of privacy, trust, and job displacement. These mixed attitudes necessitate a thoughtful outlook on the future of AI adoption in healthcare with advancements in AI, there is emerging potential to transition from a 'one-size-fits-all' treatment approach to individualized care plans that deploy sophisticated algorithms to predict patient diagnoses and prognosis, advancing the notion of personalized medicine. AI is also projected to expedite diagnosis, thus reducing waiting times and overall medical costs, an aspect much appreciated by consumers worldwide. Despite the promising attributes, reports of data breaches and cyber vulnerabilities have raised questions regarding AI's safety in healthcare, according to a report in the Lancet Digital Health. These concerns need to be assuaged by deploying robust regulatory frameworks and stringent enforcement to monitor any attempts at compromising data integrity. This involves creating and implementing robust data security measures, ensuring transparency in the information used and the functioning of AI systems, and investing in a blended model that combines technology and human touch. Guaranteeing security will be paramount in building consumer trust and confidence in AI. The narrative surrounding job displacement due to AI is another consumer apprehension that needs to be addressed. Patients fear that AI may replace human contact in healthcare, which would significantly impact the patient-doctor relationship. However, a study published in the British Medical Journal suggests that AI should be seen as an additional tool or an 'augmented intelligence' that will assist healthcare providers, not replace them. In conclusion, the future of AI in healthcare hangs in a delicate balance between its masterful possibilities and the perceivable skepticism. The potential benefits of AI include improved diagnosis, personalized treatment plans, and cost reductions, which would indisputably enhance the quality of patient care. However, conquering consumer fears and anxieties regarding privacy invasion, data security, and job displacement with concrete actions and robust policies is crucial for the successful adoption of AI in healthcare. The relationship between AI and healthcare looks promising, with the increasing probability for AI to improve patient outcomes significantly. However, for AI to optimize its full potential, healthcare providers, consumers and policy makers must collaborate to overcome the barriers of consumer attitudes and foster a healthcare culture that welcomes innovation whilst respecting individual privacy and security. Only then can the efficacy of AI in healthcare be fully realized and accepted.

References

1. Obermeyer, Z. and Emanuel, E.J., Predicting the future—big data, machine learning, and clinical medicine. *New Engl. J. Med.*, *375*, 13, 1216, 2016.
2. De Fauw, J., Ledsam, J.R., Romera-Paredes, B., Nikolov, S., Tomasev, N., Blackwell, S., Ronneberger, O., Clinically applicable deep learning for diagnosis and referral in retinal disease. *Nat. Med.*, *24*, 9, 1342–1350, 2018.
3. Barr, J., *Escaping from fundamentalism*, Wipf and Stock Publishers, 2018. https://wipfandstock.com/9781532663727/escaping-from-fundamentalism/
4. Zhavoronkov, A., Ivanenkov, Y.A., Aliper, A., Veselov, M.S., Aladinskiy, V.A., Aladinskaya, A.V., Aspuru-Guzik, A., Deep learning enables rapid identification of potent DDR1 kinase inhibitors. *Nat. Biotechnol.*, *37*, 9, 1038–1040, 2019.
5. Song, J., Price, D.J., Guvenen, F., Bloom, N., Von Wachter, T., Firming up inequality. *Q. J. Econ.*, 134, 1, 1–50, 2019.
6. Sperry, M.F., Silva, H.L., Balthazar, C.F., Esmerino, E.A., Verruck, S., Prudencio, E.S., Cruz, A.G., Probiotic Minas Frescal cheese added with L. casei 01: Physicochemical and bioactivity characterization and effects on

hematological/biochemical parameters of hypertensive overweighted women-A randomized double-blind pilot trial. *J. Funct. Foods*, *45*, 435–443, 2018.
7. Basole, R.C. and Accenture, A., II, Visualizing the Evolution of the AI Ecosystem, in: *HICSS*, vol. 1–10, 2021.
8. Bullock, E.L., Woodcock, C.E., Olofsson, P., Monitoring tropical forest degradation using spectral unmixing and Landsat time series analysis. *Remote Sens. Environ.*, 238, 110968, 2020.
9. Rajpurkar, P., Irvin, J., Zhu, K., Yang, B., Mehta, H., Duan, T., Ng, A.Y., *Chexnet: Radiologist-level pneumonia detection on chest x-rays with deep learning*, 2018, arXiv preprint arXiv:1711.05225. https://arxiv.org/abs/1711.05225. https://doi.org/10.48550/arXiv.1711.05225
10. Satava, R.M., Advanced technology for battlefield medicine, in: *Health in the New Communications Age*, IOS Press, pp. 38–40, 1995.
11. Hasanuzzaman, M., Bhuyan, M.B., Anee, T., II, Parvin, K., Nahar, K., Mahmud, J.A., Fujita, M., Regulation of ascorbate-glutathione pathway in mitigating oxidative damage in plants under abiotic stress. *Antioxidants*, *8*, 9, 384, 2019.
12. Esteva, A., Kuprel, B., Novoa, R.A., Ko, J., Swetter, S.M., Blau, H.M., Thrun, S., Dermatologist-level classification of skin cancer with deep neural networks. *Nature*, *542*, 7639, 115–118, 2017.
13. Noah, D. and Fidas, G., The global infectious disease threat and its implications for the United States, in: *National Intelligence Council*, Washington DC, 2000.
14. Krittanawong, C., The rise of artificial intelligence and the uncertain future for physicians. *Eur. J. Internal Med.*, *48*, e13–e14, 2018.
15. Atkins, M.B., Curiel-Lewandrowski, C., Fisher, D.E., Swetter, S.M., Tsao, H., Aguirre-Ghiso, J.A., Merlino, G., The state of melanoma: emergent challenges and opportunities. *Clin. Cancer Res.*, *27*, 10, 2678–2697, 2021.
16. Topol, E.J., High-performance medicine: the convergence of human and artificial intelligence. *Nat. Med.*, *25*, 1, 44–56, 2019.
17. Haynes, K.W., The importance of professional values from radiologic technologists' perspective. *Radiologic Technol.*, *91*, 6, 525–532, 2020.
18. Shorvon, S. and Luxon, L., The Royal College of Physicians at 500 years: changing roles and challenges. *Lancet*, *392*, 10152, 1004–1007, 2018.
19. Arnold, D. and Wilson, What doctor? Why AI and robotics will define new health, in: *PwC*, 2017.
20. Braunack-Mayer, A., Teaching ethics with 'cholera and nothing more'. *Public Health Ethics*, *3*, 1, 78–79, 2010.
21. Williams, E., Rodrigues, P., Novak, V., *Accenture at CheckThat! 2020: If you say so: Post-hoc fact-checking of claims using transformer-based models*, 2020, arXiv preprint arXiv:2009.02431.
22. Asmundson, G.J.G. and Taylor, S., Coronaphobia: Fear and the 2019-nCoV outbreak. *J. Anxiety Disord.*, 70, 102196, 2020 Mar. doi: 10.1016/j.janxdis.2020.102196.

23. Cassar, K., Morrison, N., Gibson, K., McEnroe, S., Wittens, C., Davies, A., Choke, E., Efficacy of cyanoacrylate glue ablation of primary truncal varicose veins compared to existing endovenous techniques: a systematic review of the literature. *Surg. J.*, *6*, 02, e77–e86, 2020.
24. Panch, T., Mattie, H., Atun, R., Artificial intelligence and algorithmic bias: implications for health systems. *J. Global Health*, *9*, 2, 543–603, 2019. https://www.ncbi.nlm.nih.gov/pmc/articles/PMC6875681/. https://doi.org/10.7189%2Fjogh.09.020318
25. Gerotziafas, G., Zhang, Z., Wan, J., Liu, P., Huang, C., Wang, Y., Bikdeli, B., Prevention and treatment of venous thromboembolism associated with coronavirus disease 2019 infection: a consensus statement before guidelines. *Thrombosis Haemostasis*, *120*, 06, 937–948, 2020.
26. Majeed, F., Asif, M.W., Hassan, M.A., Abbas, S.A., Lali, M., II, Social media news classification in healthcare communication. *J. Med. Imaging Health Inf.*, *9*, 6, 1215–1223, 2019.
27. Royal Society of Medicine (Great Britain), *Journal of the Royal Society of Medicine*, vol. 80, Academic Press, 1987. https://journals.sagepub.com/toc/jrs/80/1
28. Greenwald, G. and MacAskill, E., NSA Prism program taps into user data of Apple, Google and others. *Guardian*, *7*, 6, 1–43, 2013.
29. Nemitz, P., Constitutional democracy and technology in the age of artificial intelligence. *Philos. Trans. R. Soc. A Mathematical Phys. Eng. Sci.*, *376*, 2133, 20180089, 2018.
30. Chui, M., Manyika, J., Miremadi, M., *Where machines could replace humans-and where they can't (yet)*, 2016.
31. Akter, S. and Wamba, S.F., Big data and disaster management: a systematic review and agenda for future research. *Ann. Oper. Res.*, *283*, 939–959, 2017.
32. Scott, L.J., Redmond, N.M., Garrett, J., Whiting, P., Northstone, K., Pullyblank, A., Distributions of the National Early Warning Score (NEWS) across a healthcare system following a large-scale roll-out. *Emerge. Med. J.*, 36, 5, 287–292, 2019.
33. Jordan, M.I., Artificial intelligence—the revolution hasn't happened yet. *Harvard Data Sci. Rev.*, *1*, 1, 1–9, 2019. https://doi.org/10.1162/99608f92.f06c6e61
34. Bostrom, N., *Superintelligence*, Oxford University Press, UK, 2014.
35. Jungmann, S.M. and Witthöft, M., Health anxiety, cyberchondria, and coping in the current COVID-19 pandemic: Which factors are related to coronavirus anxiety? *J. Anxiety Disord.*, *73*, 102239, 2020.
36. Floridi, L., Cowls, J., Beltrametti, M., Chatila, R., Chazerand, P., Dignum, V., Vayena, E., An ethical framework for a good AI society: Opportunities, risks, principles, and recommendations. *Minds Mach (Dordr)*, 28, 4, 689–707, 2018, doi: 10.1007/s11023-018-9482-5.
37. Arya, V., Bellamy, R.K., Chen, P.Y., Dhurandhar, A., Hind, M., Hoffman, S.C., Zhang, Y., *One explanation does not fit all: A toolkit and taxonomy of ai explainability techniques*, 2019, arXiv preprint arXiv:1909.03012.

38. Goodman, K.W., Meslin, E.M., Ethics, Information Technology, and Public Health: Duties and Challenges in Computational Epidemiology, in: *Public Health Informatics and Information Systems. Health Informatics*. J. Magnuson, P. Fu, Jr. (eds), Springer, London, 2014.
39. Topol, E., *Deep medicine: how artificial intelligence can make healthcare human again*, Hachette, UK, 2019.
40. Haugsten, E.R., Vestergaard, T., Trettin, B., Experiences Regarding Use and Implementation of Artificial Intelligence–Supported Follow-Up of Atypical Moles at a Dermatological Outpatient Clinic: Qualitative Study. *JMIR Dermatol.*, *6*, e44913, 2023.
41. Kong, Y.C., Bhoo-Pathy, N., O'Rorke, M., Subramaniam, S., Bhoo-Pathy, N.T., See, M.H., Yip, C.H., The association between methods of biopsy and survival following breast cancer: A hospital registry based cohort study. *Medicine*, *99*, 6, 2020.
42. Brynjolfsson, E. and McAfee, A., *The second machine age: Work, progress, and prosperity in a time of brilliant technologies*, WW Norton & Company, 2014. W.W. Norton & Company, New York, https://en.wikipedia.org/wiki/The_Second_Machine_Age
43. Challen, R., Denny, J., Pitt, M., Gompels, L., Edwards, T., Tsaneva-Atanasova, K., Artificial intelligence, bias and clinical safety, in: *BMJ quality & safety*, 2019.
44. Price, W.N., Gerke, S., Cohen, I.G., How much can potential jurors tell us about liability for medical artificial intelligence? *J. Nucl. Med.*, *62*, 1, 15–16, 2021.
45. Hafner, K. and Markoff, J., *Cyberpunk: outlaws and hackers on the computer frontier, revised*, Simon and Schuster, 1995. https://katiehafner.com/books-new/cyberpunk/
46. Rao, S.V., McCoy, L.A., Spertus, J.A., Krone, R.J., Singh, M., Fitzgerald, S., Peterson, E.D., An updated bleeding model to predict the risk of post-procedure bleeding among patients undergoing percutaneous coronary intervention: a report using an expanded bleeding definition from the National Cardiovascular Data Registry CathPCI Registry. *JACC Cardiovasc. Interventions*, *6*, 9, 897–904, 2013.
47. Castelvecchi, D., Can we open the black box of AI? *Nat. News*, *538*, 7623, 20, 2016.
48. Mahdawi, A., What jobs will still be around in 20 years? Read this to prepare your future. *Guardian*, *26*, 2017.
49. Ross, C., Orr, E.S., Sisic, M., Arseneault, J.M., Simmering, M.G., Orr, R.R., Personality and motivations associated with Facebook use. *Comput. Hum. Behav.*, *25*, 2, 578–586, 2009.
50. Rodriguez, C. and Torres, J., Central coordination and profitability in large Latin American business groups. *J. Bus. Res.*, *119*, 599–609, 2020.
51. Adadi, A. and Berrada, Peeking inside the black-box: a survey on explainable artificial intelligence (XAI). *IEEE Access*, *6*, 52138–52160, 2018.
52. Ipsos, M.O.R., II, Public views of machine learning, in: *Findings From Public Research Engagement Conducted on Behalf of the Royal Society*, Ipsos MORI, The Royal Society, London, 2017.

53. Bello-Orgaz, G., Jung, J.J., Camacho, D., Social big data: Recent achievements and new challenges. *Inf. Fusion*, *28*, 45–59, 2016.
54. Kostkova, P., Brewer, H., De Lusignan, S., Fottrell, E., Goldacre, B., Hart, G., Tooke, J., Who owns the data? Open data for healthcare. *Front. Public Health*, *4*, 7, 2016.
55. Powles, J. and Hodson, H., Google DeepMind and healthcare in an age of algorithms. *Health Technol.*, *7*, 4, 351–367, 2017.
56. Makin, S., The emerging world of digital therapeutics. *Nature*, *573*, 7775, S106–S106, 2019.
57. Jeffery, M., Fisher, D., Granot, M., Kadyan, A., Pho, A., Vasquez, C., Strategic IT transformation at Accenture. *Kellogg Sch. Manage. Cases*, 1–18, 2017. https://www.emerald.com/insight/content/doi/10.1108/case.kellogg.2016.000321/full/html
58. Luxton, D.D., Recommendations for the ethical use and design of artificial intelligent care providers. *Artif. Intell. Med.*, *62*, 1, 1–10, 2014.
59. Petkovic, K., Swallow, A., Stewart, R., Gao, Y., Li, S., Glenn, F., Zhu, Y., An integrated portable multiplex microchip device for fingerprinting chemical warfare agents. *Micromachines*, *10*, 9, 617, 2019.
60. Kumar, S., Smith, S.R., Fowler, G., Velis, C., Kumar, S.J., Arya, S., Cheeseman, C., Challenges and opportunities associated with waste management in India. *R. Soc. Open Sci.*, *4*, 3, 160764, 2017.
61. Dietvorst, B.J., Simmons, J.P., Massey, C., Algorithm aversion: people erroneously avoid algorithms after seeing them err. *J. Exp. Psychology Gen.*, *144*, 1, 114, 2015.
62. Botrugno, C., Information technologies in healthcare: Enhancing or dehumanising doctor–patient interaction? *Health*, *25*, 4, 475–493, 2021.
63. Bickmore, T.W., Trinh, H., Olafsson, S., O'Leary, T.K., Asadi, R., Rickles, N.M., Cruz, R., Patient and consumer safety risks when using conversational assistants for medical information: an observational study of Siri, Alexa, and Google Assistant. *J. Med. Internet Res.*, *20*, 9, e11510, 2018.
64. Ngai, E.W., Xiu, L., Chau, D.C., Application of data mining techniques in customer relationship management: A literature review and classification. *Expert Syst. Appl.*, *36*, 2, 2592–2602, 2009.
65. Brinker, S. and Baldwin, J., Martech 2030, in: *Five trends in marketing technology for the decade of the augmented marketer. Marketing Technology Media, LLC*, 2020.
66. Van Lier, B., AI and Ethics at the Interface of Humans and Systems. *Cutter Bus. Technol. J.*, 32, 4, 34–40, 2019.
67. Ransbotham, S., Kiron, D., Gerbert, P., Reeves, M., Reshaping business with artificial intelligence: Closing the gap between ambition and action. *MIT Sloan Manage. Rev.*, *59*, 1, 2017, 2017.
68. Joshi, T., Joshi, T., Sharma, P., Mathpal, S., Pundir, H., Bhatt, V., Chandra, S., *In silico* screening of natural compounds against COVID-19 by targeting Mpro and ACE2 using molecular docking. *Eur. Rev. Med. Pharmacol. Sci.*,

2020 Apr;24, 8, 4529–4536. doi: 10.26355/eurrev_202004_21036. PMID: 32373991.
69. Li, X., Xia, H., Lu, L., ECA-CBAM: Classification of Diabetic Retinopathy: Classification of diabetic retinopathy by cross-combined attention mechanism, in: *2022 the 6th International Conference on Innovation in Artificial Intelligence (ICIAI)*, pp. 78–82, 2022.
70. Meyer, M.N., Practical tips for ethical data sharing. *Adv. Methods Practices Psychol. Sci.*, *1*, 1, 131–144, 2018.
71. Marr, B., The incredible ways artificial intelligence is now used in mental health, in: *Forbes. [Google Scholar]*, 2019.
72. Gupta, L. and Jain, S., Artificial Intelligence is Not a Foe But a Friend in the Healthcare Sector, in: *International Advanced Computing Conference*, Springer International Publishing, Cham, pp. 112–122, 2021, December 2021.
73. Möllering, G., The nature of trust: From Georg Simmel to a theory of expectation, interpretation and suspension. *Sociology*, *35*, 2, 403–420, 2001.
74. Hubble, M.A., Introduction: In MA Hubble, in: *The heart & soul of change. What works in therapy*, B.L. Duncan and S.D. Miller (Eds.), American Psychological Association, Washington, DC, 1999.
75. Goudge, J. and Gilson, L., How can trust be investigated? Drawing lessons from past experience. *Soc. Sci. Med.*, *61*, 7, 1439–1451, 2005.
76. Delgado-Bermúdez, A., Llavanera, M., Fernández-Bastit, L., Recuero, S., Mateo-Otero, Y., Bonet, S., Yeste, M., Aquaglyceroporins but not orthodox aquaporins are involved in the cryotolerance of pig spermatozoa. *J. Anim. Sci. Biotechnol.*, *10*, 1, 1–12, 2019.
77. Greenberg, L.S., Elliott, R., Watson, J.C., Bohart, A.C., Empathy. *Psychotherapy Theory Research Practice Training*, *38*, 4, 380, 2001.
78. Derksen, F., Bensing, J., Lagro-Janssen, A., Effectiveness of empathy in general practice: a systematic review. *Br. J. Gen. Pract.*, *63*, 606, e76–e84, 2013.
79. Gordon, R., *The rise and fall of American growth: The US standard of living since the civil war*, Princeton University Press, 2017.
80. BELKACEMI, S., Artificial Intelligence" AI" and its impact on global Economy. *J. Financial Accounting Managerial Stud.*, 9, 2, 2022. https://www.pwc.com/gx/en/news-room/docs/report-pwc-ai-analysis-sizing-the-prize.pdf
81. Rajkomar, A., Dean, J., Kohane, I., Machine learning in medicine. *N. Engl. J. Med.*, *380*, 14, 1347–1358, 2019.
82. Dutta, P., Paul, S., Majumder, M., *An efficient SMOTE based machine learning classification for prediction & detection of PCOS*, 2021.
83. Dutta, P., Paul, S., Kumar, A., Comparative analysis of various supervised machine learning techniques for diagnosis of COVID-19, in: *Electronic devices, circuits, and systems for biomedical applications*, Academic Press, pp. 521–540, 2021.

84. Longoni, C., Bonezzi, A., Morewedge, C.K., Resistance to medical artificial intelligence. *J. Consum. Res.*, *46*, 4, 629–650, 2019. https://doi.org/10.1093/jcr/ucz013
85. Norman, C.D. and Skinner, H.A., eHealth literacy: essential skills for consumer health in a networked world. *J. Med. Internet Res.*, *8*, 2, e506, 2006.
86. Dutta, P., Paul, S., Obaid, A.J., Pal, S., Mukhopadhyay, K., Feature selection based artificial intelligence techniques for the prediction of COVID like diseases, in: *Journal of Physics: Conference Series*, vol. 1963, IOP Publishing, p. 012167, 2021.
87. Hackl, W.O., Hoerbst, A., Section Editors of the IMIA Yearbook Section on Clinical Information Systems, Clinical information systems research in the pandemic year 2020. *Yearb. Med. Inf.*, *30*, 01, 134–140, 2021.
88. Wyckoff, M.H., Greif, R., Morley, P.T., Ng, K.C., Olasveengen, T.M., Singletary, E.M., Berg, K.M., 2022 International consensus on cardiopulmonary resuscitation and emergency cardiovascular care science with treatment recommendations: Summary from the basic life support; advanced life support; pediatric life support; neonatal life support; education, implementation, and teams; and first aid task forces. *Pediatrics*, *151*, 2, e2022060463, 2023.
89. Neudert, L.M., Knuutila, A., Howard, P.N., *Global attitudes towards AI, machine learning & automated decision making*, 2020. Working paper 2020.10 Oxford, UK: Oxford Commission on AI & Good Governance, 10pp. https://oxcaigg.oii.ox.ac.uk/publications/global-attitudes-towards-ai-machine-learning-automated-decision-making-2/#continue.
90. Dutta, P., Paul, S., Cengiz, K., Anand, R., Kumar, A., A predictive method for emotional sentiment analysis by deep learning from EEG of brainwave dataset, in: *Artificial intelligence for neurological disorders*, Academic Press, pp. 25–48, 2023.
91. Ting, D.S.W., Pasquale, L.R., Peng, L., Campbell, J.P., Lee, A.Y., Raman, R., Tan, G.S.W., Schmetterer, L., Keane, P.A., Wong, T.Y., Artificial intelligence and deep learning in ophthalmology. *Br. J. Ophthalmol.*, 103, 2, 167–175, 2019 Feb. doi: 10.1136/bjophthalmol-2018-313173.
92. Dutta, P., Paul, S., Majumder, M., *Intelligent SMOTE Based Machine learning Classification for Fetal State on Cardiotocography Dataset*, 2021.
93. Ganie, S,M,, Dutta Pramanik, P.K., Mallik, S., Zhao, Z., Chronic kidney disease prediction using boosting techniques based on clinical parameters. *PLoS One*, 18, 12, e0295234, 2023 Dec 1. doi: 10.1371/journal.pone.0295234.
94. Chu, C.H., Donato-Woodger, S., Khan, S.S. *et al.*, Age-related bias and artificial intelligence: a scoping review. *Humanit. Soc. Sci. Commun.*, 10, 510, 2023. https://doi.org/10.1057/s41599-023-01999-y
95. https://www.pewresearch.org/internet/2019/06/13/artificial-intelligence-and-the-future-of-humans/
96. https://onlinelibrary.wiley.com/doi/abs/10.1111/jrh.12278
97. https://academic.oup.com/innovateage/article/1/1/igx025/4622148

98. https://www.oii.ox.ac.uk/blog/the-ethics-of-ai-in-healthcare-a-mapping-review/
99. Mittelstadt, B., A Right to Reasonable Inferences: Re-thinking Data Protection Law in the Age of Big Data and AI, 2019.
100. Char, D.S., *et al.* Implementing Machine Learning in Healthcare - Addressing Ethical Challenges. *New England J. Med.*, 378, 14, 1366–1368, 2018.
101. Adadi, A. and Berrada, M., Peeking Inside the Black-Box: A Survey on Explainable Artificial Intelligence (XAI). *IEEE Access*, 6, 52138–52160, 2018. doi: 10.1109/ACCESS.2018.2870052.
102. Holzinger, A., *et al.*, Explainable AI (XAI) for Healthcare – A Context-Aware Approach. *J. Healthc. Eng.*, 2019, 1–13, 2019. doi: 10.1155/2019/8196923.
103. Malta, A., *et al.*, Artificial Intelligence in Healthcare: A Systematic Review of the Literature. *J. Healthc. Eng.*, 2019, 1–13, 2019. doi: 10.1155/2019/8196922.
104. Al-Haija, Q. A., *et al.*, Secure Electronic Health Record System Based on Public Key Cryptography. *J. Med. Syst.*, 44, 10, 2108, 2020. doi: 10.1007/s10916-020-01772-4.
105. Veletsianos, G. and Miller, C., Trust, credibility, and responsibility in AI for education and healthcare. *British J. Educ. Technol.*, 51, 6, 1335–1349, 2020. doi: 10.1111/bjet.13004.
106. Sanders, E.B.N. and Stappers, P.J., Co-creation and the new landscapes of design. *Co-design*, *4*, 1, 5–18, 2008.
107. Haven, K., *Story proof: The science behind the startling power of story*, Bloomsbury Publishing, USA, 2007.
108. Wakefield, J., II, Bell, S.P., Palmer, B.M., Inorganic phosphate accelerates cardiac myofilament relaxation in response to lengthening. *Front. Physiol.*, *13*, 980662, 2022.
109. Ross, C. and Swetlitz, I., IBM pitched its Watson supercomputer as a revolution in cancer care. It's nowhere close. *Stat*, 2017. https://www.statnews.com/2017/09/05/watson-ibm-cancer/
110. Darcy, A., Beaudette, A., Chiauzzi, E., Daniels, J., Goodwin, K., Mariano, T.Y., Robinson, A., Anatomy of a Woebot®(WB001): agent guided CBT for women with postpartum depression. *Expert Rev. Med. Devices*, *19*, 4, 287–301, 2022.
111. Parush, A., Mastoras, G., Bhandari, A., Momtahan, K., Day, K., Weitzman, B., Sohmer, B., Cwinn, A., Hamstra, S.J.,Calder, L., Can teamwork and situational awareness (SA) in ED resuscitations be improved with a technological cognitive aid? Design and a pilot study of a team situation display. *J. Biomed. Inform.*, 76, 154–161, 2017 Dec. doi: 10.1016/j.jbi.2017.10.009.
112. Gochoo, M., Alnajjar, F., Tan, T.H., Khalid, S., Towards privacy-preserved aging in place: a systematic review. *Sensors*, *21*, 9, 3082, 2021.
113. Tom, E., Keane, P.A., Blazes, M., Pasquale, L.R., Chiang, M.F., Lee, A.Y., Force, A.A., II, Protecting data privacy in the age of AI-enabled ophthalmology. *Transl. Vision Sci. Technol.*, *9*, 2, 36–36, 2020.

114. Zou, J. and Schiebinger, L., *AI can be sexist and racist—it's time to make it fair*, 2018. https://www.nature.com/articles/d41586-018-05707-8#:~:text=Download%20PDFWhen%20Google. Nature 559, 324-326 (2018). doi: https://doi.org/10.1038/d41586-018-05707-8
115. Dignum, V., *Responsible artificial intelligence: designing AI for human values*, 2017.
116. Jobin, A., Ienca, M., Vayena, E., The global landscape of AI ethics guidelines. *Nat. Mach. Intell.*, 1, 389–399, 2019. https://doi.org/10.1038/s42256-019-0088-2
117. Aprilicia, G., Bantas, K., Syarif, S., Kalista, K.F., Comparative Evaluation of Alpha-Fetoprotein Serum in Hepatocellular Carcinoma Patients with Non-Viral Etiology. *Jurnal Penyakit Dalam Indonesia*, 8, 1, Article 7, 2021.
118. Dutta, P., Paul, S., Shaw, N., Sen, S., Majumder, M., Heart disease prediction: A comparative study based on a machine-learning approach, in: *Artificial Intelligence and Cybersecurity*, CRC Press, pp. 1–18, 2022.
119. Ala-Pietilä, P., Bonnet, Y., Bergmann, U., Bielikova, M., Bonefeld-Dahl, C., Bauer, W., Van Wynsberghe, A., *The assessment list for trustworthy artificial intelligence (ALTAI)*, European Commission, 2020. https://futurium.ec.europa.eu/en/european-ai-alliance/pages/altai-assessment-list-trustworthy-artificial-intelligence
120. Camenisch, J., Groß, T., Heydt-Benjamin, T.S., Accountable privacy supporting services. *Identity Inf. Soc., 2*, 241–267, 2009.
121. Dutta, P., Pal, S., Kumar, A., Cengiz, K., *Artificial Intelligence for Cognitive Modeling: Theory and Practice*, CRC Press, 123, 2023, https://doi.org/10.1201/9781003216001
122. Condelius, A., Edberg, A.K., Jakobsson, U., Hallberg, I.R., Hospital admissions among people 65+ related to multimorbidity, municipal and outpatient care. *Arch. Gerontol. Geriatr., 46*, 1, 41–55, 2008.
123. Berwick, D.M., Nolan, T.W., Whittington, J., The triple aim: care, health, and cost. *Health Affairs, 27*, 3, 759–769, 2008.
124. Dutta, P., Paul, S., Cengiz, K., Anand, R., Majumder, M., A predictive method for emotional sentiment analysis by machine learning from electroencephalography of brainwave data, in: *Implementation of Smart Healthcare Systems using AI, IoT, and Blockchain*, pp. 109–130, Academic Press, 2023.
125. Dutta, P., Paul, S., Jana, G.G., Sadhu, A., Hybrid Genetic Algorithm Random Forest algorithm (HGARF) for improving the missing value Imputation in Hepatitis Medical Dataset, in: *2023 International Symposium on Devices, Circuits and Systems (ISDCS)*, vol. 1, pp. 01–05, IEEE, 2023.
126. Greenhalgh, T., Russell, J., Swinglehurst, D., Narrative methods in quality improvement research. *Qual. Saf. Health Care, 14*, 6, 443, 2005.

2

A Study of AI Application Through Integrated and Systematic Moral Cognitive Therapy in the Healthcare Sector

Anusriya Mukherjee[1]*, Tanupriya Mukherjee[2] and Mili Mitra Roy[1]

[1]*Department of Basic Science & Humanities, Budge Budge Institute of Technology, Kolkata, India*

[2]*Department of MBA, Budge Budge Institute of Technology, Kolkata, India*

Abstract

In recent times, the adoption of artificial intelligence (AI) in the Healthcare industry has become the subject of extensive research and debate. Hand in hand with emerging technology, Healthcare systems are becoming progressively complex while constantly adapting to socioeconomic, epidemiological, and demographic changes, contributing to intricate global healthcare objectives. Our study explores the use of Integrated, and Systematic Moral Cognitive Therapy (ISMCT) as a means of administering AI in the healthcare sector. The integration of AI in healthcare has the potential to improve prognosis with the help of an enormous database, but concerns have been raised regarding the vulnerability of a large section of end-users and the ethical implications of its use. ISMCT aims to address these concerns by providing a framework for the ethical development and deployment of AI in healthcare. The four major ethical issues that need to be addressed before AI can reach its full potential in healthcare: transparent, authorization, for utilization, reliability and full disclosure, computational fairness and confidentiality of information. The study investigates the efficacy of ISMCT in promoting ethical awareness and decision-making among healthcare professionals who work with AI technology. It also examines the impact of ISMCT on patient outcomes and satisfaction and provides valuable insights into the development of ethical frameworks for the use of AI in healthcare, which can contribute to the responsible integration of AI technology in healthcare systems.

**Corresponding author*: anusriyam@gmail.com

Pijush Dutta, Sudip Mandal, Korhan Cengiz, Arindam Sadhu and Gour Gopal Jana (eds.) Design and Forecasting Models for Disease Management (47–76) © 2025 Scrivener Publishing LLC

Keywords: Integrated and systematic moral cognitive therapy (ISMCT), AI, integration, healthcare, emerging technology, healthcare professionals

2.1 Introduction

Artificial Intelligence, commonly known as AI, can be described as the simulation of human intelligence processes by machines or, more specifically, by computer systems (Russell & Norvig, 2021) [1] which has already transformed various sectors, with healthcare being at the forefront of these significant shifts. AI has consistently advanced over the years, its strides undeniably remarkable, and today, it stands as a potential revolutionizing harbinger in the healthcare sector, as validated by its rising trend (Bresnick, 2019) [2]. As researchers harness the power of AI, its remit continues to evolve. In recent years, AI's role within the healthcare sector has become multifaceted, ranging from predicting patient outcomes to personalizing patient care, enhancing diagnostic precision, improving treatment protocols, aiding in repetitive tasks, and expanding telehealth services (Jiang *et al.*, 2017) [3]. The artificial intelligence (AI) healthcare market, estimated at $11 billion in 2021, is anticipated to become $187 billion economy by 2030, as per Statista. Because of this enormous growth, the way hospitals, pharmaceutical and biotechnology firms, medical providers, and other players in the healthcare sector operate is expected to undergo significant changes. However, it presents a multitude of ethical dilemmas and moral issues which questions the integrity of automation in healthcare. Little wonder then that the potential for AI to revolutionize healthcare has provoked considerable debate. There are so much ethical and legal concerns around patient data privacy and the reliance on AI technologies, that it is crucial that we tread with caution and ensure a balanced approach to AI's integration into the healthcare field. Traditional healthcare advocates for the values of empathy, dignity, privacy and compassionate care and the technological advancements requires strategic guidance by these ethical constructs as well (Mittelstadt *et al.*, 2016) [4]. And the solution is perhaps Cognitive Therapy by itself, a form of psychological treatment that helps both healthcare providers and patients to understand the feelings and thoughts that influence their behavior (Beck, 2011) [5]. It is a goal-oriented approach, which is now getting intertwined with AI to develop ethically informed healthcare algorithms that deliver patient care while upholding moral and ethical standards. The incorporation of moral cognition into AI applications to enhance the healthcare sector is a subject that straddles the line between technological advancement and ethical integrity (Edmonds, 2020) [6]. Thus it is imperative to explore the ethical implications of AI and ensure that its implementation aligns with moral principles. The future of healthcare lies in striking a balance between the technological advancements provided by AI and the humanistic values of moral cognitive therapy.

2.1.1 Understanding the Role of AI in Healthcare

AI's role in healthcare is well evident and cannot be understated. It offers tremendous potential for improving the efficiency, accuracy, and personalization of healthcare services. One significant contribution of AI in healthcare is its use in predicting patient outcomes. Identifying potential health risks and predicting outcomes are critical for effective patient management. Support Vector Machine (SVM), a type of AI technology, is exploited for this purpose (Boser, Guyon & Vapnik, 1992) [7]. Utilizing historical patient data, SVM helps healthcare professionals to identify patterns and correlations, aiding swift decision-making and early warning signs detection. For example, AI technologies like Google's DeepMind was pivotal in predicting the deterioration of patients with acute kidney injury up to 48 hours before it occurred (McKinney *et al.*, 2020) [8], a significant leap towards patient-centered care. Additionally, the enhancement of diagnostic accuracy is another role played by AI in healthcare. Computer-Aided detection (CAD) systems are a type of AI technology utilized to improve diagnostics (Doi, 2007) [9]. These systems analyze medical data such as MRI images, detecting abnormalities which may be missed by the human eye. For instance, studies show that the use of CAD systems significantly improved the detection and diagnosis of breast cancer (Giger, Karssemeijer & Schnabel, 2013) [10], indicating the potential for AI to reduce diagnostic errors and consequently, improve patient outcomes. Again, data-driven treatment plans and healthcare wearable devices are two of the ways AI has reinvented the patient experience. Machine learning algorithms can be used to customize treatment plans based on a patient's medical history. On the other hand, wearable technology provides real-time health monitoring, promoting a proactive approach to health management. Furthermore, the repetitive jobs that staff often handle in hospitals, such as processing paperwork or x-ray readings, can be automated with AI. Automation can free staff from routine tasks, allowing them to concentrate on patient care. A prime example of such an AI application is Moxi, a hospital robot that runs errands and assists nurses, providing an avenue for enhanced patient care. The expansion of telehealth services is another crucial role of AI in healthcare which uses AI chatbots and remote patient monitoring systems to provide health services without the necessity of having patients physically present in the healthcare facility. While we lean into the future where AI takes on a more prominent role within healthcare, the sector needs to balance the infinite benefits AI promises with important safeguards to protect patients and their data and ensure that the primary goal of healthcare—the best patient outcome, is always met. It is essential to note that AI cannot replace human decision-making in healthcare, the benefits it brings to improving patient care and outcomes are undeniable.

2.1.2 Advantages of AI in Healthcare

There is a symbiotic relationship between AI development and the advancements in machine learning algorithms. Increased data access allows for AI to learn and grow at a consistent rate, nuanced through the continuous input of data and experiences (Marr, 2018) [11]. Improved machine learning (ML) algorithms, more data access, more affordable technology, and the availability of 5G have all helped to increase the use of AI in the healthcare sector and quicken the speed of development. Furthermore, the advent of 5G technology has dramatically pushed the boundaries for AI application, enabling real-time, reliable, high-speed communication and data transfer, which is crucial for the proliferation of AI in healthcare (Higuera, 2020) [12]. Health records, clinical trials, and genetic data may all be sorted through and analyzed by AI and ML technologies considerably more quickly than by humans. Here are some of the key advantages of AI in healthcare:

Early Disease Detection: AI is capable of analyzing massive patient data sets, including photos and medical records, to look for patterns and anomalies that could point to the presence of diseases or disorders earlier than would be possible using conventional techniques. Early detection has the potential to result in more efficient treatments and better patient outcomes.

Accurate Diagnosis: Accurate diagnosis can be achieved with the help of AI-powered diagnostic tools, such as picture recognition algorithms. The possibility of human mistake has been decreased, for instance, by the development of AI algorithms to identify abnormalities in medical pictures such as X-rays, MRIs, and CT scans.

Personalized Treatment Plans: Using patient data analysis and an individual's medical history, genetics, and lifestyle, AI may develop personalized treatment plans. This therapy personalization may result in interventions that are more successful.

Medicinal Development: Artificial intelligence (AI) can analyze huge datasets to find possible medication candidates and forecast their efficacy and safety. This could hasten the process of developing new drugs and result in the identification of fresh medical remedies.

Efficiency in Administration: Beyond patient care, AI has also markedly enhanced the operational efficiency of healthcare institutions. The provision of better and more affordable technology has enabled healthcare facilities such as scheduling appointments, billing, and claims processing and reducing waiting time thereby cutting costs, improving patient flow. By utilizing AI in administrative tasks, healthcare systems can allocate more resources and time to patient care and other critical services (Khatri, Brown, & Hicks, 2019) [13].

Remote Monitoring: AI, combined with ML, has significantly improved patient care, transforming diagnosis, treatment, and monitoring processes. Through AI, vast amounts of health records, clinical trials, and genetic data can be quickly analyzed, contributing to precise diagnoses and personalized treatments. For instance, AI diagnostic tools have remarkably increased the speed and accuracy of medical image interpretations, in turn improving the success rate of treatment plans (Jensen, Fernandez, Sebag, & Unsupervised, 2020) [14].

Predictive Analytics: Analytics that anticipate disease outbreaks, patient admissions, and resource needs are possible using AI. This makes it possible for healthcare practitioners to better plan healthcare overall and allocate resources.

Natural Language Processing (NLP): From unstructured clinical notes and medical literature, NLP algorithms can glean important insights. By doing this, healthcare practitioners may make better decisions and keep current on the most recent research.

Robotic surgery: AI-driven robotic equipment can help surgeons carry out intricate surgeries with more control and precision. Shorter recovery times and fewer surgical problems may result from this.

Patient Engagement: Chatbots and virtual assistants powered by AI can engage with patients, respond to questions, and deliver health-related information. This improves patient education and involvement.

Reduced Costs: AI can aid in lowering the overall cost of healthcare delivery by automating jobs, enhancing diagnostic accuracy, and optimizing resource allocation.

Data Security: By identifying and blocking unauthorized access and data breaches, AI can improve data security. Additionally, it can help you comply with laws governing the protection of healthcare data, such as Health Insurance Portability and Accountability Act, 1996 in the US.

2.1.3 Moral Dilemmas and AI-Based Healthcare

There are significant restrictions and moral dilemmas associated with the application of artificial intelligence (AI) in different fields, including healthcare, which need to be properly addressed, so as to reach its full potential while maintaining patient privacy and safety. The following are some of the major restrictions and moral issues that AI implementation faces:

Limited Understanding and Interpretation: First and foremost, AI systems lack the capacity to grasp concepts in the same way human beings do (Bostrom, N. 2017) [15]. They're confined within the parameters defined by programmers, limiting their potential interpretation, understanding, and adaptability to unpredictable scenarios (Winfield, A. F., & Jirotka, M. 2018) [16]. Hence, the lack of true cognitive aptitude leads to various ethical challenges including possible risks and misuse (Russel, S. 2019) [17]. Many AI algorithms, in particular deep learning models, function as "black boxes," which means that humans have a difficult time understanding how they make decisions. Understanding how AI systems get their results can be difficult due to this lack of transparency, particularly in crucial applications like healthcare diagnosis.

Data Fairness and Bias: Artificial intelligence (AI) systems learn from previous data, which might be prejudiced and reflect current cultural preconceptions. The bias present in AI algorithms emerges from the data used to train them (Baeza-Yates, R. 2018) [18]. Even unintentional, slight inclinations can exacerbate existing societal inequalities, leading to biased decisions against certain groups (O'Neil, C. 2016) [19]. If this prejudice is not effectively addressed, it may lead to unfair consequences like incorrect diagnosis or unequal access to medical care.

Fairness: It is a critical challenge to ensure that AI systems are equitable and do not disproportionately affect specific demographic groups. Making judgment about what is "fair" can be subjective and cultural.

Privacy Issues: Using AI frequently necessitates access to sizable databases containing private patient data. Maintaining patient trust and adhering to laws like HIPAA depend on the privacy and security of this data (Taddeo, M., & Floridi, L. 2018) [20].

Concerns regarding surveillance: There are worries that AI might be used to watch people without permission and cause, potentially infringing their right to privacy, in situations like healthcare surveillance.

Responsibility and Accountability: Legal and Moral Obligations It might be difficult to identify who is at fault when an AI system commits a detrimental judgment or error. Is it the AI itself, the developer, or the operator? It is a constant struggle to create moral and legal foundations for responsibility.

Job Displacement and Economic Impact: In some industries, task automation by AI may result in job displacement, which may have an impact on livelihoods and labor markets. An ethical concern is preparing the workforce for these shifts.

Safety and Reliability: Protecting patients from harm requires ensuring the safety and dependability of AI systems, especially in crucial applications like autonomous medical equipment or surgical robots.

Informed Consent: Getting patients' consent to the use of artificial intelligence in their healthcare can be difficult, especially if they may not completely comprehend the technology or its ramifications.

Regulatory and Legal Challenges: It is a complicated and continuing process to develop and enforce regulations for AI in healthcare as well as to modify current legislation to the quickly growing area.

Ethical AI Development: Adhering to ethical principles during the design, development, and deployment phases is necessary to ensure that AI systems are created and used in an ethical manner.

Transparency and Explainability: Establishing trust in AI systems requires both the transparency of AI decision-making and the ability to explain AI-driven decisions to patients and healthcare professionals. While AI can present a wealth of benefits, there arises serious ethical and moral concerns which need to be confronted ethically and responsibly. Much progress must occur in the ethical, societal, and legal understanding of AI, establishing clear guidelines and rules to govern these powerful technologies and their potential impacts. Ignoring this responsibility may result in unforeseen consequences that could damage the fabric of our societies. Governments, healthcare organizations, AI developers, ethicists, and the general public must all work together to address these constraints and ethical issues. Clear rules and regulations should be developed to encourage the ethical and responsible use of AI in healthcare while maximizing its advantages and lowering its hazards. To further strengthen the ethical foundations of the technology, continued research and development in AI ethics and responsible AI practices are essential.

2.2 What is Integrated and Systematic Moral Cognitive Therapy (ISMCT)?

Integrated and Systematic Moral Cognitive Therapy (ISMCT) is an evolving therapeutic paradigm, primarily focused on modifying malfunctioning moral cognitions to improve individuals' moral decision-making abilities and ameliorate behavior. This psychological approach integrates cognitive-behavioral therapy techniques, with a profound focus on moral cognition, which underscores our decision-making process and shapes our behavioral response towards social situations (Jerome, Liss & Liss, 2021) [21]. ISMCT is premised on the idea that moral standards substantially drive an individual's behavior; hence any deviation or dysfunction in the moral cognitive process can result in abnormal or undesirable actions. Murphy, Lilienfeld, Skeem, and Edens (2016) [22] further elaborated that cognitive distortions or dysfunctional thoughts often contribute to criminal behavior, substance abuse, and other behavioral problems. ISMCT, therefore, aims to correct these distortions by integrating various therapeutic techniques. Particularly, the systematic aspect of this form of therapy refers to the step-by-step implementation of cognitive-behavioral techniques in conjunction with moral cognitive interventions. This systematic approach allows for greater efficacy in the intervention, primarily when used over an extended period (Miltenberger *et al.*, 2017) [23]. The intervention plan usually encompasses cognitive restructuring, problem-solving, empathy training, and role-playing exercises that aim at modifying the malfunctional moral thoughts, beliefs or values that trigger maladaptive behaviors (Rotter *et al.*, 2019) [21]. The specific composition of ISMCT interventions can vary based on the individual needs of the patient. However, the commonly used techniques included are: direct cognitive interventions to identify and rectify dysfunctional beliefs or thoughts, behavioral strategies to develop desired behaviors, and emotional techniques to manage emotional distress or instill positive emotions (Kendall *et al.*, 2011) [24]. Research support for the effectiveness of ISMCT is substantial. Several empirical studies demonstrate promising results, indicating that by using ISMCT, people have improved behavioral response due to enhanced moral cognitive functions (Schell *et al.*, 2018) [25]. Thus, ISMCT might be a viable therapeutic tool for addressing a variety of maladaptive behaviors driven by dysfunction in moral cognition.

2.2.1 Integrating Moral Cognitive Therapy with AI

Moral Cognitive Therapy is a groundbreaking, integrative therapeutic approach that merges Cognitive Behavioural Therapy (CBT) with moral and ethical philosophy. Put simply, Moral Cognitive Therapy (MCT) is an alignment of our actions, decisions, and thoughts with our moral standards and Guidelines. Looking at behavioral theories, we can trace the genesis of MCT back to Albert Bandura's work [26] in the mid-20th century. Bandura's theory of social learning suggests humans are complex beings capable of profound self-regulation in making choices that are both rational and moral. This view proposed, rather distinctly, that cognitive processes are significantly influential components of learning, which, when integrated with Bandura's moral disengagement theory, forms the basis for MCT. MCT fundamentally assumes that our cognitive processes, perceptions, and interpretations of our ecosystem significantly shape our experiences and actions. Accordingly, the essence of MCT lies in its remedial approach focused on unearthing and addressing cognitive distortions and dysfunctional thoughts that may influence one's moral and ethical decision-making. Essentially, MCT aims at correcting 'faulty thinking,' allowing individuals to align actions and decisions with their moral values. Various therapeutic methods come under the umbrella term 'MCT,' and these methods can be clubbed into two distinct categories. The first, moral reasoning, revolves around the perception of moral issues, judgment-making, and effective problem resolution. The second, moral behavior, involves aiding the client in applying these principles to real-world scenarios, therefore fostering their responsibility and empathy. The application of MCT takes place in a variety of therapeutic settings, from individual therapy to group facilitation. Techniques utilized are varied, often involving the deconstruction of cognitive distortions, the framework of SMART goals, and interactive exercises that promote critical thinking and problem-solving skills. A fundamental characteristic of MCT, significantly distinguishing it from traditional forms of cognitive therapy, is the acknowledgement and addressing of the moral and ethical structures that underpin each person's behavioral traits. A significant challenge regarding the efficacy of MCT is the formulation of a standard therapeutic model applicable across diverse client demographics, given each person's unique ethical and moral fabrics. Nevertheless, early studies into MCT's efficacy suggest that it's beneficial in treating several psychological disorders, including obsessive-compulsive disorder (OCD), anxiety, and depression. MCT has proven instrumental in accelerating progress in therapeutic treatment plans by integrating individually tailored moral structures within cognitive behavioral therapeutic models [27, 28].

2.2.2 Alignment of Moral Cognitive Therapy Principles with AI Applications

The phenomenon of interaction between the domains of artificial intelligence (AI) and psychology has precipitated a remarkable convergence of principles. Specifically, the intersections between Moral Cognitive Therapy (MCT) and AI applications have become more apparent, prompting intense deliberation in academic circles and industry alike. MCT has a well-established foundation in the realm of psychology and psychotherapy; it aims at assisting individuals in enhancing their behavioral responses through the modification of cognition. Conversely, in the technological domain, AI has emerged as an epoch-making tool whose capabilities transcend rudimentary data processing. Its far-reaching applications include healthcare, transportation, security, and also therapeutic services. The linkage between MCT principles and AI applications is therefore intriguing and demands analytical exploration. Garrard (2011), in his seminal work, elucidates the principles of MCT that are fundamentally fixated on the construction and modification of an individual's moral cognition. Garrard posits that moral cognition is an essential function that enables individuals to discern right from wrong, facilitating ethical decision-making [29].

AI applications, on the other hand, are typically geared towards achieving task efficiency and accuracy by independently analyzing and drawing conclusions from extensive data pools. However, this requires carefully constructed guidelines and principles in order to align with human values, making it relevant to moral cognition. Consequently, the potential of AI applications in MCT becomes palpable when we consider the systematic adjustment of moral cognition and the reflective decision-making capacity of an AI system. In fact, AI systems are known to emulate cognitive processes, including complex decision-making. As noted by Russell, Norvig, & Davis (2020), the algorithms utilized in AI can adapt through machine learning to not only enhance performance but also to 'learn' from real-world experiences, much like human cognition [30]. The intertwining of humanistic principles with AI applications is exhibited in the notion of 'Artificial Moral Agents' as discussed by Allen, Wallach, and Smit (2005). They affirm that AI systems must reflect sound ethical decisions based on predefined rules and regulations. Therefore, these machines serve as concrete exemplars of how the principles of moral cognition can be intertwined with AI applications [31]. Moreover, the Ethics Guidelines for Trustworthy AI, published by the European Commission High-Level Expert Group on AI, illuminates important ethical considerations, some of them congruent with MCT. These considerations accentuate the necessity of transparency, human oversight, privacy, diversity, non-discrimination, and societal well-being in the functioning of AI systems [32]. On examining these guidelines, there is an implicit alignment with MCT principles. The stress on transparency and human oversight parallels the inherent reflective nature of moral cognition, wherein an individual gauges his actions critically before implementation. Moreover, equality and societal well-being are intrinsic to moral cognition, where concern for others' rights and welfare is a fundamental tenet. Nevertheless, challenges of objectivity in moral judgment persist. Moral cognition is inherently subjective, influenced by personal beliefs and societal norms, whilst AI, on the other hand, operates on purely logical terms. But can we then initialize the AI systems variables with the nuances of human-like understanding and empathy? The intriguing question is not just about whether AI can emulate the principles of MCT but rather if can it fulfill the intended purpose of MCT: ameliorating human behavior through improving moral cognition. This aspiration might presuppose a future where AI appliances in MCT could lead to augmented human moral cognition, allowing humans to cultivate compassion and ethical discretion, consequently elevating their decision-making abilities.

2.2.3 Benefits of Integrated and Systematic Moral Cognitive Therapy

The field of psychology has witnessed a burgeoning expansion in the exploration of various therapies for addressing mental health issues. Notably, Integrated and Systematic Moral Cognitive Therapy (ISMCT) stands out as a significant departure from traditional cognitive behavioral therapy, especially concerning moral cognitive processes. This essay presents an in-depth analytical discourse on the benefits of ISMCT and its implications for contemporary mental health intervention. Conceptually, ISMCT amalgamates morality into cognitive therapy, focusing on rectifying cognitive distortions that encompass moral judgments and behaviors (Reynolds, 2018) [33]. This integrated approach includes moral principles in patients' cognitive responses while also considering the individual's systemic cognitive frameworks. The integration of moral principles into cognitive therapy facilitates imparting these principles in patients' decision-making process leading to more constructive behavioral responses, which can be directly attributed to the ISMCT practice. This approach is beneficial as it offers an additional dimension to the cognitive reframing and restructuring processes. It accords patients a moral perspective that transcends beyond perfunctory cognitively depressed notions, and allows them to consider the implications their actions may have on others (Reynolds, 2018) [33]. Adherents of ISMCT posit that this therapeutic intervention advances the moral development of an individual, which triggers cognitive change over time. Core beliefs and attitudes according to Bergner, 2017, that are misaligned with the societal moral standards are systematically challenged and realigned through this approach. This realignment encourages an increased commitment to pro-social values, which has been linked to reduced antisocial behavior among the treated patients. Moreover, ISMCT can be instrumental in reducing the recidivism rate among offenders with a history of crime. Moral cognitive therapy can significantly influence offenders' cognitive processes, diminishing their likelihood to reoffend. The therapy aids in inculcating the rudiments of ethical reasoning, thereby equipping them with the cognitive tools required to make morally correct decisions in the face of adversity and temptation.

Legal and Criminological Psychology. ISMCT has also proven beneficial in treating individuals with mental health issues that are tied to moral and ethical troubles. This therapy aids in addressing feelings of guilt, remorse and regret by helping these individuals recognize and acknowledge the moral implications of their actions, and subsequently making amends in line with these revelations Nevertheless, ISMCT is not without its criticisms and controversies, as there remains a challenge in quantifying the intricacies of morality within clinical assessments. However, this can be addressed by employing a sensitive and detailed individualized approach in the administration of the therapy.

2.2.4 Applications of AI-Integrated Moral Cognitive Therapy in Healthcare

AI-Assisted Diagnosis and Treatment Planning: Recent years have seen progressive advancements in AI's ability to interpret medical images and reports, leading to increased precision in patient diagnosis. Machine learning, a subset of AI, enables the analysis of large datasets at a speed and precision that outpaces the capacity of individual clinicians (Thakur,

Sangwan & Kumar, 2019) [34]. For instance, in medical imaging such as radiology, AI algorithms can compare images to a vast database of prior scans, identifying patterns and anomalies with an accuracy equivalent to, or in some cases surpassing, human radiologists (Rajpurkar *et al.*, 2018) [35]. Simultaneously, developments in Natural Language Processing (NLP) allow AI to sift through written reports and patient records, identifying key diagnostic details that may otherwise be overlooked. Research has shown that these AI tools, when integrated within medical systems, can significantly improve patient diagnosis (Esteva *et al.*, 2019) [36].

Smart Treatment Planning and Personalization: Subsequent to diagnosis, AI's role extends to treatment planning and personalization - tasks previously undertaken solely by clinicians. Machine learning algorithms can help develop treatment plans that are drawn from extensive databases of information, including data from previous cases, clinical studies and patient records (Bini, Mariano & Federico, 2017) [37]. For instance, in oncology, AI tools, like IBM's Watson for Oncology, analyze numerous data points to recommend patient-specific treatment plans (Lisboa & Taktak, 2006) [38]. Some AI technologies employ predictive modeling to project the likely outcomes of various treatment protocols. Thus, medical professionals can switch from a reactive treatment approach to a proactive one, markedly increasing the chances of successful treatment (Li, Shah & Cai, 2019) [39].

Engagement and Ethical Considerations: While the advantages of AI assistance in medical diagnosis and treatment are considerable, it is important to consider the ethical and social implications. Engagement with patients is a fundamental part of the healthcare journey, and the clinical relationship cannot be fully automated. The comfort, understanding, and trust inherent in a human clinician's interaction with a patient are critical elements in the healthcare process (Luxton, 2014) [40]. Moreover, medical errors made by an AI system can have potentially disastrous consequences. Furthermore, privacy issues arise with the handling of sensitive patient data. Ensuring secure, blame-free handling of AI errors and patient data is crucial in establishing trust in AI-assisted medical systems (Parasidis *et al.*, 2018) [41].

Enhancing Patient Care and Personalization: Cognitive therapy is an umbrella term for various therapeutic approaches that focus on altering thought patterns to effect change in behavior and emotion (Hofmann *et al.*, 2012) [41]. On the other hand, AI deals with the development of computer systems capable of performing tasks typically requiring human intelligence, including learning and decision-making. Integrating AI into cognitive therapy results in significant advantages, including enhanced efficiency, convenience, and effectiveness. Primarily, introducing AI in therapy augments efficiency by automating certain tasks [42]. The time therapists spend on routine tasks like administrative paperwork, report generation, and appointment scheduling can be significantly reduced by intelligent software. This extra time can be used to focus on therapeutic interactions, thus enhancing the quality of patient care. Moreover, AI integrated therapy empowers personalization of care [43]. Through advanced analytics, AI systems can predict individual responses to treatment based on their clinical records and genetic information. In the case of cognitive therapy, AI can better understand the patient's cognitive patterns and tailor the most effective therapy approach for them. This kind of personalization significantly enhances treatment outcomes. Furthermore, the role of AI in Moral Cognitive Therapy is pivotal. Moral Cognitive Therapy focuses on helping individuals make values-based decisions and align their actions with their moral compasses. AI integration can significantly augment the assessment and intervention process of moral cognition, leading to a more individualized and effectual therapy [44]. However, it is essential to understand that the integration of AI into healthcare comes with ethical considerations. Concerns regarding data security, patient privacy, and consent need to be addressed and systems should be developed in compliance with regulatory standards (Char, D., Shah, N. H., & Magnus, D., 2018) [45].

AI-Enabled Decision Support Systems: AI-Enabled DSS refers to computer-based systems which utilize artificial intelligence to support and improve decision-making processes (Pomerol, 1997) [46]. They incorporate machine learning, data mining, and predictive modeling to analyze information and ascertain the best course of action. One distinct characteristic of AI-Enabled DSS is its ability to learn over time. Just like a human, these systems continually improve with exposure to new data. The ability to process large volumes of data and make predictions based on patterns further aids in complex decision making. Next, with the aid of the latest advancements in natural language processing, AI-Enabled DSS can now interpret and respond to human speech. It allows for an exceedingly interactive and user-friendly environment in decision-making processes (Power, 2008) [47]. Lastly, the introduction of AI has made DSS more autonomous. Resultantly, human interference is reduced significantly, decreasing the likelihood of error, bias, or subjectivity. Through AI capabilities, decisions are based on pure analysis of data compared to simple heuristic rules. However, as with any technology, AI-Enabled DSS also has its limitations. Analysis of AI-Enabled DSS must also consider the limitations attached. For instance, these systems largely depend on the quality of data. Poor quality or insufficient data can lead to inaccurate predictions, adversely affecting decision-making (Davenport and Harris, 2007) [48]. Also, AI lacks human intuition and emotional intelligence which may lead to decisions devoid of human values (Brynjolfsson and McAfee, 2014) [49]. Despite these limitations, AI-Enabled DSS holds extraordinary promise. The implications of AI-Enabled DSS are profound. It could reshape industries, making companies more competitive due to better decision-making. However, businesses must equally question the ethics of implementing AI decisively. Thus, AI-Enabled Decision Support Systems is a paradigm-shifting technology that can augment and ultimately transform the decision-making processes prevalent in our global establishments. Whilst embodying the potential of complications and grey areas, the positive impacts are undeniable. It is these impacts, when harnessed correctly, that display the immense capability of AI to helm an era of profound transformation.

2.3 The Role of AI in Healthcare: A Fine Balance Between Ethics and Innovation

Research suggests that AI reduces diagnostic errors, leading to effective treatment plans, which significantly improves the patient's chances of recovery. AI tools like chatbots and wearables provide round-the-clock guidance and monitoring, refining the delivery of personalized care. Robotic applications streamline complex surgeries, maintaining precision and reducing recovery times (Liu, Y., Wu, X., Sang, Y., Zhao, C., Wang, Y., Shi, B. and Fan, Y. 2024) [50]. Innovations in AI, however, continually raise profound ethical issues. The risk of algorithmic bias, privacy infringement, and the 'black box' problem constitute some of the primary concerns. By polarizing healthcare towards quantifiable data, AI could overlook the humanistic elements of care, leading to ethical discrepancies in healthcare provision. AI applications are trained on big data, which could introduce bias into healthcare services. If the training data is not representative of the entire population, then predictions produced will lack fairness. The susceptibility of algorithmic bias results in disadvantaged diagnosis or treatment for minorities, demonstrating the ethical downsides of AI deployment (Obermeyer, Ziad, and Mullainathan, Sendhil, 2019) [51]. Another ethical concern entails data security and privacy. Healthcare data is sensitive, and its exposure could result in serious violations of patient privacy. AI's voracious appetite for data generates legitimate concerns regarding responsibility, ownership, and control, putting the ethics of AI into question.

The 'black box' conundrum is a cause for concern too. This problem pertains to the difficulty in understanding or explaining how an AI makes its decision. Lack of transparency can incite mistrust amongst the medical fraternity and patients, hence posing an ethical barrier (Whittlestone, Jess, *et al.*, 2019) [52]. So finding the balance between ethics and innovation is an imperative. Resolving the ethical issues within AI is achievable by creating strict regulatory policies centered on privacy, transparency, justice, and accountability. Policymakers, technologists, and healthcare providers must coalesce to maintain the ethical integrity of AI in healthcare. The application of AI should reflect societal values and norms with the emphasis on securing the welfare of the patients. Establishing audit trails for AI's decision-making process could help demystify the 'black box challenge' introducing necessary transparency (Clifton. L. Phua *et al.*, 2020) [53]. Cross-functional teams, comprising data scientists, clinicians and ethicists, can aid in testing the AI before implementation, ensuring unbiased predictions and absence of algorithmic discrimination. Thereby, weeding out the potential for 'algorithmic racism'. In the sphere of privacy, enhanced encryption techniques must be enforced alongside stringent data anonymization regulations in storing, sharing, and processing patient data. Besides, patients should possess the right to control how their data is used. Ultimately, balancing ethics and innovation in AI requires a continuous dialogue among stakeholders, policymakers, and end-users. Despite numerous advancements, AI cannot replace humans in healthcare; it is just a tool that aids human intelligence to deliver better healthcare.

2.3.1 Humanizing Healthcare: Towards an AI-ISMCT

Artificial Intelligence (AI) stands at the cutting-edge of advanced technology, marking a paradigm shift in revolutionizing healthcare beyond mere algorithms. Ethical mechanisms mixed with AI's promising capabilities can ensure a more empathetic, personalized, and effective treatment pathway that puts patients' needs at the center. AI's future in healthcare is brighter than ever. Still, the onus is upon us to ensure it advances ethically and responsibly while maintaining a human touch in its functionalities. By amplifying empathy, improving patient care, and supporting healthcare professionals, AI does not only provide algorithmic solutions but goes much further, ushering in an era of humanized healthcare embedded in systematic moral cognitive therapy. The following are the ten notable points that can drive AI towards a more humanistic approach while employing an Integrated Systematic Moral Cognitive Therapy (ISMCT) [54].

Strategic Data Analysis: AI is maturing towards data interpretation rather than just data generation. It uses predictive analytics for disease diagnosis, thus humanizing healthcare.
Contextual Adaptability: AI's ability to adapt to genetic, environmental, and lifestyle differences of patients contextualizes treatments, pushing healthcare beyond the 'one-size-fits-all' approach.
Emotional Intelligence: Development initiatives are incorporating emotional intelligence in AI, enabling an empathetic and personalized interaction with patients.
AI Psychiatry: AI's use in mental health diagnoses aids in systematic moral cognitive therapy with greater efficiency and reliability.
Intelligent Imaging: AI-based imaging systems surpass algorithmic pattern recognition, promoting early disease identification and personalized treatment strategies.
Cognitive Neurology: AI contributes to understanding cognitive neurology, aiding in controlling diseases like Alzheimer's, thus going beyond algorithms towards systematic moral cognitive therapy.
Predictive Medicine: AI's use in predicting disease evolution forms a significant aspect of humanizing healthcare; its early diagnosis minimizes the risk factors associated with illnesses.
Ethical Guidelines: AI healthcare maintaining ethical guidelines ensures fair treatment accessibility, patient safety, and patient data privacy, hence contributing to integrated systematic moral cognitive therapy.
Decision-Making Process: AI aids physicians in their decision-making process, giving comprehensive data-driven insights, thereby ensuring optimal medical care.
TeleHealth: AI's potential can be harnessed to facilitate remote medical consultations, increasing healthcare access and enabling individuals to manage their health proactively.

While the integration of AI in healthcare holds immense potential, transitioning from algorithm-driven responses to a more human, morally cognitive approach is challenging yet vital. Such perspective transformation humanizes the delivery of healthcare, fostering trust in AI-enhanced health systems. It heralds the shift from reactive medical actions to preventive care and promotes positive health outcomes.

2.3.2 Synergized AI and ISMCT

AI-powered with Integrated Systematic Moral Cognitive Therapy (ISMCT) in healthcare has started to delineate new cures for psychological-related disorders and improve clinical outcomes globally. The deployment of AI in healthcare is not a phenomenon of the future but a marker of the contemporary world [55]. Considering this intricate undertaking, the prowess of AI is in its capacity to enhance precision, efficiency, and outcomes of ISMCT, thereby transforming healthcare and, by extension, saving countless lives. In India, AI-ISMCT technology is instrumental to the efficient scaling of mental health services, as these systems can reach rural areas where traditional therapy services are barely available, offering enormous potentials to bridge the existing health equity gap (Mina Chandra, *et al.*, 2021) [56]. In a country, where traditional personal therapy services per person ratio is the lowest worldwide, AI synergizing with ISMCT is shifting the traditional treatment approach to a better, widespread and more personalized psychological treatment (Baweja *et al.*, 2017) [57]. The AI-ISMCT therapies have demonstrated effectiveness in managing global mental health problems such as depression, anxiety, stress, and the potential ability to meet the unmet mental health needs both globally and in India.

2.3.3 Case Study and Success Stories

***C1: AI-Powered Virtual Assistants in Mental Health Counseling*:** The advent of AI technology has led to the development of AI-powered virtual assistants that offer substantial opportunities for counseling. This presents a shift from traditional face-to-face counseling to a more advanced digital form. AI-powered virtual assistants have shown potential in assisting with mental health counseling by providing consistent and quick responses, personalized to the user's needs and communication style (Shattenkirk, 2022) [58]. Essentially, AI-powered virtual assistants harness advanced machine learning algorithms to provide interactive responses. They are designed to provide a safe, non-judgmental environment for individuals seeking help, an essential aspect of mental health counseling (Chicca *et al.*, 2020) [59]. Let's use 'Woebot', an AI-powered virtual assistant, as an example. Woebot was created in 2017 by a team of Stanford psychologists with the fundamental aim to make mental health care more accessible (Fitzpatrick *et al.*, 2017) [60]. Woebot employs the principles of Cognitive Behavioural Therapy (CBT) and interacts in a text-chat format. Its effectiveness was tested against a control group and demonstrated a significant reduction in symptoms of depression and anxiety. Thus, the introduction of AI-powered virtual assistants like Woebot is revolutionizing mental health care delivery. An analysis of how Woebot operates highlights the benefits of using AI. Woebot maintains continuous interaction with users, offering automatic responses whenever needed ("Woebot - Your Self-Care Expert", n.d.) [61]. This immediate response is valuable as it lays the ground for interactivity, fostering a sense of connection and rapport. Another benefit lies in its 24/7 availability. This feature ensures that support and help are readily available - critical in crisis situations. Even with these advantages, AI-powered virtual assistants are not without controversy. Critics argue that they cannot replace the human touch essential in therapy. They also point out the ethical concerns regarding data privacy given the sensitive information shared. Another concern is that AI bots can misunderstand or fail to recognize immediate, life-threatening situations, leading to inappropriate responses (Seabrook, 2021) [62]. On the whole, AI-powered virtual assistants present a transformative tool in mental health counseling. While considering its future, it is important to strike a balance. It should co-exist as a supplement to human therapists without aiming for replacement. For now, AI assistants have demonstrated their monumental potential in breaking down barriers in mental health support, providing accessible care available anytime and anywhere.

***C2: AI-Integrated Moral Cognitive Therapy for Chronic Pain Management*:** Chronic pain, a prolonged discomfort that weathers human strength, is a profound issue affecting around 28 million adults in the UK alone (Pulido-Martos, 2020) [63]. Addressing this global phenomenon requires an innovative, sustainable and versatile solution. In light of the recent advances in artificial intelligence (AI) in healthcare, this chapter weaves a compelling narrative of moral cognitive therapy and how AI integration can further revolutionize chronic pain management. An interesting application of AI in health management is its role in chronic pain management, where AI systems pave the way for more precise diagnosis and targeted treatments (Brigham, 2020) [64]. Through algorithms that predict response to treatments based on past medical history, AI can tailor a pain management strategy personalized to each patient. Moral cognitive therapy (MCT), a subset of cognitive-behavioral therapy (CBT), marries morality and psychological treatment in a therapeutic alliance. Rooted in the principles of empathy and understanding, MCT provides a philosophical framework for patients, aiding them to comprehend their pain, construct a positive outlook, and develop healthy coping strategies (Tong, 2004) [65]. Already, studies have confirmed the efficacy of CBT in chronic pain management, with sessions exploring the psychological realm of pain and fostering healthier attitudes towards it (Hann, 2007) [66]. It is here, in this confluence of psychological understanding and empathetic care, where AI can truly offer a seismic shift towards better chronic pain management. Recently, a case was encountered wherein a patient, who had been living with chronic lower back pain for several years, was open to the idea of incorporating AI interventions into their MCT plan. Despite the initial mistrust, the use of AI system, through interactive dialogue systems and machine learning, facilitated an individualized and holistic treatment approach.

The AI system worked in synergy with MCT to identify patterns of negative thinking, incorrect beliefs or behavioral responses contributing to psychological distress and pain prolongation. By linking patterns from over thousands of patient-data sets, the AI system effectively helped the patient to re-evaluate their thoughts and develop healthier attitudes towards pain and discomfort (Laing, 2020) [67]. Furthermore, the AI system was equipped to provide companionship through the therapeutic process, mirroring qualities of empathy often found in human-human bonds. Equipped with machine learning algorithms, the AI system was able to discern distress through speech recognition and provide comfort, paving the way for a better patient experience. In essence, this case study accentuates the immense scope of AI integration in the medical landscape and, more significantly, chronic pain management. As demonstrated, melding AI with MCT can offer a tailored, precise, and empathetic approach to chronic pain management. This not only has the potential to revolutionize treatment plans but also to reshape our understanding and perception of chronic pain itself. Nonetheless, it is pivotal to remember that these interventions do not replace human care; rather, they enhance it, offering a frontier of possibilities that elegantly knits technology and empathy together. As we continue to push the boundaries of AI, it is apparent that the fusion of AI-integrated moral cognitive therapy holds promising results for chronic pain management. While this burgeoning landscape of AI healthcare is indeed fascinating, it also warrants careful ethical navigation to ensure an optimal balance between technological innovation and human empathy, the very essence of healthcare.

C3: AI-Driven Precision Medicine for Oncology: This case study examines Tempus, a technology company pioneering AI-driven precision medicine to redefine cancer treatment. By deploying machine learning and analytic tools to decipher molecular and clinical data, Tempus makes personalized cancer care a feasible reality. Offering advanced genomic sequencing, molecular data analysis, and therapeutic data, Tempus helps clinicians make real-time, data-driven decisions to understand a patient's tumor at the molecular level, tailoring the best therapeutic approach and ultimately improving patient outcomes [68]. Genomic assessment is an integral part of precision medicine, but the extensive genetic variations within patients demand an intelligent system to process, analyze, and derive meaningful interpretations. Tempus uses its AI-based system to analyze a patient's genetic data (alongside their clinical history) and cross-references it with a vast genomic database. This enables the identification of mutations, deletions, and anomalies associated with cancer, thereby mapping out the specific 'genetic landscape'. This data-driven approach allows clinicians to design targeted therapies with potential to increase treatment efficacy [69]. However, AI and precision medicine are not devoid of challenges. Concerns about equity in access, data privacy, and algorithm transparency continue to haunt these revolutionary models. Critics also worry that an over-reliance on AI could eventually diminish the human touch in medicine [70]. To ensure ethical AI functionality within precision medicine, stringent legislative frameworks and transparent data handling protocols are becoming critical (Gerke, Minssen & Cohen, 2020) [71]. The integration of AI in precision medicine redefines clinical oncology by incorporating patient-specific genetic, lifestyle, and environmental variables for precise diagnosis and treatment. As depicted in the Tempus case study, utilizing AI's capabilities allows clinicians to unravel the complexity of each patient's genetic makeup, informing therapeutic interventions. However, substantial challenges, such as data privacy and the potential loss of 'human touch,' require urgent attention to ensure the ethical and equitable deployment of these technologies. There is no doubt that AI-driven precision medicine is carving a new path in the battle against cancer. As it continues to develop and intertwine with digital health fronts, the technology will only continue to improve living standards, providing hope for cancer patients worldwide.

2.4 Advancing Research in AI-Integrated Moral Cognitive Therapy

The dynamism of artificial intelligence (AI) and its intertwinement with various sectors, notably healthcare, has sparked revolutionary transits. More recently, healthcare settings incorporate AI technology in psychotherapy protocols, particularly in cognitive therapy. This essay explores the emerging dimension, AI-integrated Moral Cognitive Therapy, and reviews an array of pioneering research propelling the field. Robots engineered with AI technology have been utilized to maximize the quality of therapeutic interventions (Bemelmans, Gelderblom, Jonker and de Witte, 2012) [72]. Unlike humans, AI boasts an unrivaled capacity to offer consistent, bias-free treatment customized to individual clients (Syrdal, Dautenhahn,

Woods & Walters, 2013) [73]. This feature is particularly indispensable in treating patients with cognitive impairments and emotional disorders who require consistent interaction (Peters, Calvo and Ryan, 2018) [74].

AI-integrated Moral Cognitive Therapy (AI-MCT), an emergent subdivision within the AI-therapy interface, extends the capacity of AI by marrying moral and cognitive aspects of therapy. Moral Cognitive Therapy (MCT), at its essence, explores client emotions, thoughts, and actions with the aim of aligning these three components to a personal moral blueprint that advocates prosocial behavior (Prochaska & Norcross, 2010) [75]. Integrating AI in MCT could pave the way to limitless horizons of therapeutic accuracy, consistency, and personalization. Impressively, artificial intelligence has demonstrated promise in deciphering non-verbal cues and emotional states, pivotal for effective MCT. Researchers at the University of Ohio leveraged machine learning algorithms to analyze facial expressions, successfully differentiating between genuine and simulated distress (Bartlett *et al.*, 2005) [76]. Similarly, an MIT team developed an AI-controlled 'emotionally-aware' robot capable of scrutinizing human emotions and responding empathetically, potentially enhancing therapeutic interaction in MCT (Bickmore & Picard, 2004) [77]. At a more complex level, research on AI's capacity for moral cognition indicates potential towards fully-realized AI-MCT. Stanford University's Allen AI, exhibited insightful ethical reasoning based on situational analysis and moral instruction (Etzioni, Etzioni, & Lehman-Wilzig, 2020) [78]. These advancements showcase emerging research area that could revolutionize the intersection of AI and moral therapy.

While the potential that these technologies present for AI-MCT is undeniable, ethical and technological concerns have been raised, illuminating critical barriers to broad application. Privacy issues, risk of dependency and the potential ethical dilemma of an AI making moral decisions for a human stand as considerable challenges to AI-MCT (Kraemer, 2011) [79].

2.4.1 Collaborative Efforts Between Healthcare Professionals and AI Developers

The collaboration between healthcare professionals and AI developers has revolutionized the healthcare industry. Today, AI algorithms can extract, analyze health-related data, and use the analysis to make precise decisions regarding diagnoses and treatments. They employ predictive analytics to determine the risk factor of diseases, which facilitates early interventions. Moreover, AI has revolutionized surgical procedures with the introduction of robotic surgical systems for precision during operations. The central benefits of this collaboration are accuracy, efficiency, and customized medical solutions. For instance, AI can interpret medical images, ascertain critical factors and provide accurate diagnoses more rapidly than human professionals. Furthermore, AI application in healthcare improves efficiency by optimizing administrative processes such as scheduling appointments, billing, and maintaining patient records. Customizing medical solutions have become possible through personalized medicine - a realm that AI significantly impacts by predicting genetic susceptibility to certain diseases. Albeit the advantages, this synergistic approach faces hurdles such as data privacy issues, the high cost of AI systems and the risk of over-reliance on technology. Healthcare providers and patients express concerns over the security of classified health information as AI databases are susceptible to breaches. Additionally, the high cost of AI systems may lead to a digital divide, where the advanced healthcare technologies are only accessible to affluent societies. The collaboration of Healthcare Professionals and AI Developers has made significant strides in the healthcare industry. However, the challenges experienced require efficient mitigation strategies to drive this promising collaboration to convert the potential of AI-supported healthcare into reality.

2.4.2 Implications for Policy and Regulatory Frameworks

AI, simply put, involves the development of machines capable of performing tasks that typically require human intelligence – decision making, problem-solving, or learning. Despite its virtues, the adoption and widespread use of AI technologies attract considerable challenges. These challenges majorly revolve around privacy, security, accountability, and ethical concerns, among others, engendering the necessity to develop robust policy and regulatory frameworks to govern AI. The applications of AI have the potential to profoundly assert their dominance on society. A compelling issue is privacy – advanced AI technologies, like facial recognition, have raised concerns over the erosion of privacy. For instance, in an environment fraught with surveillance technologies, individual's day-to-day activities could potentially be perpetually monitored. While AI potentially augments global security levels, the privacy trade-off needs to be controlled through effective regulation. Furthermore, security is a towering concern. Researchers argue that powerful AI systems can easily fall into malevolent hands, leading to large-scale harmful outcomes. The automation of cyber-attacks is a pertinent example, where the advent of AI could essentially level up attack avenues. Consequently, policy mechanisms must be in place to mitigate such security threats associated with AI. Another key dimension is the ethical aspects of AI. While AI technologies undoubtedly carry the potential to revolutionize numerous sectors, ethical concerns emerge when machines start taking actions autonomously, specifically when those actions have moral implications. For instance, autonomous vehicles that need to make expedient and ethical decisions in the event of a looming accident. The dire necessity for policies that balance the benefits of AI while addressing ethical concerns is vividly clear. Presuming the growing impact of AI on society, the exigency for a robust regulatory framework cannot be understated. Developed regulations can govern AI applications, establish standards for data handling, and place pertinent accountability measures to ensure AI is used responsibly. Essentially, this would ensure a proper balance between leveraging AI's potential while preventing potential risks.

2.5 Conclusion

In conclusion, as AI continues to evolve, it can offer transformative contributions to society, while simultaneously presenting a myriad of challenges. Ensuring that AI is used safely and effectively mandates a comprehensive review of our current policy and regulatory frameworks. Policymakers must therefore anticipate and address the multifaceted implications of AI adoption. The significance of forward-thinking, adaptive regulations and policies that evolve with the technology itself cannot be overstated if we intend to keep pace with AI's rapid evolution.

References

1. Russell, S., The history and future of AI. *Oxford Rev. Econ. Policy*, *37*, 3, 509–520, 2021.
2. Bresnick, E., Intensified Play: Cinematic study of TikTok mobile app. *Univ. South. Calif.*, *4*, 4, 1–12, 2019.
3. Jiang, Y. and Jiang, Z.P., *Robust adaptive dynamic programming*, John Wiley & Sons, 2017. ISBN:9781119132646. Online ISBN:9781119132677. DOI:10.1002/9781119132677
4. Mittelstadt, B.D., Allo, P., Taddeo, M., Wachter, S., Floridi, L., The ethics of algorithms: Mapping the debate. *Big Data Soc.*, *3*, 2, 2053951716679679, 2016.

5. Beck, J.S., Cognitive-behavioral therapy. *Clin. Textbook Addictive Disord.*, *491*, 474–501, 2011.
6. Edmonds, D., *The murder of Professor Schlick: The rise and fall of the Vienna Circle*, pp. xiv, 313, Princeton University Press, Princeton, London, 2020. https://press.princeton.edu/books/hardcover/9780691164908/the-murder-of-professor-schlick
7. Boser, B.E., Guyon, I.M., Vapnik, V.N., A training algorithm for optimal margin classifiers, in: *Proceedings of the fifth annual workshop on Computational learning theory*, pp. 144–152, 1992.
8. McKinney, S.M., Sieniek, M., Godbole, V., Godwin, J., Antropova, N., Ashrafian, H., Shetty, S., International evaluation of an AI system for breast cancer screening. *Nature*, *577*, 7788, 89–94, 2020.
9. Blum, T., Doi, T., Hayakawa, M., Izubuchi, T., Yamada, N., RBC Collaboration, Determination of light quark masses from the electromagnetic splitting of pseudoscalar meson masses computed with two flavors of domain wall fermions. *Phys. Rev. D*, *76*, 11, 114508, 2007.
10. Giger, M.L., Karssemeijer, N., Schnabel, J.A., Breast image analysis for risk assessment, detection, diagnosis, and treatment of cancer. *Annu. Rev. Biomed. Eng.*, *15*, 327–357, 2013.
11. Marr, B., How much data do we create every day? The mind-blowing stats everyone should read. *Forbes*, *21*, 1–5, 2018.
12. Higuera, P.E. and Abatzoglou, J.T., Record-setting climate enabled the extraordinary 2020 fire season in the western United States. *Glob. Change Biol.*, *27*, 1, 1–2, 2020, doi: 10.1111/gcb.15388.
13. Yotti, R., Bermejo, J., Desco, M.M., Antoranz, J.C., Rojo-Álvarez, J.L., Cortina, C., García-Fernández, M.A., Doppler-derived ejection intraventricular pressure gradients provide a reliable assessment of left ventricular systolic chamber function. *Circulation*, *112*, 12, 1771–1779, 2005.
14. Layer, M., Leidich, J., Schwoch, S., Saske, B., Neubert, S., Robl, P., Paetzold-Byhain, K., Data management of process plants as complex systems: systematic literature review and identification of challenges and opportunities. *Rev. Chem. Eng.*, 40, 3, 329–349, 2023, https://doi.org/10.1515/revce-2022-0077.
15. Bostrom, N., Strategic Implications of Openness in AI Development. *Glob. Policy*, 8, 135–148, 2017, https://doi.org/10.1111/1758-5899.12403.
16. Winfield, A.F. and Jirotka, M., The case for an ethical black box, in: Guildford, UK July 19–21, 2017 *Towards Autonomous Robotic Systems: 18th Annual Conference, TAROS 2017*, pp. 262–273, Springer International Publishing, Proceedings 18, 2017.
17. Russel, S., Norvig, P., Armstrong, S., Fraser, H., Coiera, E., Wong, D., Salzberg, S., An open science approach to artificial intelligence in healthcare: a contribution from the international medical informatics association open source working group. *Yearb. Med. Inf.*, *28*, 01, 047–051, 2019.
18. Baeza-Yates, R., Bias on the web. *Commun. ACM*, *61*, 6, 54–61, 2018.

19. Rebello, C.J., O'Neil, C.E., Greenway, F.L., Dietary fiber and satiety: the effects of oats on satiety. *Nutr. Rev.*, *74*, 2, 131–147, 2016.
20. Taddeo, M. and Floridi, L., Regulate Artificial Intelligence to Avert Cyber Arms Race. *Nature,* 556, 296–298, 2018, doi: 10.1038/d41586-018-04602-6, Available at SSRN: https://ssrn.com/abstract=3198556 or http://dx.doi.org/10.2139/ssrn.3198556.
21. The Tabula Sapiens Consortium*, Jones, R.C., Karkanias, J., Krasnow, M.A., Pisco, A.O., Quake, S.R., Wang, S., The Tabula Sapiens: A multiple-organ, single-cell transcriptomic atlas of humans. *Science,* *376*, 6594, eabl4896, 2022.
22. Yang, G.Z., Bellingham, J., Dupont, P.E., Fischer, P., Floridi, L., Full, R., Wood, R., The grand challenges of science robotics. *Sci. Rob.*, *3,* 14, eaar7650, 2018.
23. Higgins, W.J., Luczynski, K.C., Carroll, R.A., Fisher, W.W., Mudford, O.C., Evaluation of a telehealth training package to remotely train staff to conduct a preference assessment. *J. Appl. Behav. Anal.*, *50*, 2, 238–251, 2017.
24. Kendall, P.C., Gosch, E., Furr, J.M., Sood, E., Flexibility within fidelity. *J. Am. Acad. Child Adolesc. Psychiatry*, *47*, 9, 987–993, 2008.
25. Der-Karabetian, A., Alfaro, M., Cao, Y., Relationship of sustainable behavior, world-mindedness, national and global identities, perceived environmental risk and globalization impact among college students in the United States. *Psychol. Cognit. Sciences-Open J.*, *4*, 1, 8–13, 2018.
26. Dutta, P., Paul, S., Kumar, A., Comparative analysis of various supervised machine learning techniques for diagnosis of COVID-19, in: *Electronic devices, circuits, and systems for biomedical applications*, Academic Press, pp. 521–540, 2021.
27. Dutta, P., Paul, S., Majumder, M., *An efficient SMOTE based machine learning classification for prediction & detection of PCOS*, 2021. 10.21203/rs.3.rs-1043852/v1.
28. Nossin, Y., Farrell, E., Koevoet, W.J., Datema, F., Somoza, R.A., Caplan, A., II, van Osch, G.J., The Releasate of Avascular Cartilage Demonstrates Inherent Pro-Angiogenic Properties *In Vitro* and *In Vivo. Cartilage, 13*, 2_suppl, 559S–570S, 2021.
29. Garrard, G., *Ecocriticism* (2nd ed.), Routledge, 2011. https://doi.org/10.4324/9780203806838
30. Dutta, P., Paul, S., Obaid, A.J., Pal, S., Mukhopadhyay, K., Feature selection based artificial intelligence techniques for the prediction of COVID like diseases, in: *Journal of Physics: Conference Series*, vol. 1963, IOP Publishing, p. 012167, 2021.
31. Dutta, P., Paul, S., Cengiz, K., Anand, R., Kumar, A., A predictive method for emotional sentiment analysis by deep learning from EEG of brainwave dataset, in: *Artificial intelligence for neurological disorders*, Academic Press, pp. 25–48, 2023.

32. Ganie, S.M., Dutta Pramanik, P.K., Mallik, S., Zhao, Z., Chronic kidney disease prediction using boosting techniques based on clinical parameters. *PLoS One*, *18*, 12, e0295234, 2023.
33. Thorsson, V., Gibbs, D.L., Brown, S.D., Wolf, D., Bortone, D.S., Yang, T.H.O., & Chuah, E., The immune landscape of cancer. *Immunity*, *48*, 4, 812–830, 2018.
34. Sharma, D.A., Keesari, T., Rishi, M.S., Thakur, N., Pant, D., Sangwan, P., & Kishore, N., Distribution and correlation of radon and uranium and associated hydrogeochemical processes in alluvial aquifers of northwest India. *Environ. Sci. Pollut. Res.*, *27*, 38901–38915, 2020.
35. Rajpurkar, P., Irvin, J., Ball, R.L., Zhu, K., Yang, B., Mehta, H., Lungren, M.P., Deep learning for chest radiograph diagnosis: A retrospective comparison of the CheXNeXt algorithm to practicing radiologists. *PLoS Med.*, *15*, 11, e1002686, 2018.
36. Esteva, A., Robicquet, A., Ramsundar, B., Kuleshov, V., DePristo, M., Chou, K., Dean, J., A guide to deep learning in healthcare. *Nat. Med.*, *25*, 1, 24–29, 2019.
37. Tonini, G.P. and Capasso, M., Genetic predisposition and chromosome instability in neuroblastoma. *Cancer Metastasis Rev.*, *39*, 275–285, 2020.
38. Lisboa, P.J. and Taktak, A.F., The use of artificial neural networks in decision support in cancer: a systematic review. *Neural Netw.*, *19*, 4, 408–415, 2006.
39. Dutta, P., Paul, S., Shaw, N., Sen, S., Majumder, M., Heart disease prediction: A comparative study based on a machine-learning approach, in: *Artificial Intelligence and Cybersecurity*, CRC Press, pp. 1–18, 2022.
40. Zhou, S., Zhao, J., Zhang, L., Application of artificial intelligence on psychological interventions and diagnosis: an overview. *Front. Psychiatry*, *13*, 811665, 2022.
41. Fordham, B., Sugavanam, T., Edwards, K., Stallard, P., Howard, R., Das Nair, R., Lamb, S.E., The evidence for cognitive behavioral therapy in any condition, population or context: a meta-review of systematic reviews and panoramic meta-analysis. *Psychol. Med.*, *51*, 1, 21–29, 2021.
42. Dutta, P., Pal, S., Kumar, A., Cengiz, K., *Artificial Intelligence for Cognitive Modeling: Theory and Practice* (1st ed.), CRC Press, 2023. https://doi.org/10.1201/9781003216001
43. Rajpurkar, P., Chen, E., Banerjee, O., Topol, E.J., AI in health and medicine. *Nat. Med.*, *28*, 1, 31–38, 2022.
44. Maiese, M., Mindshaping, enactivism, and ideological oppression. *Topoi*, *41*, 2, 341–354, 2022.
45. Baumgartner, C., Deserno, T., Mavrogiorgou, A., Kiourtis, A., Perakis, K., Pitsios, S., Whittaker, E., Notable papers and trends from 2019 in sensors, signals, and imaging informatics. *Yearb. Med. Inf.*, *29*, 01, 139–144, 2020.
46. Power, D.J., Heavin, C., O'Connor, Y., Balancing privacy rights and surveillance analytics: a decision process guide. *J. Bus. Anal.*, *4*, 2, 155–170, 2021.

47. Phillips-Wren, G., Pomerol, J.C., Neville, K., Adam, F., Supporting decision making during a pandemic: Influence of stress, analytics, experts, and decision aids, in: *The Business of Pandemics*, Auerbach Publications, pp. 183–212, 2020.
48. Suri, N., *Book review: Thomas H. Davenport and Jeanne G. Harris, Competing on Analytics*, 25, 3, 388–389, 2021. https://doi.org/10.1177/09722629211023024
49. Zolas, N., Kroff, Z., Brynjolfsson, E., McElheran, K., Beede, D.N., Buffington, C., Dinlersoz, E., *Advanced technologies adoption and use by us firms: Evidence from the annual business survey*, National Bureau of Economic Research, 2021, (No. w28290), doi: 10.3386/w28290.
50. Liu, Y., Wu, X., Sang, Y., Zhao, C., Wang, Y., Shi, B. and Fan, Y., Evolution of Surgical Robot Systems Enhanced by Artificial Intelligence: *A Review. Adv. Intell. Syst.*, 6, 2300268, 2024, https://doi.org/10.1002/aisy.202300268.
51. Dutta, P., Paul, S., Sadhu, A., Jana, G.G., Bhattacharjee, P., Performance of Automated Machine Learning Based Neural Network Estimators for the Classification of PCOS, in: *Doctoral Symposium on Human Centered Computing*, Springer Nature Singapore, Singapore, pp. 65–73, 2023.
52. ÓhÉigeartaigh, S.S., Whittlestone, J., Liu, Y., Zeng, Y., Liu, Z., Overcoming barriers to cross-cultural cooperation in AI ethics and governance. *Philos. Technol.*, *33*, 571–593, 2020.
53. Pimentel, M.A., Redfern, O.C., Hatch, R., Young, J.D., Tarassenko, L., Watkinson, P.J., Trajectories of vital signs in patients with COVID-19. *Resuscitation*, *156*, 99–106, 2020.
54. Sumanta, D., Avishek, S., Puspen, H., Amit, D., Arindam, S., Gour, G., Pijush, D., A Novel Binary Arithmetic Computational Method. *Res. Reviews Discrete Math. Struct.*, *10*, 1, 21–25, 2023.
55. Secinaro, S., Calandra, D., Secinaro, A., Muthurangu, V., Biancone, P., The role of artificial intelligence in healthcare: a structured literature review. *BMC Med. Inf. Decis. Making*, *21*, 1–23, 2021.
56. Amann, J., Blasimme, A., Vayena, E., Frey, D., Madai, V., II, Precise4Q Consortium, Explainability for artificial intelligence in healthcare: a multidisciplinary perspective. *BMC Med. Inf. Decis. Making*, *20*, 1–9, 2020.
57. Bohr, A. and Memarzadeh, K., The rise of artificial intelligence in healthcare applications, in: *Artificial Intelligence in healthcare*, Academic Press, pp. 25–60, 2020.
58. Chen, M. and Decary, M., Artificial intelligence in healthcare: An essential guide for health leaders, in: *Healthcare management forum*, vol. 33, SAGE Publications, Sage CA: Los Angeles, CA, pp. 10–18, 2020.
59. Guo, Y., Hao, Z., Zhao, S., Gong, J., Yang, F., Artificial intelligence in health care: bibliometric analysis. *J. Med. Internet Res.*, *22*, 7, e18228, 2020.
60. Narynov, S., Zhumanov, Z., Gumar, A., Khassanova, M., Omarov, B., Chatbots and Conversational Agents in Mental Health: A Literature Review, in: *2021 21st International Conference on Control, Automation and Systems (ICCAS)*, pp. 353–358, IEEE, 2021.

61. Cheng, Y., Xie, C., Wang, Y., Jiang, H., Chatbots and Health: Mental Health, in: *The International Encyclopedia of Health Communication*, John Wiley & Sons, Inc, 2023, doi: 10.1002/9781119678816.iehc0725.
62. Garland, A.F., Jenveja, A.K., Patterson, J.E., Psyberguide: A useful resource for mental health apps in primary care and beyond. *Fam. Syst. Health*, *39*, 1, 155, 2021.
63. Vagwala, M.K. and Asher, R., Conversational Artificial Intelligence and Distortions of the Psychotherapeutic Frame: Issues of Boundaries, Responsibility, and Industry Interests. *Am. J. Bioethics*, *23*, 5, 28–30, 2023.
64. Sabour, S., Zheng, C., Huang, M., Cem: Commonsense-aware empathetic response generation, in: *Proceedings of the AAAI Conference on Artificial Intelligence*, vol. 3611229–11237, 2022.
65. Li, L.W., Chew, A.M., Gunasekeran, D.V., Digital health for patients with chronic pain during the COVID-19 pandemic. *Br. J. Anaesth.*, *125*, 5, 657–660, 2020.
66. Fulgenzi, C.A., D'Alessio, A., Airoldi, C., Scotti, L., Demirtas, C.O., Gennari, A., Pinato, D.J., Comparative efficacy of novel combination strategies for unresectable hepatocellular carcinoma: A network metanalysis of phase III trials. *Eur. J. Cancer*, *174*, 57–67, 2022.
67. Balawender, K., Pliszka, A., Krowiak, A., Sito, M., Grabarek, B.O., Boroń, D., Does SARS-CoV-2 affect male urogenital system? *Current Pharm. Biotechnol.*, *23*, 15, 1792–1799, 2022.
68. Ruzzolini, J., Peppicelli, S., Bianchini, F., Andreucci, E., Urciuoli, S., Romani, A., Calorini, L., Cancer glycolytic dependence as a new target of olive leaf extract. *Cancers*, *12*, 2, 317, 2020.
69. Neufeld, O., Zhang, J., De Giovannini, U., Hübener, H., Rubio, A., Probing phonon dynamics with multidimensional high harmonic carrier-envelope-phase spectroscopy. *Proc. Natl. Acad. Sci.*, *119*, 25, e2204219119, 2022.
70. Chandio, A.A., Jiang, Y., Rehman, A., Rauf, A., Short and long-run impacts of climate change on agriculture: an empirical evidence from China. *Int. J. Clim. Change Strategies Manage.*, *12*, 2, 201–221, 2020.
71. Gerke, S., Minssen, T., Cohen, G., Ethical and legal challenges of artificial intelligence-driven healthcare, in: *Artificial intelligence in healthcare*, Academic Press, pp. 295–336, 2020.
72. He, Y., He, Q., Liu, Q., Technology acceptance in socially assistive robots: Scoping review of models, measurement, and influencing factors. *J. Healthcare Eng.*, 1–10, 2022, doi: 10.1155/2022/6334732.
73. Fraune, M.R., Oisted, B.C., Sembrowski, C.E., Gates, K.A., Krupp, M.M., Šabanović, S., Effects of robot-human versus robot-robot behavior and entitativity on anthropomorphism and willingness to interact. *Comput. Hum. Behav.*, *105*, 106220, 2020.
74. Peters, D. and Calvo, R., Design for wellbeing–methods and strategies for supporting psychological needs in user experience, in: *Extended Abstracts of*

the 2021 CHI Conference on Human Factors in Computing Systems, vol. 1–3, 2021, May2021.
75. Prochaska, J.O., Norcross, J.C., Saul, S.F., Generating psychotherapy breakthroughs: Transtheoretical strategies from population health psychology. *Am. Psychologist, 75*, 7, 996, 2020.
76. Solbu, A., Frank, M.G., Xu, F., Nwogu, I., Neurohr, M., The Effect of Synchrony of Happiness on Facial Expression of Negative Emotion When Lying. *J. Nonverbal Behav.*, 1–20, 2023.
77. Jeong, S., Alghowinem, S., Aymerich-Franch, L., Arias, K., Lapedriza, A., Picard, R., Breazeal, C., A robotic positive psychology coach to improve college students' wellbeing, in: *2020 29th IEEE International Conference on Robot and Human Interactive Communication (RO-MAN)*, pp. 187–194, IEEE, 2020.
78. Benjamin, L., Book Review: Climate Change and the Voiceless: Protecting Future Generations, Wildlife, and Natural Resources, by Randall S. Abate, Cambridge University Press. *Transnational Environmental Law (TEL)*, 9, 2, 379–384, 2019, doi: 10.1017/S2047102520000047.
79. Seele, P., Dierksmeier, C., Hofstetter, R., Schultz, M.D., Mapping the ethicality of algorithmic pricing: A review of dynamic and personalized pricing. *J. Bus. Ethics, 170*, 697–719, 2021.

3

A Strategic Model to Control Non-Communicable Diseases

Soumik Gangopadhyay[1]*, Amitava Ukil[2], Soma Sur[3] and Saugat Ghosh[3]

[1]*Institute of Engineering & Management (School of University of Engineering and Management, Kolkata), India*
[2]*Dinabandhu Andrews Institute of Technology & Management, Kolkata, India*
[3]*XBS, St. Xavier's University, Kolkata, India*

Abstract

Noncommunicable disease has silently evolved as a public health burden both in the developed and developing nations. Social inequities, poverty, lack of education, and risk of lifestyle diseases are mutually complementary to each other. An unhealthy lifestyle claims maximum responsibility among the factors of becoming a victim of lifestyle diseases. Zero preventive options other than lifestyle modification have enhanced the complication of management of such mass health issues. Through population-wide, policy-based health interventions, the unfinished agenda of the non-communicable disease epidemic can be thwarted. It is necessary to regularly monitor victims' subjective lifestyle choices digitally. This chapter is based on a conceptual model that can be used to bridge the gaps between healthcare, policy, and disease prevention. Such a model can assist with the recording and reporting of lifestyle-related data to analyze the victim's pattern of lifestyle adoption in place of the current healthcare alternative.

Keywords: Non-communicable diseases, lifestyle-related diseases, interventions, lifestyle modification, risk factors, strategic review centre, information and registration centre, integration centre, tertiary care hospitals

**Corresponding author*: gangopadhyaysoumik@gmail.com

Pijush Dutta, Sudip Mandal, Korhan Cengiz, Arindam Sadhu and Gour Gopal Jana (eds.) Design and Forecasting Models for Disease Management (77–98) © 2025 Scrivener Publishing LLC

3.1 Introduction

Non-communicable diseases (NCDs), which are enduring illnesses accompanied by a lengthy period, are now considered an epidemic that is spread globally. Each year, diabetes, chronic obstructive pulmonary illnesses, and cardiovascular diseases cause massive fatalities. While diabetes causes the highest amount of fatality, 7.6 million, the fatality caused by chronic obstructive pulmonary illnesses is 4.2 million, and 1.3 million is the amount of fatality caused by cardiovascular diseases. Between 1980 and 2008, the prevalence of obesity nearly doubled worldwide. An estimated 2.8 million deaths worldwide are attributed to obesity each year. Global Burden on Diseases 2019 (GBD 2019) estimates that because of NCDs, each year, more than 37 million people worldwide lose their lives, nine million people of which are below the 60 years age bracket. Among these fatalities, 90% of the cases are from nations that are characterized by low- and middle-income national economies. Lifestyle diseases are turning into a persistent issue because they are the main cause of death in particularly low-income and middle-income countries. Currently, both countries of low-income and middle-income countries are registering more than 80% fatality due to diabetes or cardiovascular diseases and approximately 90% for chronic obstructive pulmonary disease or cancer. Nearly 44% of all fatalities attributed to lifestyle disorders happen before the age of 70. In the end, death rates in low- and middle-income countries were greater (48%) than in high-income countries (26%) among them. Low- and middle-income countries have a 29% death rate from NCDs among those under 60, compared to only 13% in high-income nations.

3.1.1 India and NCDs

India accounts for 21% of the world's illness burden. According to the state-level disease burden initiative's 2017 report, India has mortality rates for communicable and non-communicable diseases of 110 per lakh and 475 per lakh, respectively. Poor dietary habits and the rapid urbanization of rural India have contributed positively to the inorganic rise in NCD cases in India [1, 2]. The problem is becoming more urgent as contagious illnesses lose ground to noncommunicable diseases (NCDs). Finally, NCDs are boosting the market value of tertiary healthcare as they have sparked an explosion in the number of private clinical facilities in India [3]. According to the National Country Profile 2019, 53% of Indian deaths before the COVID-19 pandemic were attributed to NCDs. The average out-of-pocket cost for a hospital stays in India due to NCDs is USD 149, which has resulted in catastrophic costs. There is no alternative to emergency tertiary care treatment when it comes to the silent and unexpected onset of NCDs. Table 3.1 shows the changing trend of dominance of disease in India. Due to high-tech interventions, NCD treatments are lengthy and expensive, which causes poverty due to out-of-budget expenses in the medical head [4]. According to [5], between the years 1995 and 2004, a steep increase in outpatient consultation traffic (8%) and significantly elevated (13%) hospitalization rates due to NCD-related tertiary care have

Table 3.1 Top ten reasons of death and disability in India.

Types of disease	Growth between 2009 - 2019	Rank- 2009	Rank- 2019	Reasons of having diseases	Growth from 2009 to 2019	Rank - 2009	Rank - 2019
Ischemic Heart Diseases	40.83%	1	1	Malnutrition	-39.80%	1	1
COPD	35.40%	4	2	Air pollution	-7.80%	2	2
Stroke	28.40%	5	3	High blood pressure	23.60%	4	3
Diarrheal Diseases	-21.20%	2	4	Tobacco	13.50%	5	4
Neonatal Disorders	-34%	3	5	High fasting plasma glucose	61.0%	7	5
Lower Respiratory Infection	-13.4%	6	6	Dietary risks	28.90%	6	6
Tuberculosis	-10.9%	7	7	High BMI	78.10%	10	7
Diabetes	54.2%	11	8	WaSH	-42.80%	3	8
Cirrhosis	14.30%	8	9	High LDL	29.70%	8	9
Falls	35.10%	13	10	Alcohol use	19.30%	9	10
Road injuries	4.90%	9	12	NA	NA	NA	NA
Self-harm	2.50%	10	14	NA	NA	NA	NA

Source: https://researchguides.uic.edu/c.php?g=756165&p=5916108 [11].

Table 3.2 Burden of major NCDs in India.

List of states	GSDP per capita (2018-2019)	Share of population in 60+ age group (2021)	Diabetes	Chronic respiratory diseases	Cardio vascular diseases	Cancer
	Rs.	Percentage	Cases Per one lakh population			
Delhi	394216	9.30%	4995	5598	4596	377
Haryana	260286	9.80%	4922	6314	4216	270
Karnataka	232874	11.50%	6137	5797	4606	410
Kerala	225484	16.50%	6894	6642	6955	606
Telengana	225047	11.00%	5400	5081	4392	236
Gujarat	224896	10.20%	4695	5610	4348	268
Uttarakhand	220257	10.60%	5204	6581	4336	250
Maharashtra	216169	11.70%	5287	5711	5117	312
Tamilnadu	215049	13.60%	8429	4775	5225	356
Himachal Pradesh	211325	13.10%	5105	7043	5237	323
Punjab	172149	12.60%	6537	5107	5338	293
Andhra Pradesh	168083	12.40%	5638	6149	5048	266
NE States (except Assam)	127334	8.80%	3915	4706	3569	233
Rajasthan	123343	8.60%	3727	6498	3691	204
West Bengal	119637	11.30%	4673	6202	5367	269
Jammu & Kashmir	109769	9.50%	4082	5515	4087	169
Odisha	109416	11.80%	5004	5296	4594	221
Chhattisgarh	108458	8.80%	5198	5003	3995	279
Madhya Pradesh	99025	8.50%	4276	5333	3720	246

(Continued)

Table 3.2 Burden of major NCDs in India. (*Continued*)

List of states	GSDP per capita (2018-2019)	Share of population in 60+ age group (2021)	Diabetes	Chronic respiratory diseases	Cardio vascular diseases	Cancer
Assam	94385	8.20%	4254	5033	3695	206
Jharkhand	82430	8.40%	4280	4512	3439	201
Uttar Pradesh	74402	8.10%	4186	6057	3367	214
Bihar	47541	7.70%	3327	4899	3258	155

Source: National health profile, 2019 [12].

resulted in an appalling rise in out-of-pocket spending that directly contributes to urban poverty in India. Table 3.2 has evidenced the burden of major non-communicable diseases across Indian states. Additionally, untreated cases of diabetes and cardiovascular disease and unequal access to primary healthcare exacerbate the difficulties. Additionally, the ratio of available emergency support facilities to actual demand is ideally skewed to stop the unfinished agenda of lifestyle diseases [6]. Figure 3.1 explained

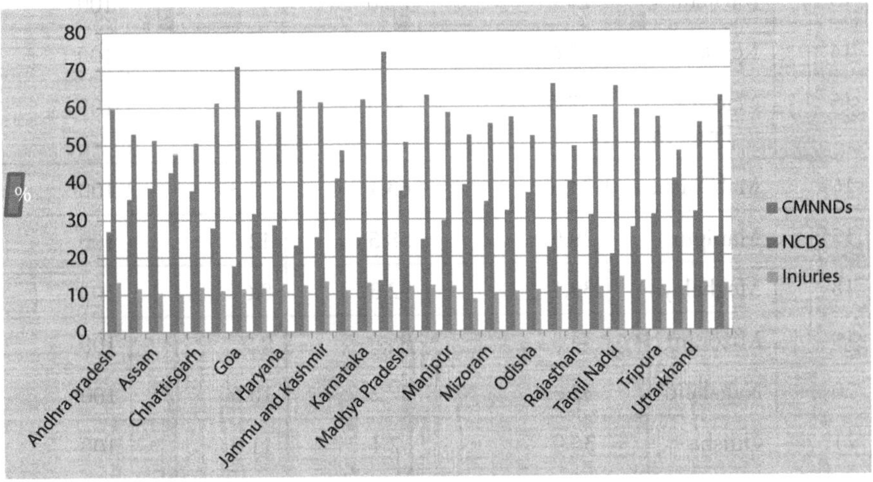

Figure 3.1 State-wise comparative status of communicable and NCDs in India. Source: India: Health of the Nation's states, ICMR, PHFI and IHME, 2017 [13].

Table 3.3 Proportion of total disease burden in 2016.

Sl. no.	Indian states	CMNNDs (%)	NCDs (%)	Injuries (%)	Total (%)
1	Andhra Pradesh	27.0	59.7	13.3	100
2	Arunachal Pradesh	35.5	52.9	11.6	100
3	Assam	38.5	51.2	10.3	100
4	Bihar	42.6	47.6	9.8	100
5	Chhattisgarh	37.7	50.4	11.9	100
6	Delhi	27.8	61.2	11	100
7	Goa	17.6	70.9	11.5	100
8	Gujarat	31.6	56.7	11.7	100
9	Haryana	28.5	58.8	12.7	100
10	Himachal Pradesh	23.1	64.5	12.4	100
11	Jammu and Kashmir	25.3	61.3	13.4	100
12	Jharkhand	40.8	48.3	10.9	100
13	Karnataka	25.1	62.0	12.9	100
14	Kerala	13.6	74.6	11.8	100
15	Madhya Pradesh	37.5	50.5	12	100
16	Maharashtra	24.6	63.1	12.3	100
17	Manipur	29.5	58.5	12	100
18	Meghalaya	39.1	52.3	8.6	100
19	Mizoram	34.6	55.5	9.9	100
20	Nagaland	32.2	57.2	10.6	100
21	Odisha	36.9	52.1	11.0	100

(*Continued*)

Table 3.3 Proportion of total disease burden in 2016. (*Continued*)

Sl. no.	Indian states	CMNNDs (%)	NCDs (%)	Injuries (%)	Total (%)
22	Punjab	22.4	66.0	11.6	100
23	Rajasthan	39.9	49.3	10.8	100
24	Sikkim	30.9	57.5	11.6	100
25	Tamil Nadu	20.4	65.3	14.3	100
26	Telengana	27.6	59.2	13.2	100
27	Tripura	31.0	57.0	12.0	100
28	UttarPradesh	40.5	47.9	11.6	100
29	Uttarakhand	31.7	55.5	12.8	100
30	West Bengal	24.8	62.7	12.5	100

Source: India: Health of the Nation's states, ICMR, PHFI, and IHME, 2017 [14].

the state-wise comparative status of communicable and NCDs in India. According to Tripathy *et al.* (2016) [7], the cost of treating NCDs in a private hospital is three to five times greater than in a public hospital. In India, 'out-of-pocket' payments still account for 75% of healthcare costs. In India, the private health sector is in charge of 40% of all hospital beds, 60% of all outpatient treatment, and 70% of all hospitals. Therefore, private healthcare's contribution to raising awareness is crucial for a successful intervention to stop the unnatural spread of NCDs.

The threat to the national economy has increased due to a few additional NCD outcomes, including an increase in ALS (average length of hospital stay) and an unpredictable extra load on GDP which amounts to 5–10% [8]. Table 3.3 has described the proportion of disease burden in 2016. These additional outcomes have also jeopardized the very foundation of sustainable health. Additionally, Indians under the age of 70 experience a greater substantial loss of disability-adjusted life years (DALYs) than people in other developing nations. In India, the financial toll of NCDs is equally significant [9]. As it already accounts for two-thirds of the morbidity burden, according to the World Bank (2010) [10], there has been a loss of 6% of the Indian GDP and both preventable disease and premature death are found to be responsible for this staggering huge loss.

Therefore, rather than relying on reactive tertiary care, a strong proactive intervention-oriented Indian health policy relating to NCDs is required [15].

To combat this new endemic, the Indian community must, however, develop a policy-based response [16].

3.2 Survey of Literature

3.2.1 Factors Contributing to the Growth of NCDs

Level 1 is the result of the sum of the hazards from the environment, the workplace, behavior, and metabolism, according to the Global Burden of Diseases. Twenty criteria are included in Level 2, including high BMIs, maternal and infant malnutrition, and air pollution. Level 3 specifically includes the risk of child growth failure and particulate matter pollution. Stunting in children and household air pollution from solid fuels has been categorized as Level 4 risks. Longevity is negatively impacted by hereditary factors and the presence of chronic lifestyles [17]. Economic success also results in NCD [18, 19]. In the end, a person's choice of lifestyle leads to NCD [19]. According to Klijs *et al.* (2011) [20], smoking, consuming alcohol, and obesity shorten the disability years of life. As daily physical habits and longevity of life are positively associated, stress, more eating, less exercise, and sedentary lifestyles are further blamed for chronic lifestyle-related disorders [21]. Occupational stress also negatively affects women's ability to lead healthy lives [22–24]. The key effects of occupational stress include emotional response, social isolation, and insufficient sleep, which, together with individual stress, contribute to the development of chronic health issues [25]. It has a major impact on satisfaction related to that work-life [26]. Moreover, it is to be blamed for the issues related to a person's mental health issues [27]. Even tension between job and personal obligations can be a source of stress and result in musculoskeletal disorders [28]. For people who work, overwork is a substantial risk factor for cardiovascular disorders [29]. Although it differs with varied degrees of working patterns, it has an impact on the subject's mental health [30]. Tiredness caused by work-related issues can also result in hypertension, metabolic syndrome, and diabetes II [31]. To increase a person's life expectancy, it is crucial to assist them in the process of changing their adaptation to stress [32].

Additionally, the availability of processed foods has led to an increase in the diet's consumption of salt, sugar, trans fats, and saturated fat. According to [33–35], there is a correlation between diet quality and the risk of cardiovascular illnesses. People still need to be made aware of the important role that wholesome food plays in ensuring human survival [36]. According to [37], the "perceived value of ill health" frequently motivates curative "health-seeking behavior." Therefore, it can be claimed that poor eating habits, intense competition, a lack of regular exercise, stress from the workplace, and so on, have made diseases that are non-communicable more prominent [38].

3.2.2 Lifestyle Modification – A Strategic Role in Mitigation of NCD

Tertiary care using cutting-edge technologies to treat NCDs is expensive and comes with no assurances. By addressing the risks linked with the unhealthy nature of diet, use of tobacco, a lack of physical exercise, and problematic alcohol consumption, the biggest health intervention to avoid NCDs can be accomplished. For private treatment facilities, these methods are affordable and even profitable (UN Secretary-General Report, 2011) [39]. According to Miller *et al.* (2013) [40], proactive preventative culture is a relatively better option than reactive treatment for reducing lifestyle diseases that can help reduce healthcare costs and have the fundamental potential to increase quality-associated life years.

Compared to medication, lifestyle changes are more successful at reducing or preventing NCDs [41, 42]. However, the low likelihood that people will adopt this strategy is a significant barrier to its effective application [43, 44]. Therefore, nonclinical and auxiliary sectors should play a bigger part [45]. The discrepancy in awareness is a factor in the prevalence of NCDs [46]. However, the consciously assessed perceived value of a healthy lifestyle is also a major factor in the subjective acceptance of a healthy lifestyle [47]. People with less education have a higher incidence of this rate [48]. Additionally, both subjective and objective bias results in unequal information on the methods of lifestyle adjustment to avoid NCDs [46].

In addition, research has shown that lower-income groups respond to health promotion programs less favorably than higher-income groups do, and their dropout rates are considerably higher [47]. Health education and communication have a strong potential to improve health outcomes and are especially helpful for healthcare services that lack the resources for human resources [48]. Additionally, encouraging people to adopt a healthy lifestyle through the use of board games is a beneficial method [49]. However, comprehensive health policy is increasingly stifling efforts to promote healthy lifestyle adoption through health education [50].

The most efficient way to raise knowledge of a healthy lifestyle and lower risk factors for NCDs is through social media marketing, individual counseling, workshops, educational materials, etc. [51]. But getting every

Indian to live a healthy lifestyle will be a huge challenge. Additionally, since cost-benefit analyses cannot be quantified, raising knowledge about NCD prevention (primary, secondary, and tertiary prevention) and encouraging people to adopt healthy lifestyles are merely social imperatives [52].

3.2.3 Policy to Control NCDs

It is necessary to switch from acute treatment-based care models to chronic, lifelong care if the growing trend of NCDs is to be reversed. Wishful thinking and deliberate involvement in clinical and non-clinical administration are necessary conditions. Strong monitoring and governance are needed to achieve primordial, primary, and secondary prevention, which is necessary to reduce prevalence [53]. The World Health Organization's Global Status Report on NCDs 2010, which was published in Geneva in 2010, also includes recommendations for "best buys" and "good buys" that are based on health promotion strategies. A fruitful partnership with linked sectors will enable the effective completion of all the prioritized actions.

As recommended by the South East European Health Network (SEEHN) (New WHO European Health Policy - Health 2020), "Health in all policies" (HiAP) has emerged as a top strategy. Table 3.4 has recorded the plan of intervention to control NCD between 2012-2016. According to Saleem *et al.* (2021) [54], the present studies conducted on NCDs have invigorated

Table 3.4 2012–2016 plan of intervention to control NCD.

Priority areas for action	Priority aspects of interventions	Supporting aspects of interventions
1. Governance for NCD 2. Evaluation, Monitoring, Surveillance and Research 3. Prevention of diseases and Health Promotion 4. Health Services	1. Promotion of healthy consumption (Fiscal & marketing policies) 2. Replacements of trans-fats in foods with poly-unsaturated fats 3. Reduction of salt intake 4. Cardiometabolic risk assessment & management 5. Early detection of cancer	1. Promotion of active mobility 2. Promotion of health in settings

Source: www.euro.who.int/ncd-actionplan.

on three major factors, the monitoring of NCD cases, surveillance of the cases and finally tracking down such cases. This entire process, according to Saleem *et al.* (2021) [54] is to be done with the help of a "Geographic Information System (GIS)" which will facilitate the monitoring of an increase in lifestyle diseases based on geographic regions. It is crucial to carefully examine the victims of lifestyle diseases' lifestyle choices by tracking and tracing.

Hence, consciousness regarding a healthy lifestyle will inspire Indians to adopt healthy lifestyles and may also serve the commercial purpose of a private hospital. This initiative needs a wide variety of health information to classify the segmented needs of residents of different geographies, that are unable to be collected manually [55]. It is possible to collect such data seamlessly with the help of Geographic Information System [56]. However, Data privacy, Lack of adequate infrastructure for data capture [57], problems of data portability, and differences in different health-seeking behaviors in different communities and regions are the hazards that may need special attention from the policymakers [58].In a country with a large population, like India, a strategic strategy can be created by creating health support centers, which are supported by a public-private partnership and require coordination between public and private tertiary care facilities.

3.3 Proposed Model

The result of a priceless healthcare service is your current health status. Before making treatment arrangements, it is important to monitor the rising burden of NCDs and lifestyle disparities depending on demographics with the use of a non-communicable disease policy [59]. Due to ignorance or incorrect inter-sectoral integration between public and private healthcare providers, there are either underreported or over-reported instances of a healthy lifestyle. For information regarding NCD mitigation, prevention, treatment, or health risk analysis to be made available to a person on demand through a digital or telephonic medium and to avoid both situations, a centralized control of information regarding victims needs to be erected in our country, India through a collaborative effort of two parties, the central government of India and the tertiary care hospitals.

Information, education, and communication (IEC) is the current management strategy for a few diseases in India [60]. The concept's overarching goal is founded on a holistic method. The current "Trace-Treat" approach to treating NCDs needs to be changed to "Trace-Track". To guide potential patients, a pyramidal management concept may be devised to research and monitor the victims' way of life. It can use both "Top-to-bottom" and "Bottom-to-top" strategies for gathering data and engaging in conversation [61]. The private clinical health establishment must take part in this project with an equal amount of interest and have the chance to exploit the model even for their financial gain.

3.3.1 Registration and Information Centre (RIC)

At the operational level, all of the public and private branches of tertiary care institutions in various Indian states would serve as the principal information suppliers. To receive hospital care, every NCD victim must first register their information (name, mobile number, address, occupation, income, ADAHR card number, status of NCDs, and lifestyle) at the local hospital [62]. Each individual can have a special identification code developed for them to help with treatment support, health information, lifestyle adjustment guidance, and to help them track better in the future. Each patient with a positive diagnosis may be required to register in any national information center, as a patient which will be supported by information centers of the state. Even a private hospital might use it as a marketing tool. However, only complete success can be guaranteed by the coordination of clinical and non-clinical parties providing health care services [63]. People need to be made more aware of the negative effects of NCD on their personal, social, and financial lives to be inspired to change their lifestyles. RIC will serve as a conduit between undiagnosed, diagnosed, and unaffected people and the Indian health authority [64].

The NCD support center can even direct patients toward wellness options and illness prevention strategies. Since private hospitals have direct access to those who are truly harmed, the centers can also utilize hospital branding to promote lifestyle modification approaches or health or disease awareness [65]. At a predetermined time, they will deliver the NCD-related data they have gathered to the IIC.

3.3.2 Integration Centre (IIC)

The National Centre for NCD Information will open a secondary and support center at the various state levels under the jurisdiction of the Health Ministry, supplying essential information and statistical data [66]. It might instead work by acting as a bridge between the national center and those at the bottom of the hierarchy. They may also serve as centers for patient counseling for NCDs. They will deliver the NCD-related data they have gathered to SRC [67].

3.3.3 Strategic Review Centre (SRC)

Any office under the supervision of the Indian Central Health Ministry in New Delhi can create this center. Health disruptions and other comments made during the conversation may be taken into consideration for upcoming healthcare planning and/or research [68]. Figure 3.2 is an author designed proposed model to control non-communicable diseases in India. A suggested PPP (public-private partnership) health model can serve as a private hospital promotion center, patient support center, research center on the NCD epidemic, and national NCD information hub. Private hospitals and the government initiative will therefore split the expense of operating such an information hub. These ideas may be useful [69].

- To track the prevalence and incidence of NCDs in India by processing real-time data more quickly, making information more readily available, and running the system efficiently. Reporting cases quickly and easily allows for better responses and helps guarantee effective operation.
- Creating NCD support organizations supported by a public-private tertiary hospital care strategy
- To support and enhance analytical techniques such as descriptive, diagnostic, prescriptive, and predictive
- Raising people's awareness of the negative effects (personal, societal, and economic) of NCDs in order to encourage

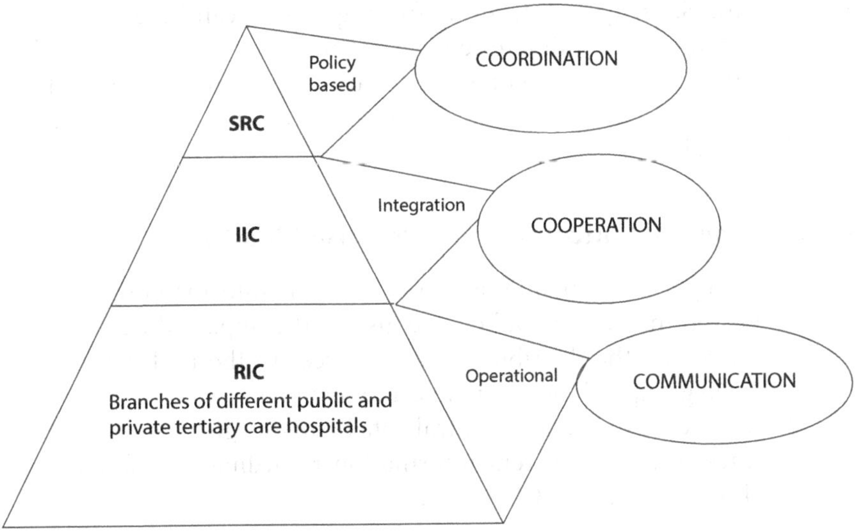

Figure 3.2 The proposed model of NCD control. Source: Author's derived model.

	Interested	Disinterested
Conscious	INDUCE (Enhances the effectiveness)	INFUSE (Develops the Responding Capacity)
Unconscious	RESURGE (Develops Real-TimeResponse Capacity)	CONDUCE (Highly vulnerable, Develops Capacity to Recover)

Figure 3.3 Proposed model of health promotion classification strategy.

them to take possible precautions and/or receive regular treatment for these conditions

- To educate people about lifestyle change methods to prevent, reduce, or avoid NCDs

A classification based on victim awareness (consciousness) and interest in a healthy lifestyle can be constructed, as shown by the matrix in Figure 3.3, using data gathered through the aforementioned centers [70]. Figure 3.3 is a proposed model of health promotion strategy based on differentiated need of health awareness. Using the first strategy (induce), relevant information can be provided to those who are concerned about their health and are interested in changing their lifestyles [71]. Strategy 2 (Resurge): A potential threat can be updated to a soul that is interested but unaware. Strategy 3 (Infuse): A fresh approach can be developed to persuade an aware but uninterested general audience to engage in healthy behavior. Instances where people are unconscious and disinterested can be made more accessible for treatment and facilities under Strategy 4 (Conduct) [72].

3.3.4 Expected Outcome of the Proposed Model

1. It might serve mainly as a resource for info for both the person making the policy and also to the impacted person. Therefore, the distribution of resources to the underprivileged group will be acceptable and justified.
2. It serves as the nation's central data bank to register and keep track of all the current information regarding the lifestyle choices made by NCD victims.

3. It serves as a database of people with NCDs who have been affected by epidemics or endemic conditions and can be used for screening and monitoring purposes.
4. Because this is a comprehensive attempt to track the minor instances that are underreported but don't enter the vicinity, it can lighten the stress on clinical administration. As a result, the government can truly plan its healthcare facilities, including emergency facilities.
5. Because private hospitals have a direct chance to engage with patients, they have a platform to advertise their businesses.
6. Therefore, given the current structure, the government initiative and private hospitals will bear the financial burden. This kind of public-private collaboration can serve as a repository for data that can be used for profit.

3.4 Conclusion

Social growth and a country's health development are interdependent. As a byproduct of growth, lifestyle diseases are on the rise in modern India. A paradigm shift in national health is indicated by the successfully strengthened silent rise of NCDs, a possible threat arising from risks related to our environment, biological shortages, and inequality caused by social affairs. We are heading into a physically weakened India nonetheless. Before the threat of NCDs causes an unanticipated health crisis, now is the time to take action rather than just speak. The government is working hard to develop a national health system that is effective, efficient, and equitable. A comprehensive strategy started at all levels of Indian healthcare can successfully instill a positive mindset to protect Indians from the harmful effects of NCDs. The need for long-lasting, caring health is urgently needed to safeguard the rights of the next generation, who should all have access to good health. The only effective protection available against the unfinished business of NCDs is more insurance coverage and an understanding of the detrimental effects of risk-increasing variables caused by lifestyles. Some "intermediate risk factors" have an influence that is hidden from view and of which people are unaware. Therefore, it is important to educate individuals about the negative effects of NCDs (social, personal, and economic) to encourage them to adopt healthy lifestyles. NCD support centers can even help patients adopt healthy lifestyle choices. It can also serve as a point of disease surveillance. Specialty and super-specialty hospitals that treat NCDs have a crucial role to play in raising awareness of lifestyle changes that can be made to ward off NCDs. Even so, it can be used in the hospital's marketing mix. Designing effective techniques to encourage patients' perceptions of the value of lifestyle alteration is where the true difficulty lies. Physical monitoring is inferior to digital monitoring in terms of utility and immediate value. However, only complete success can be guaranteed by the coordination of clinical and non-clinical parties providing health care services. This facility can serve as a data storage facility, and the collected data may even be used for commercial endeavors. Since it can also calculate the likely level of support for health services, it can serve as a prescription for policy.

References

1. Atreya, A., Kamath, A., Varma, M., Nair, S., Behavioral risk factors and health promotional activities in adults at risk of cardiovascular disease in a south Indian tertiary care hospital. *Int. J. Collab. Res. Intern. Med. Public Health*, 2, 8, 280–290, 2010.
2. *Action plan for implementation of the European strategy for the prevention and control of noncommunicable diseases 2012–2016*, WHO Regional Office for Europe, Copenhagen, 2011 (www.euro.who.int/ncd-actionplan, accessed on 18 March 2022).
3. Bertram, Y.M., Sweeny, K., Lauer, A.J., Chisholm, D., Sheehan, P., Rasmussen, B., Upreti, R.S., Dixit, P.L., George, K., Deane, S., Investing in non-communicable diseases: an estimation of the return on investment for prevention and treatment services. *Lancet Taskforce NCDs Economics*, 5, Series, 1–8, 2018.
4. Bhattacharya, S., Major Risk of Morbidity in Indian Urban Women: Causes, Prevention and Protection. *IUP J. Risk Insur.*, 8, 4, 75–79, 2011.
5. Bhojani, U., Beerenahalli, T.S., Devadasan, R., Munegowda, C.M., Devadasan, N., Criel, B., *et al.*, No longer diseases of the wealthy: Prevalence and health-seeking for self-reported chronic conditions among urban poor in southern India. *BMC Health Serv. Res.*, 13, 306, 2013.
6. Bojar, I., Humeniuk, E., Owoc, A., Wierzba, W., Wojtyła, A., Exposing women to workplace stress factors as a risk factor for developing arterial hypertension. *Ann. Agric. Environ. Med.*, 18, 1, 175–82, 2011.
7. Briney, A., *Overview of Public Health and GIS ~ GIS Lounge [Internet]*, GIS Lounge, Geography Realm, Santa Clara, USA, 2014, https://www.gislounge.com/overview-public-health-gis/.
8. Bugajska, J., Jędryka-Góral, A., Widerszal-Bazyl, M., Orłowska-Baranowska, E., Sagan, A., Michalak, J.M., Zużewicz, K., Konarska, M., Job strain, overtime, life style, and cardiovascular risk in managers and physical workers. *Int. J. Occup. Saf. Ergon.*, 17, 1, 25–32, 2011.

9. Calderon, C.L., Balague, J.M., Cortada, A.S., Health promotion in primary care: how should we intervene? A qualitative study involving both physicians and patients. *BMC Health Serv. Res.*, 11, 62–70, 2011.
10. Charansonney, O.L., Physical activity and aging: a life-long story. *Discov. Med.*, 64, 12, 177–85, 2011.
11. Crouch, R., Wilson, A., Newbury, J., A systematic review of the effectiveness of primary health education or intervention programs in improving rural women's knowledge of heart disease risk factors and changing lifestyle behaviours. *Int. J. Evidence-Based Healthcare*, 9, 3, 236–45, 2011.
12. Dahlgren, G. and Whitehead, M., *European strategies for tackling social inequities in health: levelling up part 2.Copenhagen*, WHO Regional Office for Europe, WHO, Copenhagen, 2007, available at : http://www.euro.who.int/__data/assets/pdf_file/0018/103824/E89384.pdf (accessed 28 February 2020).
13. Dobe, M., Health promotion for prevention and control of Non-communicable Diseases: Unfinished Agenda. *Indian J. Public Health*, 56, 3, 80–186, 2012.
14. Ebina, R., Kawasaki, F., Taniguchi, I., Togari, T., Yamazaki, Y., Sparks, M., The effectiveness of health communication strategies in health education in Kushima, Japan. *Global Health Promot.*, 17, 1, 05–15, 2010. https://doi.org/10.1177/1757975909356628.
15. Edme, J.L., Facq, J., Frimat, P., Vezina, M., Relationship between psychosocial factors at work and incidence of perceived health problems in the GERICOTS cohort. *Rev. Epidemiol Sante Publique*, 59, 4, 20–30, 2011.
16. Engelgau, M.M., Karan, A., Mahal, A., The Economic impact of Non-communicable Diseases on households in India. *Global Health*, 8, 9, 9–18, 2012.
17. Gangopadhyay, S. and Chakraborty, S., A model of strategic communication for clinical premises based on perceived risk of NCDs: a study of Kolkata, India. *Int. J. Bus. Innov. Res.*, 25, 3, 353–364, 2021.
18. Gangopadhyay, S. and Ukil, A., Being Resilient to Deal with Attrition of Nurses in Private COVID-19 Hospitals: Critical Analysis with Respect to the Crisis in Kolkata, India, in: *Healthcare Informatics for Fighting COVID-19 and Future Epidemics. EAI/Springer Innovations in Communication and Computing*, Springer Nature CAM, 2022, https://doi.org/10.1007/978-3-030-72752-9_18.
19. Gangopadhyay, S., Ukil, A., Chatterjee, S., Emerging Paradigm of Smart Healthcare in the Management of COVID-19 Pandemic and Future Health Crisis, in: *Intelligent Healthcare: Infrastructure, Algorithms, and Management*, pp. 455–478, Springer Nature Singapore, 2022, https://doi.org/10.1007/978-981-16-8150-9_20.
20. Glasgow, R.E., Lichtenstein, E., Marcus, C.A., Why don't we see more translation of health promotion research to practice? Rethinking the efficacy-to-effectiveness transition. *Am. J. Public Health*, 93, 1261–1267, 2003.
21. Feigin, V.L., Stark, B.A., Johnson, C.O., Roth, G.A., Bisignano, C., Abady, G.G., Hamidi, S., Global, regional, and national burden of stroke and its risk

factors, 1990–2019: a systematic analysis for the Global Burden of Disease Study 2019. *Lancet Neurol.*, *20*, 10, 795–820, 2021.
22. https://researchguides.uic.edu/c.php?g=756165&p=5916108.
23. *Global status report on noncommunicable diseases 2010*, World Health Organization, Geneva, 2011 (http://www.who.int/nmh/publications/ncd_report2010/en/, accessed 20th March 2012).
24. *GOI, National Health Profile 2019 report*, Central Bureau of Health Intelligence, Directorate of Health Service,Ministry of Health &Family Welfare, India, 2019, Retrieved Jun 20, 2021, http://www.cbhidghs.nic.in/showfile.php?lid=1147.
25. Hämmig, O., Knecht, M., Läubli, T., Bauer, G.F., Work-life conflict and musculoskeletal disorders: a cross-sectional study of an unexplored association. *BMC Musculoskelet. Disord.*, 16, 12, 60–69, 2011.
26. Hazra, J., The changing pattern of diseases in West Bengal, India. *Singapore J. Trop. Geogr.*, 10, 2, 144–159, 1990.
27. Hooda, K.S., Out-of-pocket payments for healthcare in India: who have affected the most and why? *J. Health Manage.*, 19, 1, 1–15, 2017.
28. Inoue, M., Tsurugano, S., Yano, E., Job stress and mental health of permanent and fixed-term workers measured by effort-reward imbalance model, depressive complaints, and clinic utilization. *J. Occup. Health*, 53, 2, 93–101, 2011.
29. Institute for Health Metrics and Evaluation (IHME), *GBD Compare Data Visualization*, IHME, University of Washington, Seattle, WA, 2020. Available from http://vizhub.healthdata.org/gbd-compare. (Accessed 16th October 2021).
30. Johansson, H., Weinehall, L., Emmelin, M., If we only got a chance Barriers to and possibilities for a more health-promoting health service. *J. Multi. Healthcare*, 3, 1–9, 2009.
31. Karasek, R., Collins, S., Clays, E., Bortkiewicz, A., Ferrario, M., Description of a large-scale study design to assess work-stress-disease associations for cardiovascular disease. *Int. J. Occup. Med. Environ. Health*, 23, 3, 293–312, 2010.
32. Klijs, B., Mackenbach, J.P., *et al.*, Obesity, smoking, alcohol consumption and years lived with disability: a Sullivan life table approach. *BMC Public Health*, 11, 378, 2011.
33. Krishnaswamy, K., Perspectives on nutrition needs for the new millennium for South Asian regions. *Biomed. Environ. Sci.*, 14, 1–2, 66–74, 2011.
34. Kundu, M.K., Hazra, S., Pal, D., Bhattacharya, M., A review on Noncommunicable Diseases (NCDs) burden, its socio-economic impact and the strategies for prevention and control of NCDs in India. *Indian J. Public Health*, 62, 4, 302–304, 2018. https://doi.org/10.4103/ijph.IJPH_324_16.
35. Lancon, C., Depression and health state. *Encephale*, 36, 5, 112–16, 2010.

36. Litaker, D., Flocke, S.A., Frolkis, J.P., Stange, K.C., Physicians' attitudes and preventive care delivery: insights from the DOPC study. *Prev. Med.*, 40, 556–63, 2005.
37. Malta, D.C., Bernal, R.T.I., DeSouza, MFM., *et al.*, Social inequalities in the prevalence of self-reported chronic non-communicable diseases in Brazil: national health survey 2013. *Int. J. Equity Health*, 15, 153, 2016. https://doi.org/10.1186/s12939-016-0427-4.
38. Miller, G., Daly, M., Roehrig, C., Tradeoffs in Cardiovascular disease prevention, treatment and research. *Healthcare Manage. Sci.*, 16, 2, 87–100, 2013.
39. Mondal, S., Lucas, H., Peters, D., Kanjilal, B., Catastrophic out-of-pocket payment for healthcare and implications for household coping strategies: evidence from West Bengal, India. *Econ. Bulletin Access Econ.*, 34, 2, 1303–1316, 2016.
40. Mote, B.N., A regional epidemiology of India's "NCD's risk factors" focusing particularly on Maharashtra: A call for "Health promotion" once again. *Int. J. Med. Public Health*, 6, 1, 26–30, 2016.
41. Mozaffarian, D., Peter, W.F., Wilson, M.D., Kannel, M.D., Contemporary reviews in cardiovascular medicine. *Circulation*, 117, 3031–3038, 2008.
42. Musa, G.J., Chiang, P.H., Sylk, T., *et al.*, Use of GIS Mapping as a Public Health Tool-From Cholera to Cancer. *Health Serv. Insights*, 2013, 6, 111–16, 2013. https://doi.org/10.4137/HSI.S10471.
43. Naghii, M.R., Almadadi, M., Zarchi, A.A., Regular physical activity as a basic component of lifestyle modification reduces major cardiovascular risk factors among male armored force personnel of Shabestar army installation in Iran. *Work*, 40, 2, 217–227, 2011.
44. Nakao, M., Special series on "effects of board games on health education and promotion" board games as a promising tool for health promotion: a review of recent literature. *BioPsychoSocial Med.*, 13, 5, 2019. https://doi.org/10.1186/s13030-019-0146-3.
45. Nolan, C., Deehan, A., A. and Wylie, R.J., Practice nurses and obesity: professional and practice-based factors affecting role adequacy and role legitimacy. *Prim. Health Care Res. Dev.*, 13, 353–63, 2012.
46. Nutbeam, D., Health education and health promotion revisited. *Health Educ. J.*, 78, 6, 705–709, 2019. https://doi.org/10.1177/0017896918770215.
47. Okamoto, E., Is health a right or an obligation. *Int. J. Integrated Care*, 7, Jan-Mar, 1–2, 2007.
48. *Prevention and control of noncommunicable diseases*, Report of secretary general, UN general assembly 66th Session, United Nations Publication, New York, 19th May, 2011 (A/66/83).
49. Ramaraj, R. and Alpert, J.S., Indian poverty and cardiovascular disease. *Am. J. Cardiol.*, 102, 1, 102–106, 2008.
50. Reineholm, C., Gustavsson, M., Ekberg, K., Evaluation of job stress models for predicting health at work. *Work*, 40, 2, 229–237, 2011.

51. Ribera, A.P. and McKenna. J. and Riddoch, C., Attitudes and practices of physicians and nurses regarding physical activity promotion in the Catalan primary health-care system. *Eur. J. Public Health*, 15, 569–575, 2005.
52. Robertson, A., Lobstein, T., Knai, C., *Obesity and socio-economic groups in Europe: Evidence review and implications for action*, European Commission, Brussels, 2007, http://ec.europa.eu/health/ph_determinants/life_style/nutrition/documents/ev20081028_rep_en.pdf (accessed 7 February 2012).
53. Roman, B., Carta, L., Martínez-González, A.M., Serra, M.L., Effectiveness of the Mediterranean diet in the elderly. *Clin. Interventions Aging*, 3, 1, 97–109, 2008.
54. Saleem, S.M., Aggarwal, C., Bera, O.P., Rana, R., Singh, G., Bhattacharya, S., Non-communicable disease surveillance in India using Geographical Information System-An experience from Punjab. *Indian J. Community Health*, 33, 3, 506–511, 2021. https://doi.org/10.47203/IJCH.2021.v33i03.017.
55. Sinha, R., Anderson, D.E., McDonald, S.S., Greenwald, P., Cancer risk and diet in India. *J. Postgrad. Med.*, 49, 3, 222–228, 2003.
56. Sun, W., Wu, H., Wang, L., Occupational stress and its related factors among university teachers in China. *J. Occup. Health*, 53, 4, 280–286, 2011.
57. Syed, M.A., Alayne, M.A., Mushtaque, C., Abbas, B., Gender, socio-economic development and health-seeking behaviour in Bangladesh. *Soc. Sci. Med.*, 51, 361–71, 2000. https://doi.org/10.1016/S0277-9536(99)00461-X.
58. Amin Syed, T., *Hospital and Health Services Administration (Principles and Practice)*, pp. 94–113, Oxford University Press, New Delhi, 2010.
59. Thakur, J.S., Prinja, S., Garg, C.C., Mendis, S., Menabde, N., Social and Economic Implications of Noncommunicable diseases in India. *Indian J. Community Med.*, 36, Suppl, 13–22, 2011.
60. The new European policy for health – Health 2020: vision, values, main directions and approaches, *Provisional agenda item 6(a). Sixty-first session of the Regional Committee for Europe*, Baku, Azerbaijan, 12–15 September 2011, WHO Regional Office for Europe, Copenhagen, 2011, (http://www.euro.who.int/__data/assets/pdf_file/0007/147724/wd09E_Health2020_111332.pdf, accessed 25 March 2012).
61. Tourlouki, E., Matalas, A.L., Panagiotakos, D.B., Dietary habits and cardiovascular disease risk in middle-aged and elderly populations: A review of evidence. *Clin. Interventions Aging*, 4, 319–330, 2009.
62. Tripathy, P.J., Prasad, M.B., Shewade, D.H., Kumar, V.M.A., Zachariah, R., Chadah, S., Tonsing, J., Harries, D.A., Cost of hospitalisation for non-communicable diseases in India: are we pro-poor ? *Trop. Med. Int. Health*, 21, 8, 1019–1028, 2016. https://doi.org/10.1111/tmi.12732.
63. Tripathy, P.J., Thakur, J.S., Jeet, G., Chawla, S., Jain, S., Prasad, R., Urban rural differences in diet, physical activity and obesity in India: are we witnessing the great Indian equalisation? Results from a cross-sectional STEPS survey. *BMC Public Health*, 16, 816, 2016. https://doi.org/10.1186/s12889-016-3489-8.

64. Tyrovolas, S. and Panagiotakos, D.B., The role of Mediterranean type of diet on the development of cancer and cardiovascular disease, in the elderly: A systematic review. *Maturitas*, 65, 2, 122–130, 2010.
65. Veit, H.Z., 'So few fat ones grow old': diet, health, and virtue in the golden age of rising life expectancy. *Endeavour*, 35, 2–3, 91–98, 2011.
66. Weidner, G., Sustainability in Medicine: A Case for Prevention of Chronic Noncommunicable Diseases. *Environmentalist*, 32, 3, 353–359, 2012.
67. WHO, World Health Report, *Health Systems Financing: The Path to Universal Coverage*, WHO Press, Switzerland, 2010.
68. World Health Organization. Regional Office for South-East Asia. WHO country cooperation strategy: India 2012-2017. World Health Organization. Regional Office for South-East Asia, 2012, https://iris.who.int/handle/10665/161136.
69. Winter, A.C., Hoffmann, W., Meisinger, C., Evers, S., Vennemann, M., Pfaffenrath, V., Fendrich, K., Baumeister, S.E., Kurth, T., Berger, K., Association between lifestyle factors and headache. *J. Headache Pain*, 12, 2, 147–55, 2011.
70. Yadav, A.K., Paltasingh, K.R., Jena, P.K., Incidence of Communicable and Non-communicable Diseases in India: Trends, Distributional Pattern and Determinants. *Indian Econ. J.*, 68, 4, 593–609, 2020. https://doi.org/10.1177/0019466221998841.
71. Yeung, S.S. and Yuan, J., Low back pain among personal care workers in an old age home: work-related and individual factors. *AAOHN J.*, 59, 8, 345–53, 2011.
72. Zhang, Q., Lan, Y.J., Chen, H.M., Association of occupational stress with quality of working life. *Sichuan Da Xue Xue Bao Yi XueBan*, 42, 4, 540–543, 2011.

4

Image Compression Technique Using Color Filter Array (CFA) for Disease Diagnosis and Treatment

Indrani Dalui[1]*, Avisek Chatterjee[2], Surajit Goon[1] and Pubali Das Sarkar[1]

[1]*Computer Application Department, Eminent College of Management & Technology, Barasat, Kolkata, India*
[2]*Computer Application Department, BCDA College of Pharmacy & Technology, Barasat, Kolkata, India*

Abstract
This study demonstrates that the Color Filter Array (CFA) data can be efficiently compressed using existing image compression algorithms such as JPEG. In addition, it tested the overall performance of CFA compression methods and demosaicing algorithm selection. These CFA compression methods exploit how CFA images do not contain repetitive data introduced by demosaicing. This work established a relationship in the discovery of different CFA compression techniques and described how the choice of demosaicing algorithm affects visual quality. Using a simple bilinear demosaicing method, CFA image compression produces better images compared to the traditional compression scheme.

However, this may not happen if we choose advanced demosaicing methods that minimize color artifacts. In addition, Wavelet Packet Tree generation is used to exchange still and color images. This method checks the difference between the deconstructed nodes' entropy and the decomposed node and then chooses the node to decompose. A versatile quantification threshold was also discussed in the research. The proposed algorithm ensures a good compression result. This study is useful for the diagnosis of a disease and its treatment. Digital health records are used in the health sector where patient information and medical reports and records are digitally stored for convenient use. The traditional image processing system requires more space for its data storage. Image compression using CFA data helps to store more medical images and reports like X-rays, MRI, USG, etc.

**Corresponding author*: indrani@ecmt.in

Pijush Dutta, Sudip Mandal, Korhan Cengiz, Arindam Sadhu and Gour Gopal Jana (eds.) Design and Forecasting Models for Disease Management (99–114) © 2025 Scrivener Publishing LLC

without compromising the quality of the image which is helpful for accurate diagnosis of disease and its treatment.

***Keywords*:** Color filter array (CFA), electronic health record (EHR), PSNR, demosaicing, compression

4.1 Introduction

Image compression is a method for limiting the size of the graphic image in bytes without degrading the character of the image to an inappropriate level.

It is an interesting work that helps us to do something with an image easily and fluently. Since digital images have very much volume, efficient image compression techniques are essential for their archiving and transmission. This project describes a new technique for further decomposition of approximation and detail components when the sum of the information content of the separate detail and approximation components is less than the information content of the decomposed component.

4.1.1 Color Filter Array

A color filter is a mosaic of color filters in front of the image sensor. Most digital cameras use this to filter out the red, green, and blue components of the light falling on them. Figure 4.1 illustrates the spatial arrangement of the Color Filter Array (CFA). This group is a commonly used arrangement called Bayer CFA. There are two green and one red or one blue filter in two Bayer patterns.

Thus, in this research work, Base is taken into consideration for developing an image editor has been constructed. The main segments of compression are redundancy and irrelevance reduction. So, we go for expelling

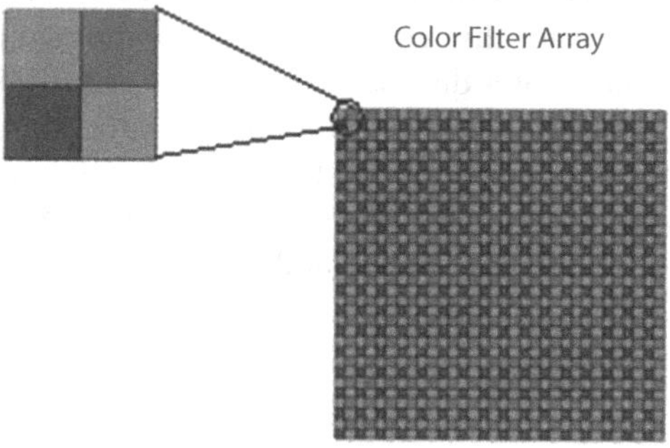

Figure 4.1 Colour filter array (CFA).

duplication from the image. By analyzing an image using this editor the easiest method to comprehend the compression of an image is obtained. Image can be compressed with comparatively better image quality. The image becomes more user-friendly by this editor. Here we can partition the image into many regions so it can be easy to find out whether there is any noise affecting the image.

4.1.2 Electronic Health Record (EHR)

(EHR) Electronic Health Record Standards in India were first released in 2013 by the Ministry of Health & Family Welfare (MoH&FW). The collection of different medical records in digital form generated during clinical investigations is called an Electronic Health Record (EHR). The benefits of gathering medical records include better and evidence-based treatment, faster and more accurate diagnoses that ensure better treatment at lower costs, avoiding repeated unnecessary medical tests, accurate analysis of reports and predictions, easy exchange of data, long-term preservation of health records, etc. The importance of maintenance of medical records is increasing day by day for the above-mentioned benefits.

To support clinical operations, the health record system contains documents and files in different formats, which serve as official records of different analytic and prescription data or information. In the digital health record system, the following categories of documents are generally stored.

1. Images (in series or single form), Waveforms (like those in an ECG or EEG), audio (like those in a digital stethoscope), etc.
2. Captured or Scanned Records:
    - Image: JPEG lossy or lossless, 300 dpi, 1024 x 768 pixel minimum size and resolution
    - Audio/Video: Audio-Visual Objects coding
    - Scanned Documents: Electronic Document File for Document Management.

The following is the paper's structure. In the next section 4.2, we discussed some relevant research on image compression approaches. Afterward, in Section 4.3. We describe the detailed design of the proposed model. In Section 4.4. We discussed the detailed implementations of the proposed work and 4.5. We discussed the experimental results. In section 4.6. We demonstrate the comparison of past work and our work and lastly in section 4.7. Gives the conclusion and future scope of the entire work.

4.2 Related Works

In this proposed work researcher have to recognize the real color images from graphics images and to compress graphics image utilizing a wavelet transform augmentation under JPEG2000 standard that will improve the compression execution. This technique can be effectively adjusted in image compression applications without changing the sentence structure of compressed stream of JPEG2000. The exhibition of JPEG2000 corrupts when it compresses the images having low - depth color images, such as graphics images.

In this paper [1] the display of JPEG2000 is undermined when it compresses the images having low-depth color images. Using JPEG2000 greatly improves the performance of JPEG2000 Part-2 down sampling factor styles. We used the entropy to recognize the genuine color image or graphics images. The proposed strategy can be effectively adjusted in graphics or drawing-type image compression applications without changing the sentence structure of the compressed JPEG2000 bit stream. This strategy can be extended to compress the video sequences under the JPEG2000 standard.

In paper [2], Andrew J. Penrose and Neil A. Dodgson build up a lossless image compression technique for a color video that exploits the spectral, spatial, and temporal redundancy inherent in image data. They demonstrate that a versatile plan is crucial to guarantee full misuse of these redundancies for compression purposes. The consequences of the proposed plan are observed to be ideal when contrasted with lossless image compression gauges and this presentation is accomplished while as yet allowing a computationally simple decoder. The paper [3] demonstrates the several kinds of multi-viewed image sets and the corresponding coding scheme and, gives a helpful correlation of the pressure proficiency of those two classes of multi-viewed picture encoders over a variety of test images. Portrayal quality and application explicit prerequisites are considered to rule for the encoder to be utilized. In paper [4], the orthonormal MLFRIT (multilevel Finite Ridgelet Transform) performs on the input images and the resultant ridge let coefficients are assembled properly and exposed to vector quantization. The results of the suggested compression algorithm have been thoroughly discussed for a two-level MLFRIT. Similar to DWT, Multilevel FRIT iteratively decomposes the input image into approximations and details. A progressive compression technique has been developed for efficient lossy image compression using MLFRIT (multilevel Finite Ridgelet Transform) and VQ (Vector Quantization). The ridgelet coefficients obtained from various levels of the multilevel Finite Ridgelet Transform are vector-quantized based on the level and subband in which they appear. The results of the suggested approach for a two-level MLFRIT verify the technique's applicability for progressive compression at low bit rates. The approach may be applied to bigger photographs with more than two degrees of breakdown. The type of reconstruction is additionally determined by the codebook's dimensions. The downside of this strategy is the overhead associated with codebook storage. Ridgelet compression, which is effective for certain types of image, notably those with straight edge segments, creates sparser image representation than other transformations.

The aim of this paper [5] is to assess the effect on image classification with compression quality factors and image JPEG compression for SPOT multispectral data and compression quality factors. Every compressed image is categorized using classification methods of Maximum Likelihood Classification (MLC) and ranking method of ISO-DATA group (ISOCLUS). When the classified output uses the original image as a benchmark, the aftereffects of the controlled and exclusive feature study show a decrease in classification efficiency relative to the compression scale. Finally, some recommendations are made regarding the optimal JPEG compression factor for image classification.

The use of image compression techniques to decrease the quantity of image data has become mandatory rather than essential. If the image quality inevitably degrades during the compression process, it is necessary to determine the ideal mode to balance the degraded image quality and the compression ratio In regard of the experiments in this study, the ISOCLUS unsupervised classification technique's number of clusters is influenced by the compression effect, which distorts spectral content. The number of resulting clusters generally decreases as the degree of compression increases. A small percentage increase is also observed as a result of an unforeseen blocking effect. In addition, the change of cluster numbers after compression also depends on the scene. In the paper [6], there are various topics on the structure of image compression algorithms. The primary concentration in their work is the improvement of proficient entropy coding algorithms, advancement of enhancement procedures for context modeling, and utilization of strategies in the structure of an algorithm for the compression of medical images. In the paper [7], proposes an improved compression approach that employs slightly plane cutting and adaptable Huffman with a (LZW) Lempel-Ziv-Welch word reference. When using LZW compression to compress an image, the kind of image and number of colors must be considered, and LZW compression will be applied in specific circumstances. The constraint will be reduced by using bitmap cutting for grayscale and color images. The proposed approach outperforms the usual LZW technique by 55.6% for color and 102% for grayscale images. Test findings for the proposed method outperformed existing approaches for a wide variety of image types. The proposed method enhances LZW compression in the following ways. To begin, convert the grayscale images into eight binary (monochrome) images using bit-plane cutting. The colored images are often addressed by the tristimulus red, green, and blue signals, each of which is a grayscale image that will be sliced into eight binary (monochrome) images via bit-plane cutting. The information images should be divided into groups based on color or semantics. The resulting binary images have extra bits, which increases the compression ratio. Because the number of hues will be reduced to two: black (value 0) and white (value 1). Coding techniques improve compression ratios. It is challenging to identify implementable ways to boost the LZW compression ratio, especially when dealing with pictures with varying colors and large sizes. We have chosen 8-bit plane slicing to add redundant data and adaptive Huffman to reduce the number of bits of yield code as methods for enhancing LZW. Similarly, instead of (256) characters, it uses two characters (0) and (1) to represent LZW word references.

In paper [8], recently, several image compression algorithms are offered to simply deal with CFA images captured before color addition. These CFA compression schemes use the fact that CFA images do not contain the additional data produced by demosaicing. In this study, they will consider the presentation of various CFA compression algorithms and demonstrate how the choice of demosaic algorithm influences visual qualities. In this paper [9], a strategy dependent on a co-evolutionary genetic algorithm is presented, which is utilized to develop specific wavelets for fingerprint images. These wavelets are distinguished from the hand-designed wavelet, which is now utilized by the FBI to compress fingerprints. The results show that the advanced wavelets routinely outperform the hand-designed wavelet. Using development to adapt wavelets to certain picture classes can significantly improve the quality of the compressed image. A co-evolutionary genetic method based on Enforced Sub-Populations and a mathematical approach known as Lifting are used to generate wavelets that are specifically tailored to a given class of pictures. The technique is tested in the fingerprint compression domain, allowing for a thorough comparison with other existing approaches. When the wavelets discovered by the GA are tested in a cutting-edge image coder and contrasted with standard wavelets, including the winner of the FBI's challenge to find the best wavelet for fingerprint compression, the evolved wavelets consistently outperform human design, demonstrating that evolutionary discovery can outperform human design in an important task. These findings demonstrate that developing wavelets tailored to certain image classes may greatly improve the image coder's compression efficiency. They also show how evolutionary discovery may outperform major human design efforts in a critical job. The first portion of this exercise focuses on the measurements for bi-level image coding. This paper [10] shows the initial segment of this exercise is given to the measures for bi-level image coding. The altered Huffman (MH) and Modified Relative Element Address Designate (MREAD) models are used for content-based documents, although newer standards like JBIG1 and JBIG2, presented by the Joint bi-level experts' group (JBIG), may efficiently encode handwritten letters and binary halftone images. The last piece of this exercise is dedicated to the benchmarks for continuous tone images. They will provide insights into the Joint Photographic Experts Group (JPEG) standard and its different modes, such as baseline (sequential), lossless progressive, and hierarchical. The very recent and advanced coding standard – the JPEG-2000 will be talked about in the lesson. This paper [11], presents a new algorithm for demosaicing digital images, for example, interpolation of Bayer patterns, A review of existing demosaicing algorithms has been made, and a new high-performance algorithm is proposed including noteworthy enhancements to interpolation for visual quality and PSNR. The algorithm introduces a few enhancements upon a notable Kimmel strategy: progressively complex insertion of green components, adaptive control of the number of iterations, and projection on starting data.

The Bayer Pattern must be processed to generate a high-quality color picture. Each pixel in this interpolated image contains a full-color spectrum depending on the colors of the pixels around it. The procedure is known as demosaicing or demosaicking. This paper describes the work I completed for the final assignment, which was to implement and compare various frequently used demosaicing algorithms. In this proposed method [12].

Many consumer digital color cameras employ a single light-sensitive sensor and a color filter array (CFA), each pixel element stores intensity information for one color component. In many applications, the data is interpolated into a full-color image, which is subsequently compressed. Performing color interpolation before compression adds redundancy to the data. With this study, we explore the methodologies and challenges involved with CFA data compression before complete color interpolation. The described compression methods use the same number of pixels as the sensor data. As a post-processing technique, median filtering is used to improve image quality. Furthermore, JPEG is used to compress the CFA data to ensure low complexity. Simulations have shown that image quality can be significantly improved using these new schemes in this paper [13].

In this study, S. A. Curtis and another researcher work on fractal image compression by demonstrating the repetition of modification of a Sierpinski triangle. They also show encoding and decoding techniques by optimizing the image quality. In this work [14]. In this paper Nasir Menon and his friends tried to produce a new image compression technique whose name is Lossless/Near-Lossless Image Compression It combines progressive transmission with near-lossless compression in a single framework. They are certain that their compression approach ensures that there is no pixel difference between the original and compressed images, whereas lossless techniques provide reconstructed data that perfectly matches the original. In this work [15].

Michael and his friends work on multiwavelets and multiwavelet packets. They developed algorithms capable of outperforming current image compression standards, such as the Joint Photographic Experts Group (JPEG) algorithm. This study introduces multi-wavelet packets and introduces new multi-wavelet transform and quantization methods. Extensive experimental findings show that our approaches outperform, and in some circumstances outperform, existing wavelet filters [16].

The discrete cosine transform (DCT), which Andrew works on, is a method for breaking down a signal into its fundamental frequency components. He creates some straightforward routines to compute the DCT and compress photos here. These features demonstrate Mathematica's capability to quickly prototype image processing techniques. In this paper [17].

Christine Bako explores the JPEG 2000 picture compression standard in her essay. It establishes a brand-new image-coding strategy based on cutting-edge wavelet-based compression methods. Christine also discusses how helpful it is for a wide range of applications, such as digital photography, security systems, Internet image sharing, and medical imaging. In [18].

Dinesh and al analyze the most recent picture-compressing methods in their work. This essay discusses image compression and how it relates to many aspects of image processing. Additionally, they demonstrate how Principal Component Analysis is used in image reduction. In this paper [19].

This Paper [20], Anish Kumar *et al.* work on the New Image Compression Technique which is known as MRT. They executed simulations with fixed thresholds for various image forms. By altering the threshold, the MRT-based technique has an advantage over the DCT in terms of balancing quality and percentage compression. The potential of the MRT-based image compression technology is demonstrated in the paper. By selecting the right threshold, this method's flexibility in defining the amount of compression at the sacrifice of image quality is its main advantage. This study encourages a great deal further MRT-based picture compression research.

Fabio and his friends conduct a comparison analysis of the Lifting Method and Principal Components Analysis, two different image compression methods. In the initial image of a cutting instrument, lifting and principal components analysis were used to produce a low-resolution version, which will be utilized as input cases while training neural networks. In this paper [21].

This research [22], In Public health data is still being digitized and made quickly available in the form of electronic records in developing countries like India and others. In India, the primary and community health centers (PHC/CHC), secondary health centers (district hospitals), and tertiary health centers (national level) are the three main levels at which patient care is provided. Many issues confront healthcare institutions in gathering, processing, storing, and managing this data while maintaining security and privacy. To manage patient information electronically, some secondary and tertiary healthcare organizations have so far begun to deploy health IT applications such the hospital information systems, hospital management information systems, electronic medical records (EMR), etc. Different companies used different databases and programming languages to develop these systems. The system is unique because of this method; however patient information stays inside the same hospital and is not transferable to another hospital when a patient transfers for more complex or specialized care. This is because there aren't enough useful and interoperable data sources. A secure electronic health record (EHR) standard will be created as part of the proposed work using accepted medical terminology and coding standards. The public healthcare system in India will operate more efficiently once the EHR framework is implemented. EHR enables continuous and efficient patient care across the board in the healthcare system.

[23], unanticipated arrival of the pandemic has enhanced the scope of digital healthcare (Gangopadhyay *et al.*, 2023). [24], Implementing health IT has led to the emergence of smart health choices including telemedicine, m-health, and e-health management policies for those people whose health status has worsened because it works at the intersection of "health", "health care" and "health policy". Such technology can only be implemented successfully if a reliable high-speed network is available, the availability of affordable devices, public awareness, and the wishful

thinking of policymakers. Based on the goals of sustainable development, many underdeveloped countries have adopted telemedicine to reduce the health inequalities of the population. M-Health is a parallel opportunity for better management of scarce resources. Cost, distance, or travel time affect people's perceived importance of adopting telemedicine for self-care. It was first implemented in India as a trial-and-error technique to lighten the patient load in open clinical settings. Insufficient knowledge and accessibility are the primary obstacles to the expansion of e-health in India. However, telemedicine has emerged as a therapy option in the wake of the COVID-19 epidemic, free travel restrictions, and nationwide closures aimed at preventing infections. The necessity for telemedicine as a practical and dependable therapeutic alternative has grown as a result of force rather than choice. However, growing capacity is only feasible with adequate backing in related fields.

[25], Human crisis in serving the COVID-19 patients has been followed in Kolkata-based hospitals and it may occur in future in pandemic situation. The problem can be solved by applying technology for remote health care service. Digital health care system is playing an important role in the remote health services. These organizations need to be flexible and adaptable in order to become robust. For this reason, sustaining a harmonic balance of interdependence among the hospital's four stakeholders is crucial to its competitive advantages as a technological institution. Nurse Appointment or Temporary a full-time nurse is an additional option that can provide acute relief for clinical facilities. Installation and induction of the robot service system AI healthy and economical practices can be implemented at these facilities, leading to positive changes in the future. For competitive edging, the key managing premise of such a varied situation is the 3I strategy (information, integration, and innovation).

4.3 Proposed Model

To implement an image editor that will do the following steps on an image (Figure 4.2).

- ***CFA Data:*** A raw picture file consists of lightly altered data from the image sensor of a digital camera. As an input image, we use the raw image. A demosaicing technique optimized for each type of color filter converts the unprocessed image data obtained from the image sensor to a full-color image

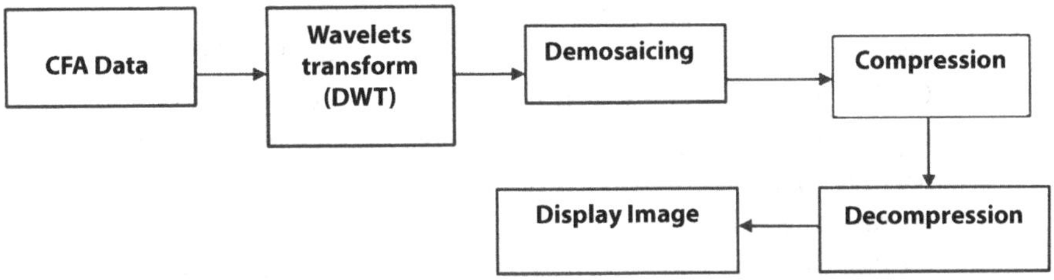

Figure 4.2 The proposed image editor.

(each pixel representing the intensity of one of the three primary colors).

- ***Wavelets transform (DWT)***: A mathematical function known as a wavelet splits a given function or continuous-time signal into discrete frequency components and then analyses each component at a scale-appropriate resolution. A wavelet-based representation of a function is called a wavelet transform. Waves are scaled and translated copies (called "daughter waves") of finite length or rapidly decaying oscillatory waveforms (called mother waves). Wavelet transformations outperform typical Fourier transforms in modeling functions with discontinuities and strong peaks, as well as accurately deconstructing and rebuilding finite, non-periodic, and/or non-stationary signals. Discrete wavelet transforms (DWTs) and continuous wavelet transforms (CWTs) are the two types of wavelet transforms. Examples of continuous-time (analog) transforms are DWT and CWT. They can depict analog signals that are continuous in time. Whereas DWTs employ a representation grid or a subset of scale and translation values, CWTs function on all feasible scales and translations. The input image is transferred using discrete wavelet transformations (DWT).
- ***Demosaicing***: To display the image, we employ a demosaicing process; therefore, we need to generate an image with a red, green, and blue pixel at each place. To achieve this, we interpolate the absent sensor values. "Demosaicing" refers to techniques for interpolating the missing sensor information. A demosaicing algorithm is a digital imaging method that uses incomplete color samples generated from an image sensor overlaid with a color filter array (CFA) to rebuild a full-color image. Another spelling for color reconstruction

or CFA interpolation is demosaicing. Because most modern digital cameras use a single image sensor overlaid with a CFA to gather images, demosaicing is part of the processing pipeline necessary to transform these images into a readable format. A Bayer image, on the other hand, can be viewed as a grayscale image.

- ***Compression***: The CFA image's color components are divided into distinct channels and then molded into a rectangle. Any image compression method can then be applied.

A CFA image is divided into two sub-images by it:

(a) Every green sample in the CFA image is contained in a green sub-image.
(b) The red and blue samples from the CFA image are contained in the non-green sub image.

Compared to a traditional RGB image compressor, the CFA data compressor deals with only 1/3 of the data.

- ***Image Display***: Then the compressed images need to be displayed. Then we can save it and use it for our purpose.

4.4 Implementation

This proposal includes a GUI for image compression. The easy-to-use Graphical User Interface (GUI) for Image Compression was created specifically for MatLab TM computing environments. Through the GUI, the user can effectively use several cutting-edge image compression algorithms and acquire knowledge of the techniques that are provided many image compression methods that were developed employing key ideas from wavelet transform, fuzzy logic, and artificial neural networks are incorporated into the GUI. The graphical user interface is meant for inexperienced users with the technical specifics of the regions, making the GUI a more interactive user interface tool. Construction of the image editor that will do the following steps on an image.

1. In this proposed approach, we must first implement the image compression GUI.

2. Then we have to browse and upload the image we want to compress with the help of browse and upload image buttons respectively.
3. After uploading the respective image we will compress the image with the help of the compress button in the image compression interface.
4. Finally we have to click the save button to save the compressed image.

4.5 Results

The images were compressed using CFA image compression techniques. The only rate that has a bit-per-pixel expression, matters in this situation because

Table 4.1 Experimental results.

Serial no.	Original image	Size	Compressed image	Size	PSNR
1	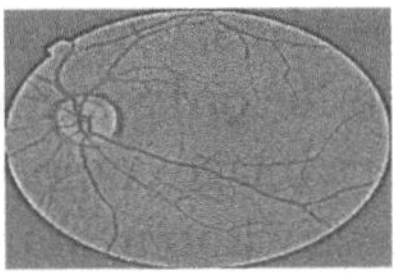	1.33 MB	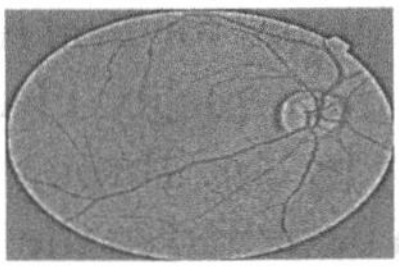	98 KB	38.09 dB
2	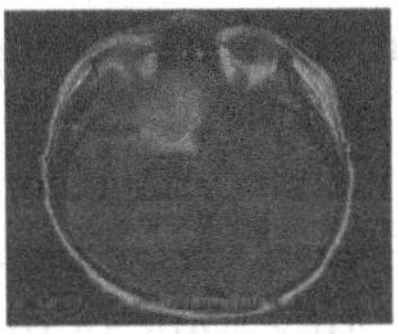	204 KB	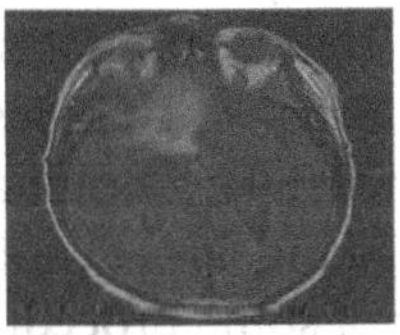	9.78 KB	30.22 dB
3	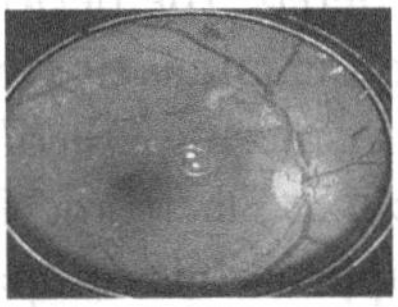	225 KB	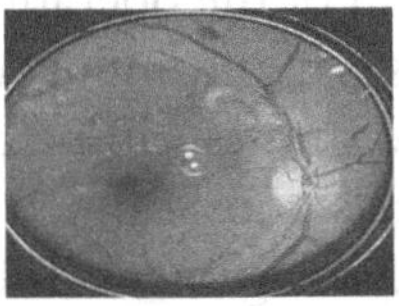	103 KB	43.59 dB
4	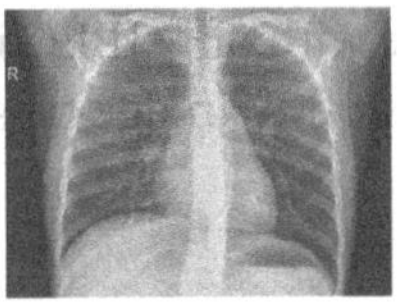	236 KB	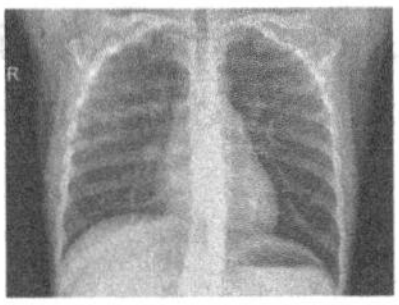	81.3 KB	41.16 dB

the reconstructed images are identical to the input images. Compared to the conventional RGB image compressor, only one-third of the data can be processed by the CFA data compressor, which is explained in Table 4.1.

Table 4.1 shows the typical PSNR values for a lossy image range from 30 to 50 dB, where higher is better. This compressor handles CFA data compression, whereas a traditional RGB image compressor handles only 1/3 of the data. In the resulting table Image No. 1 is in its original size of 1.33 MB and after compression, the size will be 98 KB and the PSNR value for this image is 38.09 dB. Image No. 2 is in its original size of 204 kb and after compression, the size will be 9.78 kb and the PSNR value for this image is 30.22 dB. Image No. 3 is an initial size of 225 kb and after compression, the size will be 103 kb and the PSNR value for this image is 43.59 dB. Image No. 4 is an initial size of 236 kb and after compression, the size will be 81.3 kb and the PSNR value for this image is 41.16 dB.

4.6 Conclusion

Therefore, we conclude that this software allows us to deliver digital medical reports such as x-rays. Thus, we conclude that this program allows us to deliver compressed digital medical reports, such as MRIs, USGs, and X-rays, without sacrificing their quality. Electronic Health Record System is getting more important day by day. Through the use of this image compression system, more data can be stored in the medical records without compromising the quality. The whole system is computerized and hence minimizes errors and manual labor. The user can easily access the software through a user-friendly interface design which hides the coding complexities.

Future research will take a simultaneous approach to image compression, deblocking, demosaicing, and denoising. Several works have been done on reversing the order of some procedures or attempting some of them simultaneously, but not all of the operations are considered. The performance of the four procedures is highly influenced by each other and is tightly tied to it. Additionally, future studies can be conducted considering video/audio like Audio-Visual Objects and Scanned Documents like Electronic Document File for Document Management. This proposed approach is compatible with only JPEG image files that can be used further programmed for various other image files. This proposed approach is compatible with only JPEG image files that can be used further programmed for various other image files.

References

1. Zhang, H., Shao, S., Tao, M., Bi, X., Letaief, K.B., Deep learning-enabled semantic communication systems with task-unaware transmitter and dynamic data. *IEEE J. Sel. Areas Commun.*, *41*, 1, 170–185, 2022.
2. Salihu, A., Snopce, H., Luma, A., Ajdari, J., Optimization of Dodgson's condensation method for rectangular determinant calculations. *Adv. Math. Models Appl.*, *7*, 3, 264–274, 2022.
3. Tang, Y., Chang, K., Huang, M., Li, B., BMISP: Bidirectional Mapping of Image Signal Processing Pipeline. *Signal Process.*, 109135, 2023.
4. Dadras, A., Leffler, K., Yu, J., *A Ridgelet Approach to Poisson Denoising*, Umeå University, Sweden, 2024, arXiv preprint arXiv:2401.16099.
5. Chen, M., Wen, L., Pan, D., Cumming, D.R., Yang, X., Chen, Q., Full-color nanorouter for high-resolution imaging. *Nanoscale*, *13*, 30, 13024–13029, 2021.
6. Bertocchi, L., Vecchio, R., Sorbello, S., Correale, L., Gentile, L., Buzzachera, C., Odone, A., Impact of the COVID-19 pandemic on physical activity among university students in Pavia, Northern Italy. *Acta Bio Medica Atenei Parmensis*, *92*, Suppl 6, 1–7, 2021.
7. Chiu, M.C., Liu, Y.S., Yang, Z.A., Fan, Y.C., Adaptive Constant Hue Algorithm Design of Color Filter Array, in: *2023 IEEE International Conference on Consumer Electronics (ICCE)*, pp. 01–04, IEEE, 2023.
8. Yip, C.C.Y., Ho, C.C., Chan, J.F.W., To, K.K.W., Chan, H.S.Y., Wong, S.C.Y., Yuen, K.Y., Development of a novel, genome subtraction-derived, SARS-CoV-2-specific COVID-19-nsp2 real-time RT-PCR assay and its evaluation using clinical specimens. *Int. J. Mol. Sci.*, *21*, 7, 2574, 2020.
9. Ayyoubzadeh, S.M. and Wu, X., *Lossless compression of mosaic images with convolutional neural network prediction*, McMaster University, Hamilton, 2020, arXiv preprint arXiv:2001.10484.
10. Wang, L., Guo, Y., Wang, Y., Li, J., Gu, S., Timofte, R., Zhang, W., NTIRE 2023 challenge on stereo image super-resolution: Methods and results, in: *Proceedings of the IEEE/CVF Conference on Computer Vision and Pattern Recognition*, pp. 1346–1372, 2023.
11. Yu, Y., Wang, Y., Yang, W., Lu, S., Tan, Y.P., Kot, A.C., Backdoor attacks against deep image compression via adaptive frequency trigger, in: *Proceedings of the IEEE/CVF Conference on Computer Vision and Pattern Recognition*, pp. 12250–12259, 2023.
12. Park, C.W., Moon, Y.H., Eom, I.K., Image tampering localization using demosaicing patterns and singular value based prediction residue. *IEEE Access*, *9*, 91921–91933, 2021.
13. Elgendy, O.A., Gnanasambandam, A., Chan, S.H., Ma, J., Low-light demosaicking and denoising for small pixels using learned frequency selection. *IEEE Trans. Comput. Imaging*, *7*, 137–150, 2021.

14. Gangopadhyay, B., Mallik, A., Ali, M.S., Halder, A., Layek, R., Karmakar, D., Chattopadhyay, D., Thermal Effects on Electrochemical Performance of Copper Oxide Nanoparticles Decorated Amine-Functionalized Graphene Oxide for Ultrahigh Energy Density Supercapacitor with Real-Life Application. *J. Phys. Chem. C, 127*, 45, 21940–21953, 2023.
15. Gangopadhyay, S., Ukil, A., Chatterjee, S., Emerging Paradigm of Smart Healthcare in the Management of COVID-19 Pandemic and Future Health Crisis, in: *Intelligent Healthcare: Infrastructure, Algorithms and Management*, pp. 455–478, Springer Nature Singapore, Singapore, 2022.
16. Gangopadhyay, S. and Ukil, A., Being resilient to deal with attrition of nurses in private COVID-19 hospitals: critical analysis with respect to the crisis in Kolkata, India. *Healthcare Inf. Fighting COVID-19 Future Epidemics*, 1, 353–363, 2022.
17. Dutta, P., Paul, S., Kumar, A., Comparative analysis of various supervised machine learning techniques for diagnosis of COVID-19, in: *Electronic devices, circuits, and systems for biomedical applications*, pp. 521–540, Academic Press, 2021.
18. Dutta, P., Paul, S., Cengiz, K., Anand, R., Kumar, A., A predictive method for emotional sentiment analysis by deep learning from EEG of brainwave dataset, in: *Artificial intelligence for neurological disorders*, pp. 25–48, Academic Press, 2023.
19. Dutta, P., Paul, S., Shaw, N., Sen, S., Majumder, M., Heart disease prediction: A comparative study based on a machine-learning approach, in: *Artificial Intelligence and Cybersecurity*, pp. 1–18, CRC Press, 2022.
20. Kumar, A. and Gopikakumari, R., Development of 2-D MRT based image compression techniques (Doctoral dissertation, Cochin University of Science And Technology), 2013.
21. Pereira, F.H., Baptista, E. *et al.*, Low-loss image compression techniques for cutting tool images: a comparative study of compression quality measures. *Exacta, São Paulo,* 8, 2, 225–235, 2010.
22. Pai, M.M.M., Ganiga, R., Pai, R.M. *et al.*, Standard electronic health record (EHR) framework for Indian healthcare system. *Health Serv. Outcomes Res. Method,* 21, 339–362, 2021.
23. Blandford, A., Wesson, J., Amalberti, R. *et al.*, Opportunities and challenges for telehealth within, and beyond, a pandemic. *Lancet Glob. Health,* 8, e1364–e1365, 2020.
24. Yeung, A., Torkamani, A., Butte, A., *et al.* The promise of digital healthcare technologies, *Front. Public Health,* 11, 1196596, 2023.
25. Alghamdi, N. and Alghamdi, S., The Role of Digital Technology in Curbing COVID-19. *Int. J. Environ. Res. Public Health,* 19, 14, 8287, 2022.

5

Research in Image Processing for Medical Applications Using the Secure Smart Healthcare Technique

Debraj Modak* and Chowdhury Jaminur Rahaman

Department of Electronics and Communication Engineering, Abacus Institute of Engineering and Management, Kolkata, West Bengal, India

Abstract

Image processing technology refers to the techniques and algorithms used to manipulate and analyze digital images. It involves using software tools and mathematical algorithms to enhance, transform, or analyze digital images to extract useful information or improve their visual quality. Through image processing techniques, digital images can be compressed, monitored, and transmitted effectively in remote settings. Digital images have a significant effect on modern society, science, technology, and art. Some techniques that can be used in Image enhancement, filtering, segmentation, object recognition, and image fusion. Image processing techniques have been playing a crucial role in advancing healthcare, particularly in the field of smart healthcare. Through the Markov Random Field model image segmentation can be identified and separated with different structures within medical images, such as MRI or CT scans. For medical image processing, Bayesian image restoration incorporates with MRF model and improves image quality. Bayesian inference estimates the posterior distribution of the underlying image given the observed noisy image and the prior distribution by using Markov Chain Monte Carlo. In the medical 4.0 era, smart medical care introduces AI and deep image recognition. This is significantly developed by learning or neural networks and provides predictable and hand-held medical services, reducing the monotonous work of doctors, and improving the accuracy, efficiency, and expediency of medical services. Artificial intelligence (AI) is introduced into the medical industry and assists the betterment of medical treatment. The World Health Organization (WHO) expresses eHealth as "the use of information and

**Corresponding author*: ddebraj.hetc@gmail.com

Pijush Dutta, Sudip Mandal, Korhan Cengiz, Arindam Sadhu and Gour Gopal Jana (eds.) Design and Forecasting Models for Disease Management (115–154) © 2025 Scrivener Publishing LLC

communication technologies (ICT) to support health and health-related fields". The WHO has officially acknowledged the critical role those digital technologies enhanced public health by shifting its emphasis from information networks.

Keywords: Image processing, MRF model, Bayesian image, MCMC, AI, deep image, ICT

5.1 Introduction

An image is typically a 2-D representation such as a photograph, drawing, or graphic. It can be captured by various means such as a camera, scanner, or created digitally using software. Images can be used to convey information, express ideas, or emotions, or simply capture a moment in time. Images can represent a wide range of things, depending on their context, content, and purpose. Some possible things that images can represent include:

Objects or scenes: Images can depict real or imagined objects or scenes, such as landscapes, people, animals, buildings, or abstract shapes.
Concepts or ideas: Images can illustrate or symbolize abstract concepts or ideas, such as love, freedom, justice, or success.
Information or data: Images can convey information or data in a visual format, such as charts, diagrams, maps, or infographics.
Identity or culture: Images can reflect or represent aspects of identity or culture, such as race, ethnicity, gender, or religion.

Images can represent a diverse range of things and can be interpreted differently by different people depending on their experiences, perspectives, and contexts. Images are processed using a combination of hardware and software tools. The process of image processing typically involves the following steps:

- Acquisition: The first step in processing an image is to acquire it using a camera, scanner, or other image-capturing device. This involves capturing the light that reflects or emits from the object being imaged and converting it into a digital format.
- Preprocessing: Once the image is acquired, it may undergo preprocessing to enhance its quality or correct any distortions or noise. This may involve operations such as filtering, smoothing, or sharpening.

- Feature extraction: In this step, the relevant features or characteristics of the image are identified and extracted using various techniques such as edge detection, segmentation, or pattern recognition.
- Image analysis: The extracted features are then analyzed to draw conclusions or make decisions based on the image content. This may involve tasks such as object recognition, classification, or tracking.
- Interpretation: The final step in image processing involves interpreting the results of the analysis to glean pertinent data from the image. This may involve making decisions or taking actions based on the image content.

Image processing techniques apply some algorithms to glean pertinent data from digital images and enhance their quality. Through mathematical calculation, it performs various operations by using hardware and software operations. Such as enhancement, filtering, compressing, restoring, and segmentation. Since 1950, Digital Image Processing has been used in many aspects such as Computer picture processing. The United States Jet Propulsion Exp. extensively surveyed the moon in 1964, transmitting frequent images from the spacecraft [1]. Since 1970, DIP has found extensive and diverse applications in a variety of fields, including medical imaging, satellite imaging, digital photography, and video processing. The Authors show the block diagram of medical imaging system in Figure 5.1. By using Image Processing Technology in medical image science, is used to improve the quality of medical images, identify diseases more accurately, and aid with doctors as immediate basis treatment. After diagnosis patient's body X-ray, an echo of sonar and medical imaging instruments collect raw data and derive the output characteristics.

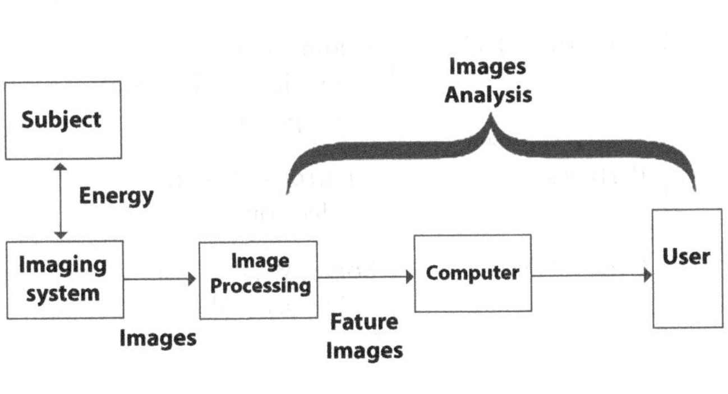

Figure 5.1 Block diagram of medical imaging system.

5.1.1 Imaging Systems

In recent technological advancements, the scope of visual observation has been greatly expanded in many fields. Imaging systems may be categorized by using different parameters. According to radiation, imaging system classification is shown in Table 5.1. Figure 5.2, shows the EM spectrum photon energy concerning frequency and wavelength. There are two different methods i.e. Direct imaging and indirect imaging, are used in various fields shown in Table 5.2. Direct imaging refers to the procedure of apprehending a visual representation of an object or scene by collecting photons or other particles emitted or reflected directly from the object or scene. In astronomy, telescopes capture the light emitted by stars and other celestial objects and form direct images of them. In medical imaging, X-rays are used to directly capture images of bones, while ultrasound waves are used to directly capture images of soft tissues. In astronomy, indirect imaging techniques such as interferometry and gravitational lensing are used to infer the presence and properties of celestial objects by analyzing the way their light is distorted or interferes with other light sources. In medical imaging, CT scans and MRI scans use indirect imaging techniques to create images by analyzing how X-rays and magnetic fields interact with the body's tissues and organs. The RGB image is a 3D function that may be stored and edited in a computer because it can be signified as a set of discrete values.

Table 5.1 Classification of radiation type in the imaging system.

Types of radiation	Parameter
EM Waves	Microwaves, Infrared, Visible Light
Quasistatic Fields	Geomagnetic, Bioelectric, Electrical Properties
Particles	Neutrons, Protons, Electrons
Other Waves	Sonar, Water, Seismic, Ultrasound

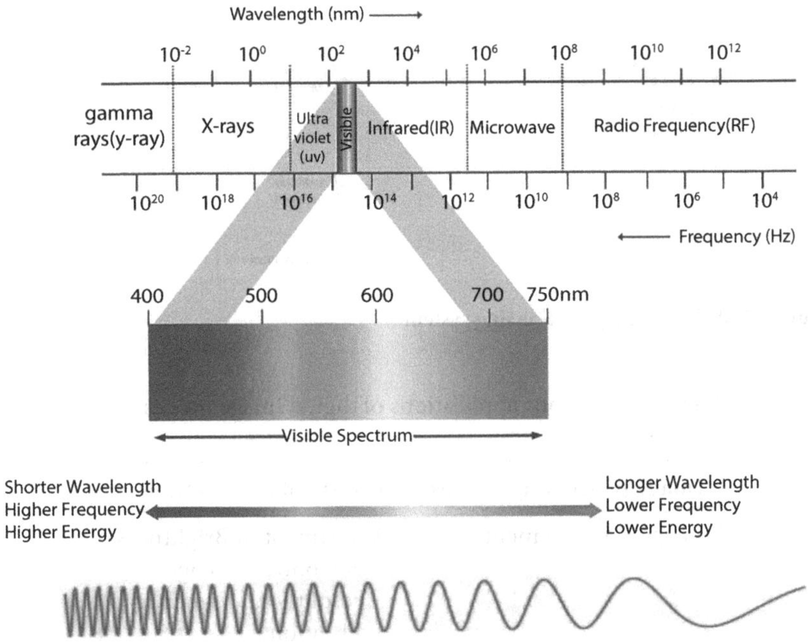

Figure 5.2 The EM spectrum of photon energy.

Table 5.2 Classification of imaging systems.

Classification of imaging		Parameters
Direct Imaging	Parallel acquisition	Human Eye, Optical large-scale telescopic and microscopic imaging instruments.
	Serial acquisition	Scanning probe Microscope, gamma-camera
Indirect Imaging		X-ray, MRI, SPECT, Film Camera

5.1.2 The Digital Image Processing System

It is a computer-based system that is designed to process digital images. It involves the use of algorithms and mathematical models to manipulate and enhance digital images. The system consists of hardware and software components, including input devices such as cameras and scanners, processing

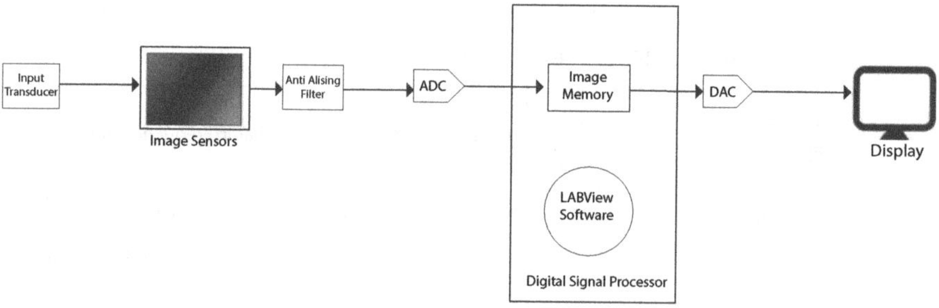

Figure 5.3 A digital image processing system.

Table 5.3 Various applications of digital image processing systems.

Image processing classes	Variety of applications
Image Enhancement	Adjustment of Brightness & Contrast, Image resolution, frequency filtration
Image Restoration	Photometric Correction, Inverse Filtering
Image Analysis	Image Segmentation, Feature Extraction
Image Compression	Lossless and Lossy Compression
Image Synthesis	Tomographic Imaging, 3-D construction

units such as CPUs and GPUs, and output devices such as printers and displays. A digital image processing system is shown in Figure 5.3. The digital image processing system can be used for a variety of applications, as shown in Table 5.3.

5.1.3 Image Enhancement

The technique of enhancing a picture's quality to make it more aesthetically pleasing and informative is known as image enhancement. Image enhancement techniques are used in the medical field to enhance the clarity of medical pictures, such as CT scans and X-rays, to facilitate diagnosis.

This process is done by convolution method using an operator i.e. called kernel denoted by $w(k,l)$. Considering $w(k,l)$ to be an array of $(2K+1*2L+1)$ coefficients, where $(k,l)=(0,0)$ i.e. the center of the kernel, is defined by,

$$g(m,n) = w(k,l) * f(m,n) = \Sigma_{k=-K}^{K} \Sigma_{l=-L}^{L} w(k,l) \bullet f(m-k,n-l) \quad (5.1)$$

Where, $g(m,n)$ is the convolution or output image. The image pixel (m,n).

5.2 Classification of Digital Images

Digital images can be classified into various categories based on different criteria, including their content, format, resolution, and application. Here are some common classification methods:

Content-based classification: This method classifies images based on their content, such as natural scenes, portraits, or abstract art.
Format-based classification: Images can be classified based on their file format, such as JPEG, PNG, TIFF, or RAW [2, 3].
Resolution-based classification: Images can be classified based on their resolution, such as low-resolution images, high-resolution images, or ultra-high-resolution images.
Application-based classification: Images can be classified based on their intended application, such as medical imaging, remote sensing, digital photography, or graphic design.
Color-based classification: Images can be classified based on their color scheme, such as black and white, grayscale, or full color.
Shape-based classification: Images can be classified based on their shape or geometry, such as circular images, rectangular images, or irregular-shaped images.
Texture-based classification: Images can be classified based on their texture, such as images with smooth surfaces, rough surfaces, or patterns.

5.2.1 Utilizations of Digital Image Processing (DIP)

5.2.1.1 Medicine

Digital image processing systems are widely used in medical imaging systems. Image Fusion and Image registering methods find applications across a broad spectrum of uses [4, 5].

5.2.1.2 *Forensics*

Digital image processing techniques are commonly used in various fields such as edge detection, security, identification of patterns and authentication purposes, and face recognition. Forensics is constructed by an individual's database information. To establish a person's identification, forensics compares the input data (images, fingerprints, eyes, etc.) with the database.

5.2.2 Medical Image Analysis

Image analysis involves analyzing and understanding the visual content of an image, including its shape, color, texture, and spatial relationships. Compression is necessary because video files can be very large and take up a lot of storage space, making it difficult to store and transfer them efficiently. Video image compression minimizes video file size by eliminating redundant or non-essential information from the video frames. There are two primary types. Such as lossless compression and lossy compression. Lossless compression algorithms preserve all the information in the original video file, without compromising quality but achieve less compression of data. Lossy compression algorithms, remove some of the information in the video file to accomplish a higher compression ratio. Considering simple image format, there classified into two parts. i.e., the background part and a significant part. At this point, the segmentation procedure just requires the setting of a threshold value, resulting in a binary image with just two grey values.

5.2.3 Max-Variance Automatic Cut-Off Method

The Gray-scale histogram's shape can be variable. The histogram is denoted by h(i) and defined as,

$$h(i) = \Sigma_{m=0}^{M-1}\Sigma_{n=0}^{N-1}\delta(f(m,n)-i), i = 0,1,\ldots..,P-1 \tag{5.2}$$

Where, $\delta(w) = \begin{cases} 1, w = 0, \\ 0, otherwise \end{cases}$

The average gray value of the entire image of Area 1 and Area 2, is defined as:

$$\mu = \mu_1\theta_1 + \mu_2\theta_2 \tag{5.3}$$

A considerable grayscale contrast exists between two regions, μ1 and μ2, separated by the threshold t. In such cases, the difference between the average grey value of the two regions (μ2) and the overall average grey level of the entire image (μ1) is also substantial.

$$\sigma_B^2(t) = \theta_1(\mu_1 - \mu)^2 + \theta_2(\mu_2 - \mu)^2 \tag{5.4}$$

Image Segmentation: The technique of dividing an image into several segments or areas according to its properties is known as image segmentation. To help with diagnosis and treatment, image segmentation in smart healthcare is used to identify and isolate particular regions of interest in medical pictures, such as tumors or lesions.

Image registration: Aligning two or more photos of the same scene or item that were shot at various times or from different perspectives is known as image registration. To facilitate precise comparison and analysis, image registration is used in the healthcare industry to align medical pictures, including X-rays, CT scans, and MRIs.

Computer-aided diagnosis (CAD): CAD is a type of software that uses image processing techniques to analyze medical images and aid in diagnosis. CAD can help detect abnormalities in medical images, such as tumors, cysts, and other diseases. The Authors represent the block diagram of Computer-aided diagnosis (CAD) in Figure 5.4.

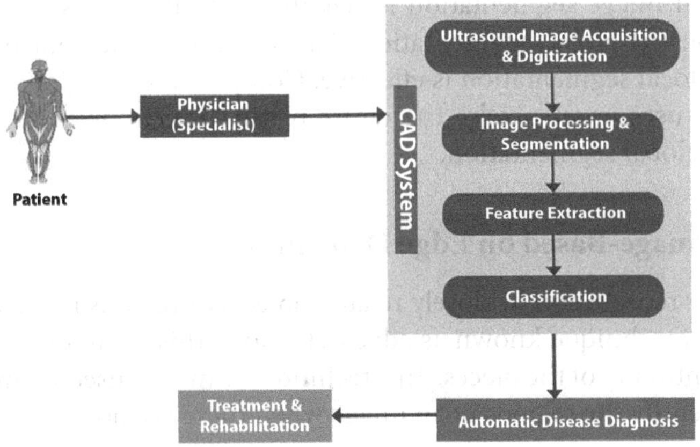

Figure 5.4 Block diagram of CAD.

5.2.4 Medical Imaging Segmentation

Medical imaging is the technique of taking pictures of various body areas for use in diagnosing or studying medical conditions. Medical imaging is growing rapidly as a result of improvements in image processing techniques, such as image recognition, analysis, and enhancement. Here, illustrates how several image processing methods, including k-means, ROI-based segmentation, and watershed approaches, are used in the area of medical imaging to apply both straightforward and advanced image analysis techniques. Sonography, magnetic scopes, thermal, isotope, radiological, and organic imaging employing electromagnetic radiation (X-rays and gamma rays) are all used in medical imaging procedures. These visuals were created utilizing efficient processors and the arithmetical and logical translation of energy to signals. Signals are transformed into digital images and represent the various types of tissue in the body. Since 1960, image-processing techniques used in many diversified fields. In the 20th century, image processing technology became simpler, faster, and more reasonable price.

The advantages of digital images are low-cost processors, processor speed, easy-to-communicate and storing ability, quick quality assessment, fast reproduction rate, and ease of manipulation. It has some drawbacks also. Such as copyright infringement, the difficulty to resize while maintaining quality, the necessity for high-capacity memory for storing high-quality files, and faster processor for high-quality data manipulation. Medical image-processing signifies the specific problem and enhance the raw medical data for some selective visualization.

Medical Image segmentation is classified into two types. (i) local segmentation (ii) global segmentation. Particularly in one split of the picture, the local segmentation is effective. Compared to the global type, this approach uses fewer pixels. The entire picture functions as a single unit after the global segmentation.

5.2.5 Image-Based on Edge Detection

Boundary recognition of closely related objects or regions is used in a segmentation technique known as edge detection. This technique establishes the discontinuity of the pieces. This technique is mostly used in image analysis to identify the regions of an image when there is a noticeable intensity shift.

Edge Detection Techniques are explained as follows:

5.2.5.1 *Robert's Kernel Method*

This method is used to distinguish between two closely spaced pixels. It is known specifically as forward differences. This approach employs the 1st order derivative and cross gradient operator for edge detection in images with high noise levels described in Equations (5.5) and (5.6) [6].

$$\frac{\partial f}{\partial x} = f(i+1, j) - f(i, j+1) \tag{5.5}$$

$$\frac{\partial f}{\partial x} = f(i, j) - f(i+1, j+1) \tag{5.6}$$

5.2.5.2 *Prewitt Kernel*

The core difference concept is the foundation of the Prewitt kernel approach. Compared to the Roberts operator, this method is superior. The fractional derivative equation of Prewitt operator is described in Equation (5.7):

$$G_x = (a_2 + ca_3 + a_4) - (a_0 + ca_7 + a_6) \tag{5.7}$$

5.2.5.3 *Sobel Kernel*

The point of difference, which generally leans toward the core pixels, may be the basis for this strategy. This method may be written as a $3X3\ matrix$ to the Gaussian kernel's first derivative. The related equation described in given at Equations (5.8)–(5.9) [6–8]:

$$G_x = (a_2 + 2a_3 + a_4) - (a_0 + 2a_7 + a_6) \tag{5.8}$$

$$G_y = (a_6 + 2a_5 + a_4) - (a_0 + 2a_1 + a_2) \tag{5.9}$$

In terms of noise reduction, the Sobel outperforms the Prewitt. This method is applied in nuclear medicine and other functional imaging modalities. The unwinding of precisely nearby cells in the examination of red blood cell pictures is regarded as a challenging problem because of the ambient noise. This has an impact on the interpretive processes and makes it challenging for the doctor to diagnose them.

5.2.5.4 *k-Means Segmentation*

A method of signal and vector valuation is called k-means clustering. This method helps to divide the image into n numbers and k no of clusters, where each data point conforms to the cluster with an equivalent mean. Cluster segmentation tends to determine the clusters with set of comments $x_1, x_2, \ldots\ldots x_n$ where each comment represents d-dimensional real vector. In order to reduce the within-cluster sum of squares, k-means clustering divides the n observations into k (≤ n) sets $S = \{S_{1,}S_{2,}\ldots\ldots S_k\}$ given in Equation (5.10) and Equation (5.11) [9].

$$arg\ min_s\ \Sigma_{i=1}^{k}\ \Sigma_{x\in S}||x-)(\mu_i||^2 = argmin_s\ \Sigma_{i=1}^{k}|S_i VarS_i \quad (5.10)$$

Where, μ_i is the mean of points in S_i

$$arg\ min_s\ \Sigma_{i=1}^{k} \frac{1}{2|S_i|} \Sigma x|y \in S_i||X-Y||^2 \quad (5.11)$$

5.2.6 Images from γ-Rays

These pharmaceutical tracers with radioactive labels are injected or consumed and then circulate and/or undergo metabolism within the body. During passing radioactive decay, the γ-rays emit from the body frame and gamma camera capture image that depict the relevant organs' functional or metabolic activity. The perfect radioisotope should only emit monochromatic, or single-energy, γ-rays. In nuclear medicine, there are three imaging modalities. projection studies, also known as planar scintigraphy, like projection radiography. A revolving gamma camera is used in tomographic image renowned as single-photon-emission-computed-tomography (SPECT), to collect projection pictures from various angles that are then utilized to build cross-sectional images. Additionally, pairs of 511 keV gamma photons are detected via positron emission tomography (PET), which occurs when positrons are destroyed. Figure 5.5 illustrates the process of extracting planar nuclear medicine images.

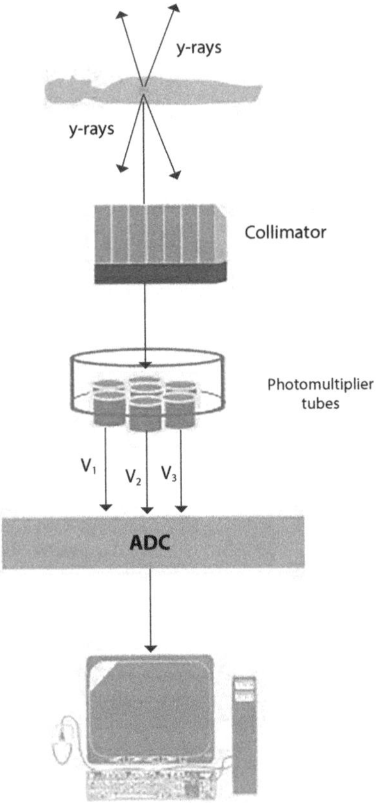

Figure 5.5 Diagram for extraction of planar nuclear-medicine image.

5.2.6.1 Non-Ionizing Radiation

Medical-Ultrasound utilizes pulse-echo technique. Some characteristics of the ultrasound technology such as creating non-invasive device, low cost, reliability, it is widely used. Unpaired nucleons, such as protons, align within a MF, and RF pulses can alter this alignment balance.

Ultrasound imaging:
Sound waves are longitudinal waves, the material's particles oscillate along the same path as pulse wave. The authors depict a schematic diagram outlining the principles of clinical ultrasound imaging in Figure 5.6. Here, sound-speed denoted by 'v', density 'ρ', compressibility 'K'. Thus:

$$v = \frac{1}{\sqrt{K_\rho}} \tag{5.12}$$

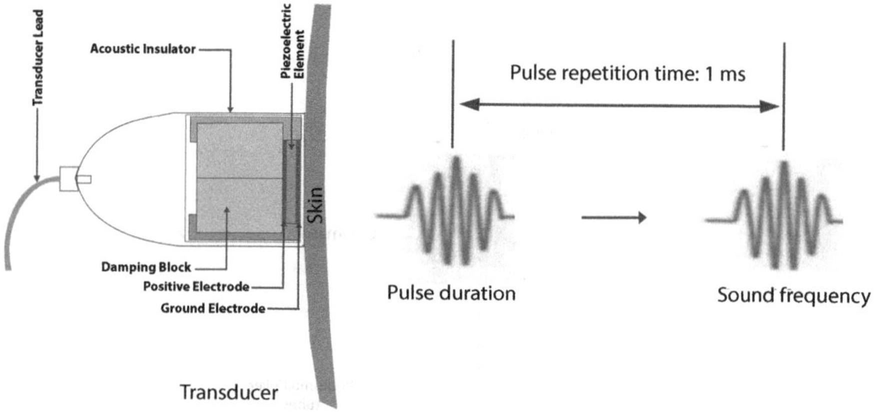

Figure 5.6 Schematic diagram of clinical ultrasound imaging.

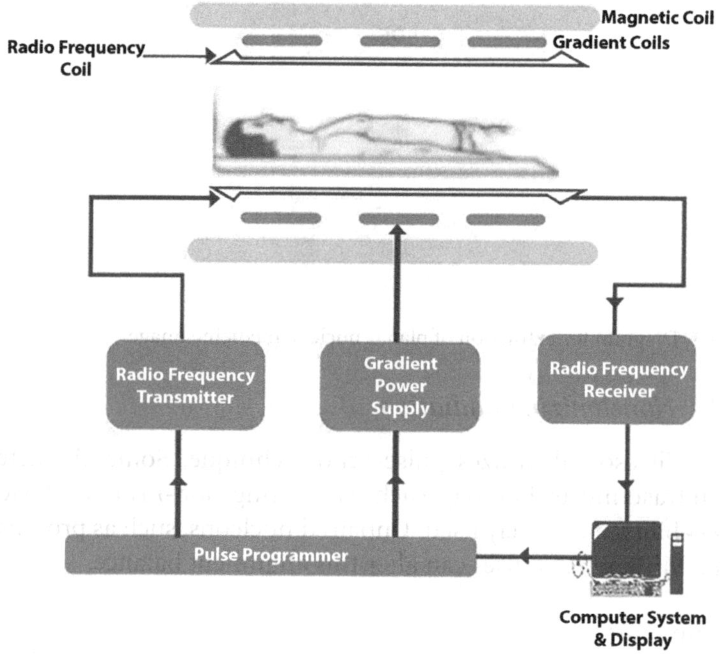

Figure 5.7 Basic components of MRI or magnetic resonance imaging.

5.2.6.2 Magnetic Resonance Imaging

Another non-ionization technique is Magnetic resonance imaging (MRI) which employs 200MHz–2 GHz and around 1–2 tesla. The superconducting magnets formed strong magnetic fields, which transmit electricity through coils of superconducting wire with virtually zero electrical resistance. MRI images provide High spatial resolution (~1 mm), anatomical and physiological information with 3D capability and brilliant soft-tissue visualization. Using several no of slice images, the device reconstructs the image. Due to superconducting magnets, MRI scanners more costly than CT scanners. The primary components of the MRI scanner are illustrated in Figure 5.7.

5.2.6.3 Segmentation Using Multiple Images Acquired by Different Imaging Techniques

Pixel classification for a single picture is based on a single feature (grey level), and segmentation is carried out in a one-dimensional (single-channel) feature space. The EM algorithm iterates between tissue categorization and inhomogeneity correction ("E" and "M" steps), inspired by Markov models and aiming to minimize thermal noise in MR images. The ensuing combinatorial optimization issue was solved by Held *et al.* [10] using the iterated conditional modes approach.

Since the identity covariance matrix is entirely unconstrained. So, like Markov Random Field (MRF) model, the smoothness constraint incorporated with covariance matrix associated by adjacent points.

$$C_{smooth} = \begin{bmatrix} 1\ 0\ 0.5\ 0 & \cdots & 0.5\ 0 \\ \vdots & \ddots & \vdots \\ 0\ 0.5\ 0\ 0 & \cdots & 0\ 1 \end{bmatrix}$$

Where C_{smooth} represents 2N X 2N matrix. In that matrix, the first two eigenvectors represent rigid translation and other eigenvectors represent greater variations in frequency.

The MRF model simulates the Spatial relationships among neighboring elements often employed in statistical classifiers [11]. The Supervised classifiers may integrate MRFs under a Bayesian prior model [10, 12] and unsupervised classifiers integrate with the K-means method [13, 14] and the EM algorithm [15, 16]. This Markov RF also suggested the for-tissue classification [17, 18]. Within fuzzy classifiers, smoothness might also be required [19–21].

5.3 Methods

The real image is thought to be made up of K tissue classes, each of which has a unique, unknowable intensity value known as the class mean, indicated by the letters v_k, k = 1, ..., K. Spatial fluctuations in the picture domain Ω may arise in the class means because of inhomogeneities or variations within the tissue itself.

$$y_j = \Sigma_{k=1}^{K} z_{jk} v_{jk} + \eta_j, j \in \Omega \tag{5.13}$$

Where y_j is represented intensity at pixel j, White Gaussian Noise denoted by, n_j with variance σ2.

The multiplicative gain field, g_j varies with mean of the function illustrated as,

$$y_j = g_j \, \Sigma_{k=1}^{K} z_{jk} v_{jk} + \eta_j, j \in \Omega \tag{5.14}$$

We define z_{jk} as an indicator function used in the MRF model that satisfy,

$$z_{jk} = \begin{cases} 1, & \textit{if pixel j is in tissue class k} \\ 0, & \textit{otherwise} \end{cases} \tag{5.15}$$

5.3.1 k-Means Approach

In the MRF model optimal segmentation incorporated by K-means algorithm [22]. The joint-posterior probability (JPP) function illustrated by summation of the two energy functions being minimized:

$$E = \frac{1}{2\sigma^2} E_1 + \beta E_2 \tag{5.16}$$

The first term represents:

$$E_1 = \Sigma_{j\in\Omega} \Sigma_{k=1}^{K} z_{jk} (y_j - g_j v_k)^2 \tag{5.17}$$

The second term represents the Markov Random Field:

$$E_2 = \Sigma_{j\in\Omega}\Sigma_{i\in N_j} z_j^T V z_i \tag{5.18}$$

The equation for E1 follows from the properties that z_2, $z_{jk} = 1$, $j_k = z_{jk}$, k=1. E_1 calculates the objective function by the K-means clustering algorithm. 'E' is minimized by the K-means, i.e. ISODATA algorithm with iterative method. We disregard the consequences of σ2 without losing generality.

The standard K-means algorithm defines three steps, the typical K-means approach, the means of every cluster data, and the gain. To evaluate the indicator functions, we compute for each pixel j ∈:

$$z_j = arg\ minz_j\left[(y_j - g_j\Sigma_{k=1}^{K} z_{jk}v_k)^2 + \lambda\Sigma_{i\in N_j} z_j^T V_{Z_i}\right] \tag{5.19}$$

The zero Gradient equation derived by:

$$v_k = \frac{\Sigma_{j\in\Omega} z_{jk} g_j y_j}{\Sigma_{j\in\Omega} z_{jk}(g_j^2)} \tag{5.20}$$

When g_j is unity, standard K-means algorithm derived by:

$$g_j = \Sigma_{n=1}^{N} f_n P_n(j) \tag{5.21}$$

Here, P_n represent the Chebyshev polynomial basis functions and f_n denotes as the coefficients of functions.

Markov Random Fields (MRFs) are a powerful tool for modeling dependencies among variables in a variety of fields, including the medical field. One simple MRF model in the medical field is a model for medical image segmentation. The MRF model captures the dependencies between neighboring pixels in the image. Specifically, the probability of assigning a particular label to a pixel depends on the labels assigned to its neighboring pixels. This is because neighboring pixels tend to have similar labels, due to the spatial continuity of the structures in the image. The MRF model can be represented as a graph, where each pixel is a node and edges connect neighboring pixels. The edges in the graph encode the spatial relationships between the pixels, and the weights on the edges represent the strength of the dependency between neighboring pixels. The MRF model provides a powerful framework for modeling dependencies between variables in the medical field, and has been used successfully in a variety of applications, including medical image segmentation, image registration, and disease diagnosis.

5.3.2 Bayesian Objective Function

Bayesian image restoration is a popular technique for medical image processing, which allows for a principled approach to incorporate prior knowledge about the image and the imaging process in order to improve image quality. In the medical field, images acquired from various imaging modalities (such as MRI, CT, PET) are often noisy, distorted, or otherwise degraded. Bayesian image restoration provides a powerful framework for improving the quality of these images, by incorporating prior knowledge about the imaging process and the structure of the image. The basic idea behind Bayesian image restoration is to estimate the underlying image from the observed noisy image by exploiting the prior knowledge of the image and the imaging process. The estimation process involves formulating a statistical model that describes the relationship between the observed image and the underlying image, and then using Bayesian inference to estimate the underlying image.

In covariance matrices, prior models are used for structure in the individual image, and determine the shape parameters $w = (w_{1,}w_2, \ldots., w_t)$ T & parameters: scale, rotation and translation referred as a, θ and $T_{x,}T_y$ respectively. The pose and shape parameter vector represented by $p = (a, \theta, T_x, T_y, w_1, w_2, \ldots\ldots w_t)T$. The nth boundary point, $(n = 0, 1, \ldots\ldots, N-1)$ is

$$x(p,n) = a\cos\theta\,[\overline{x}(n) + \Sigma_{k=1}^{t}\, Q_{2n,k} w_k] - a\sin\theta\,[\overline{y}(n) + \Sigma_{k=1}^{t}\, Q_{2n+1,k} w_k] + T_x \tag{5.22}$$

$$y(p,n) = a\sin\theta\,[\overline{x}(n) + \Sigma_{k=1}^{t}\, Q_{2n,k} w_k] + a\cos\theta\,[\overline{y}(n) + \Sigma_{k=1}^{t}\, Q_{2n+1,k} w_k] + T_y \tag{5.23}$$

where Q is represented the covariance matrix contains the first 't' eigenvectors. The authors frame the issue as a Bayesian maximum a posteriori problem, leveraging prior knowledge of the shape model to enhance boundary determination.

5.4 Segmentation and Database Extraction with Neural Networks

After data segmentation, data must be extracted & represented by various patterns. Due to the complex, noisy and difficult recognition patterns of medical images, machine learning algorithms are been used for testing and computer-aided design (CAD) is involved. Artificial neural networks (ANNs) and Bayesian belief networks (BBNs) are two of the ML approaches that are used. CAD system incorporates with ANN and BBN, determine good and reliable performance in medical purpose and optimize the database and selection of feature of medical image processing.

5.4.1 Artificial Neural Network

A computer structure called an ANN was developed as a result of research into biological neuron processing. It can simulate the human brain in certain ways and learn or recognize patterns based on insufficient or partial information. Using the back-propagation method, there are many other ANN topologies, ANN trained a multilayer feed-forward is the most common network design. Most artificial neural networks (ANNs) employed in computer-aided design (CAD) of medical images typically consist of three layers. This choice is based on empirical evidence suggesting that a single hidden layer can approximate any continuous function [23].

Figure 5.8 depicts a standard three-layer artificial neural network (ANN). This configuration comprises an input layer, where neurons are characterized by a set of features, a hidden layer responsible for internal computations, and an output layer where final decisions are generated. The equation describes the relationship between the hidden neurons $(h_j, j=1,2,\ldots,m)$ that link the input and output neurons denoted as $(X_i, i=1,2,\ldots,n)$ and $(Y_K, k=1,2,\ldots,N)$ respectively.

$$Y_k = g\left[\sum_{j=1}^{m} w_{kj}\, g\Big(\sum_{i=1}^{n} w_{ji} x_i + \theta_{in}\Big) + \theta_{hid}\right] \tag{5.24}$$

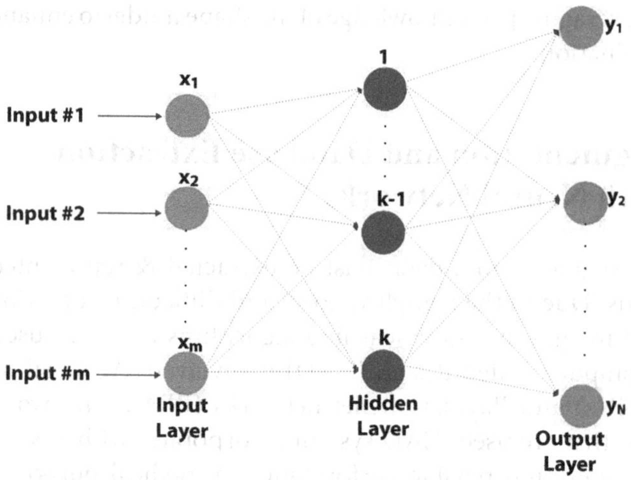

Figure 5.8 A three-layer artificial neural network.

Where $g_z = 1/(1+d-z)$, w_{ji} denoted as the weight from the input neuron i_{th} terms to the hidden neuron j_{th} terms and w_{ki} is denoted the weight from the j_{th} hidden neuron to the output neuron k_{th} terms, θ_{in} is referred as a input layer bias neuron, θ_{hid} is hidden layer bias neuron.

5.4.2 Bayesian Belief Networks

A Bayesian Belief Network (BBN) is a graphical representation of data that employs conditional dependencies to efficiently describe the joint probability distribution function in the occurrence of a problem. Computer-assisted diagnosis of breast cancer using a data-driven Bayesian belief network. Instead of having a fixed topology like a three-layer ANN, the structure of BBN may be flexibly adjusted for various purposes depending on human understanding. The network topology enables researchers to describe the dependency.

For instance, one research created and evaluated a BBN to distinguish between benign and malignant breast tumors. The network has 13 characteristics, of which 5 are derived from the clinical histories of the patients, 4 from general physical examination results, and 4 more from mammography findings. The topology of a BBN was created based on the connection (dependence) of these properties (Figure 5.9). In Bayesian image restoration, the underlying image is modeled as a random field, with a prior distribution that captures the prior knowledge about the image structure.

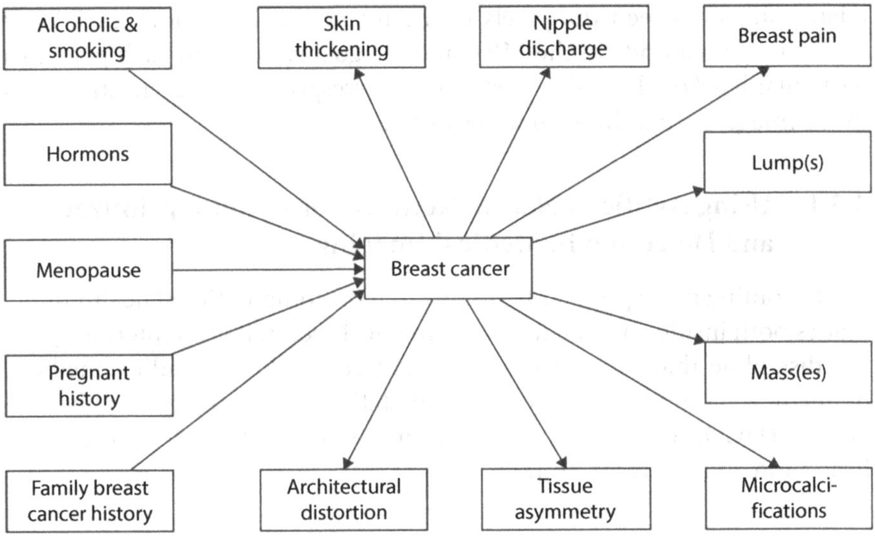

Figure 5.9 Diagram of a BBN for breast cancer diagnosis.

The observed noisy image is then modeled as a noisy version of the underlying image, with the noise modeled as a random variable with a known distribution. The Bayesian inference is to estimate the posterior distribution of the underlying image given the observed noisy image and the prior distribution. This can be done using techniques such as Markov Chain Monte Carlo (MCMC) or variational inference.

In the medical field, Bayesian image restoration has been successfully applied to a wide range of problems, including denoising, deblurring, super-resolution, and image segmentation. It has been used to improve the quality of medical images acquired from various imaging modalities, and to enhance the accuracy of medical image analysis and diagnosis.

5.5 Applications in Medical Image Analysis

Medical imaging relies heavily on classification algorithms, such as disease detection and categorization of tumors. After segmentation, the suspected area recognized and extract the pertinent information from the region. Breast cancer detection is a highly significant application. Medical image segmentation is the process of identifying and separating different structures within medical images, such as MRI or CT scans. One approach to medical image segmentation is to use MRFs to model the spatial relationships between the pixels in the image. In this model, each pixel in the image is a node in the MRF, and the goal is to assign each pixel to a particular label or class. The labels might correspond to different structures in the image, such as bone, muscle, or fat.

5.5.1 Using Artificial Neural Network for Better Optimization and Detection in Medical Imaging

It is becoming more popular to do research on computational health informatics both inside and outside of the medical sector. It is an interdisciplinary discipline that draws from many different fields, including medical, biomedical, computer technology, and statistics. Improving Health Care Output (HCO), or the standard care for patients is the main objective of health informatics research [24].

5.5.1.1 Opportunities

Big data analytics in healthcare face significant hurdles because to all these big data features (including all vs.). Different models and visualization strategies might be used to adapt to and maximize the benefits of big data analytics by utilizing parallel computing platforms [25, 26] and the following are some specific instances of opportunities that could be investigated:

Personalized care: Develop prediction models to use real time personalized therapies for patients such as genomic DNA sequence for cancer care. Before a patient has any signs of the condition, these methods could provide early identification and diagnosis.
Genomic analytics: Enhance the traditional healthcare decision-making process by incorporating genomic analysis through the development of efficient and effective gene sequencing technologies. Utilize high-throughput genomic sequencers to conduct comprehensive genome-wide association analyses (GWASs) for a better understanding of genetic factors associated with human illnesses.

5.6 Standardize Analytics Pipeline for the Health Sector

A complete health informatics processing pipeline is comprised of a series of phases that make up a big data computing architecture for the field of

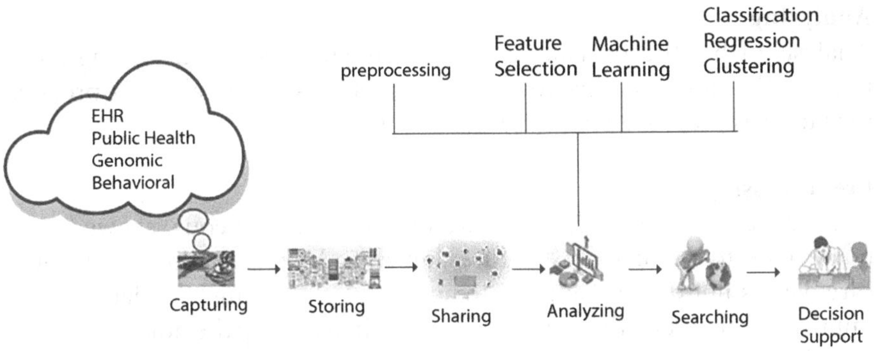

Figure 5.10 Health informatics processing pipeline.

health informatics. Every stage of the pipeline is crucial to producing accurate and worthwhile results from big data analytics. In particular, the gathering, storage, and sharing of large-scale data provide the essential input for the subsequent analysis process, in which a variety of analytical techniques are used to discover significant patterns from big data in healthcare to make quick and efficient judgments. The pipeline (as seen in Figure 5.10) will be covered in detail in this section.

Capturing

The recent adoption of EHRs by U.S. hospitals propelled the healthcare sector into a new, high-tech era with significant promise for using analytics to enhance outcomes. The EHR is renowned for its advantages in providing standardized medical data, enhancing patient care, and lowering costs. Healthcare professionals can use the EHR to obtain data about specific individuals or groups [27].

Storing Availability

The provision of excellent management and storage strategies is crucial for achieving cost-effectiveness. An ideal healthcare data storage and management system should have a number of unique characteristics.

Sharing

Acquisition of patient's diagnostics data, the primary aim is how to safely share the data across every necessary place and communicate with scientists and doctors. Nowadays, Big data has become increasingly and it is the subset of cloud computing. Cloud computing will support the synchronization of data storage with devices by assuring storage capacity, server administration, bandwidth utilities, and network power.

Analyzing

Analysis of healthcare data has frequently employed data mining techniques. The next sections cover the main stages of the analysis process, including as preprocessing, data extraction/selection.

Preprocessing

The nature of real-world healthcare data is heterogeneous, skewed, and noisy. Due to the complex nature and mixed-up data, the raw healthcare data is not directly applied to analytical algorithms. So, data quality improvement and further study and research, handling the noisy and complex dataset in large scale zone by preprocessing technique.

5.7 Feature Extraction/Selection

Feature selection, also known as attribute selection or variable selection. It is extracting and choosing a subset of significant and pertinent characteristics from a vast set of measured data. In general, incoming data may contain redundant characteristics that may be reduced utilizing dimensional reduction methods to a limited collection of important features. Big, multidimensional medical data may be ineffective and inaccurate. Healthcare informatics feature extraction techniques are frequently used by researchers to get over this problem.

5.7.1 Significance of Machine Learning for Medical Image Processing

A subset of Artificial Intelligence (AI) known as "machine learning (ML)" is teaching algorithms to make predictions or judgments based on information. Medical image analysis and diagnosis and treatment are made easier with the use of machine learning techniques in the healthcare industry. As an illustration, machine learning algorithms may be trained to spot patterns in photos of medical conditions and diagnose illnesses. Both supervised learning (also known as predictive learning) and unsupervised learning (sometimes known as descriptive learning) are common subcategories of machine learning algorithms [28].

In supervised learning, inputs and the intended outputs are given simultaneously so that a universal prediction guideline establishes connections between inputs and outputs can be acquired through learning. The two main subcategories of supervised learning examined in this review are classification and regression. The objective of unsupervised learning is to

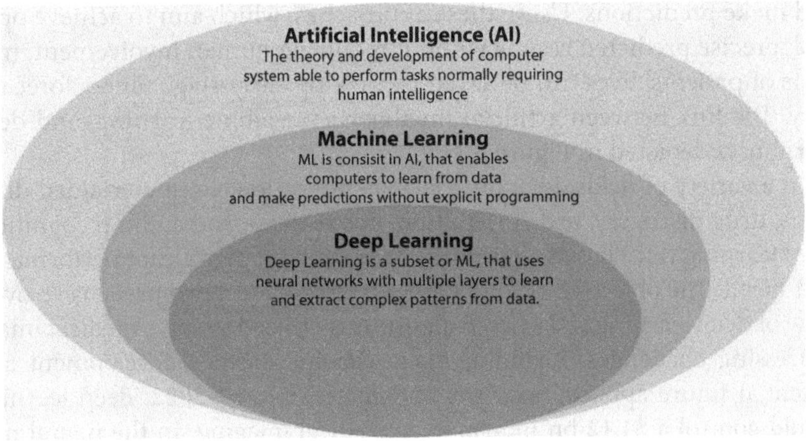

Figure 5.11 Relationship between AI, ML, and Deep Learning.

identify the unknown data by looking for similarities across records. One form of unsupervised learning strategy is clustering. This study covers the following three kinds of **clustering algorithms**: The partitioning approach is suggested because it is straightforward, quick, and able to handle huge datasets. Although it is occasionally not ideal for massive data owing to space and time constraints, the **hierarchical algorithm** emphasizes the another clustering approach with its visualization capabilities. The final approach, **Density-Based Clustering**, manage dynamic and intricate datasets. It identifying outliers and capturing arbitrary patterns in biomedical images. Similar to hierarchical clustering for extensive datasets, a subset of artificial intelligence, employing machine learning, is utilized to enhance the learning experience without direct human interaction.

5.7.2 Significance of Deep Learning for Medical Image Processing

Deep learning is another method that is frequently applied to the prediction of different diseases. It is a constituent of both artificial intelligence (AI) and machine learning (ML), which have made significant advances in fields including computer vision methods, picture captioning, image coloring, medical image study, drug detection and fraud detection methods. Large scale datasets are employed drug discovery, neural networks, and even fraud detection systems. It employed in deep learning, where feature extraction techniques are used to determine the most important qualities, and then different data mining algorithms are used to conduct data analysis and make predictions. Using these approaches, which aim to achieve optimal, precise predicted results without requiring human involvement, millions of patients' lives can be saved by making early-stage illness forecasts easy. The link between artificial intelligence, machine learning, and deep learning is depicted in Figure 5.11.

In a variety of fields, including face identification, computer-aided diagnosis, drug discovery, text recognition, lips reading, and audio recognition, deep learning (DL) based algorithms demonstrated significant performance and speed. The objective of this chapter is to provide a comprehensive overview of deep learning (DL) based algorithms utilized to solve medical image processing challenges, including their current state of development and potential future applications. It is anticipated that by 2022, deep learning would control a $1.12 bn industry for medical imaging. In the neural network, fundamental computing unit is the neuron, which was developed from research on the human brain. Neurons accept many signals as inputs, combine signals linearly and sends to the output signal via nonlinear processes.

A wide range of issues, including cancer screening, illness monitoring, and personalized therapy recommendations, are addressed by deep learning applications in the healthcare industry. Today, a huge quantity of data is available to doctors thanks to a variety of sources, including radiological imaging (X-ray, CT, and MRI scans), pathological imaging, and most recently, genetic sequencing.

Diabetic Retinopathy

One of the most prevalent retinal diseases in the world is known as diabetic retinopathy (DR). Depending on the degree of severity, it results in total blindness. It harms the microscopic inner layers of the eye as well as the tiny blood vessels of the retina. However, these diagnostic techniques are quite costly and challenging. The research made the following original contributions, among others: first, giving a comprehensive history of the DR illness and current methods of diagnostics. Second, applications for deep learning in DR are described together with several imaging approaches. Third, several DR detections use cases and real-world scenarios are investigated using deep learning approaches.

Detection of Gastrointestinal (GI) Diseases

The gastrointestinal system (GI) comprises all the organs involved in food digestion, nutrition absorption. The esophagus, stomach, large (colon or big bowel), and small (small bowel) intestines are the organs. Different conditions and disorders such as GI tract inflammation, bleeding, infections, and cancer impair food digestion and absorption [29]. In order to diagnose and analyze these disorders, image processing and machine learning are essential. They also assist clinicians in reaching quick decisions on effective and precise therapy. A common technique for automatically extracting features from endoscopic pictures is to use CNN [30]. The SVM is then used to classify and identify gastrointestinal lesions using the characteristics.

Cardiac Imaging

DL has demonstrated significant outcomes in the domain of cardiac imaging, particularly in the accurate estimation of Calcium scores. The most popular imaging modalities are CT and MRI, and left ventricle segmentation is a frequent problem for image processing. A number of various applications have been created. Since manual CAC identification in cardiac CT needs significant expert engagement, it is time-consuming and impractical for large-scale or epidemiological research. Recent research on cardiac imaging is centered on CAC calculation utilizing deep conventional neural networks based on CT angiographic images.

Tumor Detection

Using Artificial Neural network for better optimization and detection in medical image processing. A tumor, also known as neoplasm, is a mass formed when cells from any region of the body develop abnormally. Tumors come in two different varieties. Malignant tumors are malignant, while benign tumors are not. A benign tumor does not spread to other parts of the body and is not particularly hazardous as long as it stays in one place. The spread of a malignant tumor, however, is extremely damaging and affects other bodily parts. Wang *et al.* experimented, in between 32 to 70 years of the person's age, 482 mammograms were utilized [31] and observed 246 ladies who have tumors. Following median filter denoise, region growth, morphological procedures, and modified wavelet transformation are used to separate the breast tumor from the surrounding tissue. To classify and detect breast tumors, morphological and textural characteristics were then given to SVM and an extreme learning machine.

5.8 Image-Based Forecasting Using Internet of Things (IoT) in Smart Healthcare System

Image processing, AI, machine learning, and deep learning are increasingly being used in medical aid to improve the accuracy, speed, and efficiency of medical diagnosis and treatment. Smart healthcare systems built on the Internet of Things have the potential to completely change how medical care is provided. To enhance patient outcomes and quality of life, IoT (Internet of Things) may be employed in several health monitoring systems. An IoT-based smart healthcare system has the following critical attributes:

- Wearable devices: Wearable devices may be used by IoT-based smart healthcare systems to continually monitor patient health. Vital indicators such as heart rate, blood pressure, and oxygen saturation levels may be measured by wearable technology such as smart watches, fitness trackers, and health monitors. Real-time transmission of this data to medical professionals enables them to monitor patients from a distance and take appropriate action if necessary.
- Remote patient monitoring: Using smart healthcare solutions built on the Internet of Things, patients may be monitored remotely. Patients may assess their health parameters at home using IoT-enabled equipment like glucometers, blood pressure monitors, and spirometers. Healthcare professionals may access this information and remotely monitor patients while giving advice and assistance.
- Electronic health records (EHRs): Electronic health records (EHRs) can be used by IoT-based smart healthcare systems to store patient data. EHRs may give medical professionals access to patient health information whenever and wherever they need it, improving care coordination and lowering mistake rates.
- Automated alerts: Automated alerts can be used by IoT-based smart healthcare systems to inform medical professionals of possible health risks. For instance, alerts can be issued to healthcare professionals in case a patient's vital signs deviate from a certain range, allowing them to take appropriate action.
- Predictive analytics: Predictive analytics may be used by IoT-based smart healthcare systems to spot possible health problems before they become severe. Predictive analytics algorithms, for instance, may detect patients who are at a high risk of contracting a specific disease or condition by tracking patient health data over time and offering preventative therapies.

- Medication management: Patients may better manage their prescriptions by ensuring they take the proper dose at the right time with the use of IoT-enabled pill dispensers and medication tracking devices. If a patient forgets to take a dosage or takes too much medicine, notifications can be sent to the healthcare professional.
- Chronic disease management: Chronic conditions including diabetes, asthma, and heart disease may be monitored and managed with the help of IoT devices. Patients can monitor their symptoms and get notifications if anything changes. The data gathered can be used by healthcare professionals to make better choices about the patient's care.

5.9 IoT Monitoring Applications Based on Image Processing

Real-time visual data analysis is a common feature of IoT (Internet of Things) monitoring systems. Consider these instances:

Smart security systems: With real-time object and person detection and recognition, IoT cameras with image processing capabilities may inform users via their smartphones when they observe questionable activity.
Industrial automation: IoT cameras make real-time product fault detection and manufacturing line monitoring possible. Algorithms for image processing can be used to categorize errors and launch automatic reactions like halting the manufacturing line or notifying a human operator.

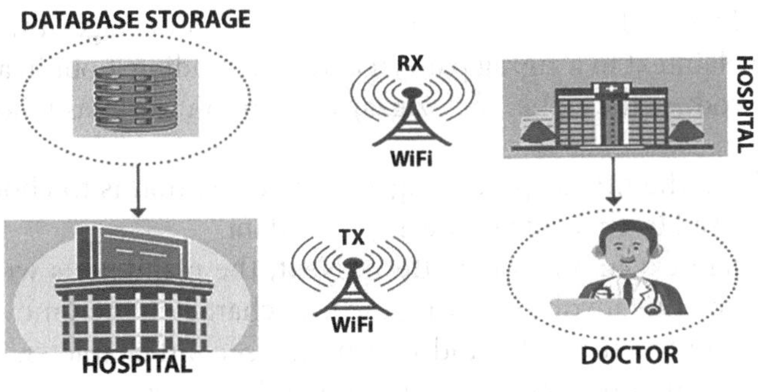

Figure 5.12 IoT monitoring system in healthcare.

Agriculture: IoT cameras may be used to keep an eye on crops and look for symptoms of pests, illnesses, and other problems. Algorithms for image processing can be utilized to pinpoint the precise issue and provide the farmer with a solution. For IoT monitoring applications, image processing is an effective tool that may offer real-time insights and automate answers to a variety of problems in many sectors. IoT is used for medical imaging purposes. As an example, images operated in the textual version of the analytical data. Images are used to conceal information. AES and RSA, which stand for Advanced Encryption Standard, were used to encrypt the data using industry-standard methods. The cipher text is hidden using a 2D discrete wavelet transform (1 level) picture. In order to hide its contents, the text scale has been suppressed. To do this, two different cover pictures with two different text sizes were produced using grayscale and color photographs. Another illustration of this idea may be found in [32], which provides pre-encrypted medical photos with an extra layer of security supplied by an integrated one-time password (OTP). The same user will get both the OTP and the encrypted picture.

The encoded OTP will then be shown when the medical picture has been encrypted and decoded. Finally, a comparison is made between the two OTPs that were previously extracted from the medical picture. To ensure the integrity of the encrypted medical picture, a verification step is required. Original medical data will be extracted following the verification process. Figure 5.12 depicts an example of an image-processing based IoT healthcare application. When sending information securely through communication channels like the cloud for processing and/or storage purposes, this application may be apparent. Medical image zigzag encryption is one of these benefits [33]. The suggested approach outperforms the opposition and might be useful for diagnosing medical photos.

Zigzag encryption is a type of encryption algorithm that can be used to secure data in an IoT healthcare system. This encryption algorithm involves writing the plaintext in a zigzag pattern and then reading it out in a different pattern to produce the cipher text. zigzag encryption works as follows:

Choose a key: The first step in using zigzag encryption is to choose a key that will be used to encrypt and decrypt the data.

Write the plaintext in a zigzag pattern: Next, the plaintext is written in a zigzag pattern. This involves writing the first character, then moving diagonally down and to the right, and writing the second character. The third character is written directly below the second character, and so on, until the entire plaintext is written in a zigzag pattern.

Read the cipher text in a different pattern: To decrypt the data, the cipher text is read out in a different pattern. This involves reading the first character, then moving diagonally down and to the right, and reading the second character. The third character is read directly below the second character, and so on, until the entire cipher text is read out in a different pattern.

Zigzag encryption can be used to secure data in an IoT healthcare system by encrypting patient data such as medical records, vital signs, and other sensitive information. This helps to protect the privacy and confidentiality of patient data, and ensure that it is only accessible to authorized users.

5.10 Significance of Computer-aided Big Healthcare Data (BHD) for Medical Image Processing

Computer-aided big healthcare data (BHD) analytics refers to the use of advanced computational and analytical techniques to analyze large volumes of healthcare data. Here are some key features of computer-aided BHD analytics:

- Data integration: Computer-aided BHD analytics involves integrating data from various sources, such as electronic health records, medical imaging, clinical trials, and genomics data. By integrating data from multiple sources, healthcare providers can get a comprehensive view of patient health and make better-informed decisions.
- Predictive analytics: Computer-aided BHD analytics uses predictive analytics algorithms to forecast future healthcare outcomes based on historical data. Predictive analytics can help healthcare providers anticipate potential health issues and take proactive measures to prevent them.
- Clinical decision support: Computer-aided BHD analytics provides clinical decision support by giving healthcare providers access to real-time patient data and insights. This data can help healthcare providers make better-informed decisions about patient care.
- Precision medicine: Computer-aided BHD analytics enables precision medicine by tailoring treatments to individual patients based on their unique genetic makeup, medical

history, and other factors. By analyzing large volumes of data, computer-aided BHD analytics can identify patient subgroups that are more likely to respond to a particular treatment and can help healthcare providers make more personalized treatment decisions.

It has been noted that the usage of big data architecture and methodologies is continually helping to manage the healthcare sector's rapid data development. Big data has enormous prospects for the healthcare industry in terms of increasing healthcare effectiveness and quality, identifying dangers to public health, controlling human health via early illness diagnosis, and supporting wiser decisions. However, some of the biggest issues facing the big data sector are data storage, data standardization, data transport, and privacy. Many studies have attempted to include security component in healthcare system. Below are shown two distinct strategies: For the healthcare information system (HIS), [34] presented the structure seen in Figure 5.13. To handle and analyze a large quantity of data for the benefit of both patients and healthcare providers, this framework was utilized. As listed below, there are five primary parts that make up HIS framework.

The cloud environment, the first element, provides diverse services and enables authorized data sharing. The second element, the electronic health record, integrates patient statistics from several sources. The third component, the security layer, efficiently manages security challenges like authentication, authorization, and patient data confidentiality. AES and RC4 are only two of the encryption techniques used at this layer to secure data. In order to prevent unauthorized users, multiple authentication methods, such as One-Time Password (OTP) and Two-Factor Authentication (2FA), have been implemented. Using OTP, it is possible to safeguard network data by creating a new password each time.

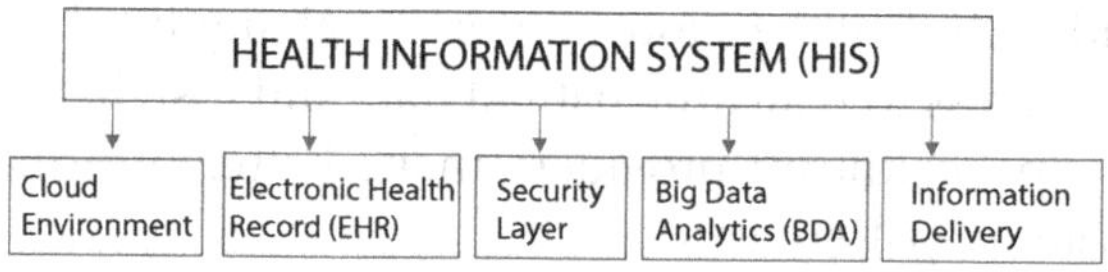

Figure 5.13 Health information system framework.

5.11 Applications of Big Data

5.11.1 Big Data Analytics in Health Sector

The complexity and timeliness of healthcare and medical data make it like big data, rather than its enormous amount and quantity. Electronic Health Reports (EHR), diagnostic reports, physician's orders, photographs of the patient, pharmaceutical records, and research data from medical publications are only a few examples of the patient-related data that constitutes health information [35]. In order to increase the standard of care, detect illnesses early enough to reduce risk factors, and manage hospital information systems more effectively, healthcare sectors must digitize the data they collect today. With the help of big data analytics, it is possible to significantly influence the quality and efficiency of the healthcare industry.

We are describing the four parameters can be effectively accomplished by using descriptive, predictive, and prescriptive big data analytical healthcare techniques.

Patient Centric Care
By making early illness diagnoses that are supported by clinical data and minimizing medicine dose to prevent side effects, patients can be relieved. In addition to lowering patient costs, this lowers hospital readmission rates.

Predictive Analysis of Diseases
Using real-time data to forecast viral diseases before they spread. This may be discovered by looking at the social media accounts of the patients who are localized in that place and are afflicted with a certain condition. By implementing the appropriate preventative steps, this further enables the medical experts to assist the victims.

Real Time Patient Monitoring
Monitoring and observing if the hospital setup complies with the standards specified by the IMC. By conducting regular inspections, the government is better able to take the required action against facilities that should be excluded.

Improving the Treatment Methods
The study of drug doses is a key component in the development of personalized patient care monitoring. Doctors can provide new patients

beneficial medications by using patient data from others who are experiencing similar symptoms. Diabetes, cardiovascular disease, liver disorders, and psychiatric disorders are a few illnesses that have a connection to chronic health problems. Through early illness detection and patient life expectancy improvement, big data analysis contributes to better healthcare. For big data applications across all industries, almost 6K papers have been published between 2018 and 2023. The security and privacy of medical information are equally vital in the modern healthcare business as data size, which is not the only serious problem. In order to solve issues connected to data privacy and optimization, the big data ecosystem in healthcare is necessary. In Figure 5.14 the authors illustrated the design the machine learning based healthcare framework. This framework consists four layers. Such as, ML based application layer, Data Security layer, Data Storage layer, Data Source layer. The distributed framework and storage optimization aid in the optimization of storage resources, which eventually improves system operating performance.

Additionally, by implementing various data protection mechanisms, the attempt is made to uphold the confidentiality and privacy of patient data. The first layer handles many data sources. The major sources of data layer are patient diagnostics records, operational database, genetic information, and medical images and datasheet like X-ray report, MRI scan, CT scan report etc. Heterogeneous data are stored in the second layer. Depending on the kind of data, it can be kept in a data warehouse or in the form of

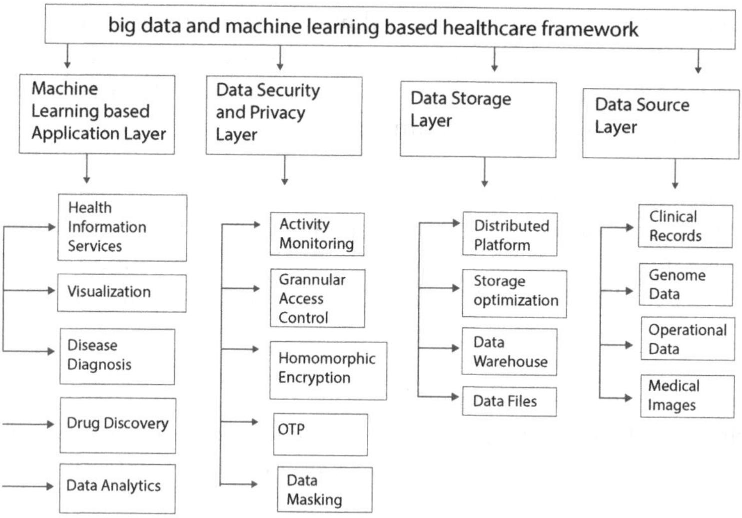

Figure 5.14 Architectural design of machine learning based healthcare framework.

data files. Finally, this layer handles storage optimization to make the most of the memory resources. The integration and analysis of scattered data offered by this system will help patients receive better care and healthcare professionals make wise judgments. Big data is seen to have a serious issue with privacy and security. In order to prevent unauthorized access to data, it must be used appropriately.

5.11.2 Computer-Aided Diagnosis in Mammography

The most crucial method for examining breast cancer is mammography. Early illness detection makes it possible to do surgery or receive the most beneficial treatment. Image processing, segmentation, and feature extraction are used in computer-assisted mammography diagnosis. Segmenting the breast area helps to focus the search for lesions and microcalcifications. Receiver operating characteristic (ROC) analysis may be used to assess various approaches, but it cannot be used to compare them unless the same image datasets were utilized. Since CAD systems are utilized in conjunction with a radiologist and not in isolation, they do not need to be flawless.

5.11.3 Tumor Imaging and Treatment

In order to diagnose and cure cancer, multi-modality imaging is crucial. It is utilized in diagnostics to find, identify, and describe the "tumor burden" [36]. Imaging serves as a guiding tool for surgical resection, radiation treatments, and evaluating therapy effectiveness upon tumor detection. Each imaging technique, including radiography, CT, SPECT, MRI, PET, and others, offers complimentary data that helps medical professionals better comprehend the anatomy and operation of the tumor and surrounding organs.

5.11.4 Molecular Imaging

The non-invasive, *in-vivo* study of molecular cellular events crucial to both healthy and unhealthy processes is known as molecular imaging. Molecular imaging simultaneously detects antibiotic resistance gene, identify, and diagnostics diseases inside the body while also evaluating the effectiveness of different treatments. Researchers aim to integrate molecular imaging with nanotechnologies (utilizing sub-100nm structures or sensors) to precisely locate and diagnostics the affected cells. Quantum-dots (Q-dots) are semiconductor-based fluorescent nanoparticles with the potential for identifying cancer biomarkers [37].

5.11.5 Surgical Interventions

Images used in medicine are utilized for surgical procedures as well as diagnostics. A robot equipped with surgical instruments and a camera is controlled by a surgeon during robotic surgery using a joystick. It is used for treatments with a low risk of complications, such removing the prostate or clearing obstructions between the kidney and ureter. Small servo motors in the robotic arms give the surgeon additional tactile (haptic) feedback. The surgeon uses optically enhanced 3-D glasses to see the patient in stereoscopic and three-dimensional (3-D) fashion.

5.12 Conclusion

The healthcare industry is very distinct from other industries. People anticipate the best level of care and services, regardless of cost, in this high priority area. Following the success of deep learning in other real-world applications, it is now offering innovative solutions for medical imaging with high accuracy and significant technique use for upcoming applications in the health industry. In this article, the author describes the Machine Learning (ML) and Deep Learning [DL] based various healthcare framework and optimized the medical image segmentation architecture.

According to the literature already in existence, big data applications in the healthcare sector revolutionize the medical sector by giving patients better health and knowledge. AI and ML algorithms can be used to analyze medical images to automatically identify patterns or abnormalities. Deep learning has made significant advancements over typical machine learning algorithms in the last several years, taking a leading role in the automation of our daily lives. It analyze large amounts of medical data, such as patient records, lab results, and medical imaging, to identify patterns that are indicative of a particular disease or condition. This can help doctors to identify and treat patients earlier, before the disease or condition becomes more serious. Many large research institutions are developing deep learning-based solutions that promote the usage of deep learning when applied to medical imaging. Image processing, AI, machine learning, and deep learning in medical aid is rapidly advancing, leading to new and innovative ways to diagnose and treat diseases and injuries. These complete system structure and significant, vibrant technology of the smart patient monitoring system are familiarized in this chapter. Here author clearly elaborated the remote monitoring imaging system and expected that it will be performed crucial role in healthcare in the future.

References

1. Kirubakaran, B. and langkumaran, M., Selection of optimum maintenance strategy based on fahp integrated with gra–topsis. *Ann. Oper. Res.*, 245, 1-2, 285–313, 2016.
2. Abdallah, Y.M., History of medical imaging. *Arch. Med. Health Sci.*, 2017, 5, 275–278, 2017. 10.4103/amhs.amhs_97_17.
3. Abdallah, Y., *An Introduction to PACS in Radiology Service: Theory and Practice*, LAP LAMBERT Academic Publishing, Germany, 2012, 978-3846588987.
4. Abdallah, Y., Increasing of edges recognition in cardiac scintigraphy for ischemic patients. *J. Biomed. Eng. Med. Imaging*, 2, 6, 40–48, 2016. http://dx.doi.org/10.14738/jbemi.26.1697.
5. Abdallah, Y., *Increasing of Edges Recognition in Cardiac Scintography for Ischemic Patients*, pp. 123–125, Lambert Publishing Press GmbH & Co. KG, Germany, 2011.
6. Batenburg, K.J. and Sijbers, J., Adaptive thresholding of tomograms by projection distance minimization. *Pattern Recognit.*, 42, 10, 2297–2305, 2009. 10.1016/j.patcog.2008.11.027.
7. Lauren, B. and Lee, L.W., *Perceptual information processing system*, Paravue Inc, U.S., July 11, 2003, U.S. Patent Application: 10/618,543.
8. Kashanipour, A., Milani, N., Kashanipour, A., Eghrary, H., Robust color classification using fuzzy rule-based particle swarm optimization. *IEEE Congr. Image Signal Process.*, 2, 110–114, 2008.
9. Abdallah, Y., Lungs detection in ventilation and perfusion scintigraphy using watershed transform. *Int. J. Electron. Commun. Comput. Eng. (IJECCE)*, 2, 3, 416–419, 2015.
10. Held, K., Kops, E.R., Krause, B.J., Wells, W.M., Kikinis, R., Muller, G.H., Markov random field segmentation of brain MR images. *IEEE Trans. Med. Imaging*, 16, 6, 878–887, 1998.
11. Li, S.Z., *Markov Random Field Modeling in Computer Vision*, Springer, - Verlag, Berlin, Heidelberg, Nov 01, 1995.
12. Kapur, T., Grimson, W., Kikinis, R., Wells, W.M., Enhanced spatial priors for segmentation of magnetic resonance imagery, in: *Proc of the First Int. Conf. on Medical Image Computing and Computer Assisted Interventions (MICCAI98)*, pp. 457–468, 1998.
13. Pappas, T.N., An adaptive clustering algorithm for image segmentation. *IEEE Trans. Signal Process.*, 40, 901–914, 1992.

14. Rajapakse, J.C., Giedd, J.N., Rapoport, J.L., Statistical approach to segmentation of single-channel cerebral MR images. *IEEE Trans. Med. Imaging*, 16, 176–186, 1997.
15. Leemput, K.V., Maes, F., Vandermulen, D., Seutens, P., Automated model-based tissue classification of MR images of the brain. *IEEE Trans. Med. Imaging*, 18, 10, 897–908, 1999.
16. Zhang, Y., Brady, M., Smith, S., Segmentation of brain MR images through a hidden Markov random field model and the expectation-maximization algorithm. *IEEE Trans. Med. Imaging*, 20, 1, 45–57, 2001. see http://www. fmrib.ox.ac.uk/fsl/fast/index.html.
17. Marroquin, J.L., Vemuri, B.C., Botello, S., Calderon, F., Fernandez, A., An accurate and efficient Bayesian method for automatic segmentation of brain MRI. *IEEE Trans. Med. Imaging*, 21, 8, 934–945, 2002.
18. Marroquin, J.L., Santana, E.A., Botello, S., Hidden Markov measure field models for image segmentation. *IEEE Trans. Pattern Anal. Mach. Intell.*, 25, 11, 1380–1387, 2003.
19. Pham, D.L., Spatial models for fuzzy clustering. *Comput. Vis. Image Und.*, 84, 2, 285–297, 2001.
20. Liew, A.W.C., Leung, S.H., Lau, W.H., Fuzzy image clustering incorporating spatial continuity. *IEE Proc.—Vis. Image Signal Process.*, 147, 2, 185–192, 2000.
21. Ahmed, M.N., Yamany, S.M., Mohamed, N., Farag, A.A., Moriarty, T.A., modified fuzzy c-means algorithm for bias field estimation and segmentation of MRI data. *IEEE Trans. Med. Imaging*, 21, 193–199, 2002.
22. Duda, R.O., Hart, P.E., Stork, D.G., *Pattern Classification*, 2nd, Wiley-Interscience, New York, 2000.
23. Hornik, K., Stinchcombe, M., White, H., Multilayer feed-forward networks are universal approximators. *Neural Netw.*, 2, 359–366, 1989.
24. Matthew, H.T., Khoshgoftaar, M., Randall, W., A review of data mining using big data in health informatics. *J. Big Data Springer*, 1, 1, 1–35, 2014.
25. Naiem, T., II, Stephen, W.B., Sivanesan, D., Big data: The next frontier for innovation in therapeutics and healthcare. *Expert Rev. Clin. Pharmacol.*, 7, 3, 293–298, 2014.
26. Susan, E.W., A review of big data in health care: Challenges and opportunities. *Clinical Cosmet. Invest. Dent.*, 6, 45–56, 2014.
27. Andy, T., *Data capture and analytics in healthcare*, Zebra Technologies: [Zebra Technologies Blog], U.S. August 02, 2014, http://blogs.zebra.com/data-capture-andanalytics-in-healthcare. Retrieved 02-08-2015.
28. Illhoi, Y., Patricia, A., Miroslav, M., Keila, P.H., Rajitha, G., Jia, F.C., Lei, H., Data mining in healthcare and biomedicine: A survey of the literature. *J. Med. Syst.*, 36, 4, 2431–2448, 2012.
29. John, A.Q., Nakasi, R., Mugagga, P.KB., Byanyima, P., Lubega, W., Andama, A., *Deep convolutional neural networks for microscopy-based point of care*

diagnostics,: Proceedings of the 1st Machine Learning for Healthcare Conference, PMLR, Los Angeles, CA, 56:271-281, 2016, arXiv preprint arXiv:1608.02989.
30. Zhu, R., Zhang, R., Xue, D., Lesion detection of endoscopy images based on convolutional neural network features, in: *Image and Signal Processing (CISP), 2015 8th International Congress on*, pp. 372–376, IEEE, 2015.
31. Kamnitsas, K., Ledig, C., Newcombe, V.FJ., Simpson, J.P., Kane, A.D., Menon, D.K., Rueckert, D., Glocker, B., Efficient multi-scale 3d cnn with fully connected crf for accurate brain lesion segmentation. *Med. Image Anal.*, 36, 61–78, 2017.
32. Rajagopalan, S., Janakiraman, S., Rengarajan, A., Rethinam, S., Arumugham, S., Saravanan, G., IoT Framework for Secure Medical Image Transmission, in: *Proceedings of the 2018 International Conference on Computer Communication and Informatics (ICCCI)*, Coimbatore, pp. 1–5, India, 4–6 January, 2018.
33. Deepika, J., Rajan, C., Senthil, T., Security and Privacy of Cloud- and IoT-Based Medical Image Diagnosis Using Fuzzy Convolutional Neural Network. *Comput. Intell. Neurosci.*, 6615411, 1–17, 2021.
34. Youssef, A.E., A framework for secure healthcare systems based on big data analytics in mobile cloud computing environments. *Int. J. Ambient Syst. Appl.*, 2, 2, 1–11, 2014.
35. Saravana, N.M.K., Eswari, T., Sampath, P., Lavanya, S., Predictive Methodology for Diabetic Data Analysis in Big Data. *Procedia Comput. Sci.*, 50, 203–208, 2015.
36. Dougherty, G., Quantitative indices for ranking the severity of hepatocellular carcinoma. *Comp. Med. Imaging Graphics*, 19, 329–38, 1995.
37. Carts, P.Y., Small wonders. *Engin. Technol.*, 1, 32–5, 2006.

6

Comparative Study on Image Enhancement Techniques for Biomedical Images

Sudip Mandal*, Uma Biswas, Aparna Mahato and Aurgha Karmakar

Department of Electronics and Communication Engineering, Jalpaiguri Government Engineering College, Jalpaiguri, India

Abstract

Image enhancement techniques engage an essential and fundamental task and interpretability of biomedical images, enabling accurate diagnosis and analysis. This review paper provides a comprehensive overview of image enhancement techniques specifically tailored for biomedical images. The paper aims to analyze and compare the effectiveness of various spatial domain techniques such as log transformation, gamma correction, contrast stretching (based on spatial domain) histogram equalization, adaptive histogram equalization, and image adjustment (based on histogram technique). The clarity and interpretability of biomedical images are greatly improved by image enhancement techniques, allowing for precise analysis and diagnosis. This review article offers a thorough overview of image enhancement methods designed specifically for biomedical imaging. The purpose of the paper is to evaluate and contrast the usefulness of various approaches and the extent to which they can be used in the biomedical field. These indicators include subjective assessments by qualified radiologists, and objective measurements like the Mean Square Error (MSE) and Peak Signal-to-Noise Ratio (PSNR), Each metric's benefits and drawbacks are highlighted, along with factors to take into account while choosing the best evaluation techniques. This review paper offers a thorough examination of image-enhancing methods for biological images, making it easier to comprehend their benefits, drawbacks, and applicability to diverse biomedical imaging modalities.

**Corresponding author*: sudip.mandal007@gmail.com

Pijush Dutta, Sudip Mandal, Korhan Cengiz, Arindam Sadhu and Gour Gopal Jana (eds.) Design and Forecasting Models for Disease Management (155–180) © 2025 Scrivener Publishing LLC

Keywords: Biomedical image enhancement, log transformation, gamma correction, contrast stretching, histogram equalization, adaptive histogram equalization, image adjustment

6.1 Introduction

Any visual representation of the architecture or physiological functions of the human body acquired using diagnostic imaging techniques is referred to as a medical picture. X-rays, CT, MRI, ultrasound, and PET are a few examples of the modalities that can be used to capture these images. Each modality has particular benefits and supplies various kinds of information.

Medical imaging is an essential aspect of modern healthcare, as it allows doctors and other medical professionals to see inside the human body and diagnose a wide range of conditions. However, medical images are often affected by various issues, such as noise, artifacts, and low contrast. Therefore, there is a need for medical image enhancement techniques to raise the quality of images and assist in accurate diagnosis and treatment.

Techniques for enhancing the quality, clarity, and interpretability of images are used in medicine. Several factors highlight the significance of image enhancement in medical research:

- *Visualization*: With the use of improved photographs, medical personnel can more clearly see and assess anatomical features, tissue properties, and probable problems. This aids in the diagnosis of illnesses, the planning of therapies, and the tracking of patient improvement.
- *Diagnosis and identification*: Enhancement techniques help in the identification and diagnosis of a variety of medical disorders by enhancing image quality. Enhanced pictures can highlight minute details, increase lesion visibility, and help detect diseases in their early stages.
- *Image Interpretation*: Medical image improvement aids in reducing distortion, noise, and artifacts that might hide key details. Clearer images make proper interpretation easier, which leads to more accurate diagnoses and better-informed choices.
- *Research and Analysis*: For experiments and investigations, researchers frequently use medical photographs. Researchers can do in-depth analysis, extract more accurate quantitative

measurements, and get superior outcomes by improving these photos.

In this work, we have explored the concept of medical image enhancement, the problems associated with medical images, and the reasons why medical images need to be enhanced. Through a review of the literature and analysis of various case studies, we will argue that medical image enhancement is a crucial aspect of modern healthcare that can improve patient outcomes and optimize clinical workflows.

A scene or item is represented in two dimensions by an image. It is a graphic depiction of an item or situation that can be captured using a variety of techniques, including photography, video recording, scanning, or digital painting. One of the main issues with medical photos is that they frequently have a poor contrast or foggy appearance, which makes it challenging for medical practitioners to analyze them correctly. This may be because of several things, including poor lighting during imaging, motion artifacts, or inherent limitations in the imaging technique. The quality and clarity of medical images have been improved by the development of numerous image-enhancing techniques to address this problem. Techniques for picture enhancement can help to increase contrast, decrease noise, and improve the image's overall visual appeal.

Furthermore, MATLAB can be used to perform image registration, which can help align images from multiple modalities, making it easier to compare and diagnose abnormalities. Overall, image processing in MATLAB can help improve the quality and accuracy of X-ray, CT scan, and MRI images, ultimately aiding in more accurate diagnoses and better patient outcomes. There are various techniques for image quality enhancement, Log transformation, power law transformation or gamma correction, piecewise linear transformation, or contrast stretching. These techniques have been implemented in MATLAB, a popular programming language and environment for image processing. Overall, the application of histogram-based image enhancement techniques can help to increase the clarity and diagnostic value of medical images, enabling doctors to make more precise diagnoses and give patients better care.

6.2 Literature Review

Image processing is a method for enhancing and manipulating digital photographs. It includes several procedures, including picture enhancement, segmentation, filtering, and pattern recognition. Techniques for image processing are widely employed in a variety of industries, including digital art, remote sensing, and medical imaging. Universally there are lots of techniques that are used in image processing and there are many research works done by many authors. Here are some of these.

Mohamed *et al.* [1] define how this image-processing method helps in the identification of any disorder and describes the classification of digital images in different fields. Olumayowa *et al.* [2] explain the different medical imaging systems and the algorithm of the proposed method. They also discussed the AMBE, PSNR, and entropy of different medical techniques. Also, they conclude the opinion score of these values got by MRI or CT scan, etc. Shruthishree and Tiwari [3] describe the two types of Image Processing, algorithms used for Image Processing at different stages, and different image processing techniques like Edge detection, Feature Extraction, etc. Sargun and Rana [4] describe the effect of image enhancement, and different types of enhancement techniques with their application in various fields. Perumal and Velmurugan [5] mainly explain the preprocessing of lung images for better visualization and also the detection of lung cancer in the initial stage. Ikhsan *et al.* [6] discuss the oldest photographic films X-Ray and their impacts. They also showed the image enhancement techniques for getting a good segmentation for X-ray images. Madaean [7] explains how MATLAB is important and the uses of image enhancement techniques in it. Gopika [8] introduces about the blood disorder. She also describes how machine learning can propose an advanced method of identifying blood disorders. Naseera *et al.* [9] state the advantages, and applications of image processing. They also show the different imaging techniques of medical tests like CT scans, Radiography, etc. Agaian *et al.* [10] explain the new quantitative measure using current magnitude reduction techniques for image enhancement. However, this is still an interesting research area in the field of image processing.

6.3 Theoretical Concepts

There are many techniques we use to enhance an image, some of the most notable techniques are:

a) Spatial domain-based methods:

- Log Transformation.
- Power Law Transformation or Gamma Correction.
- Piecewise Linear Transformation or Contrast Stretching.

b) Histogram based methods:

- Histogram Equalization.
- Contrast-Limited Adaptive Histogram Equalization (CLAHE).
- Adjustment Function.

The details of the above-mentioned methods are discussed one by one.

6.3.1 Logarithmic Transformation

Log transformation is a common technique to enhance the contrast of images with low dynamic range. The log transformation involves taking the logarithm of the intensity values of an image, which can compress the dynamic range of the image and enhance the visibility of details in dark regions. For effective contrast enhancement of dark images, this transformation technique makes use of the fundamental image enhancement of the spatial domain. The pixels are transformed to a grey level by the log transform, which is essentially a grey-level transform. The low grey level values of the input image are converted using this transform approach to a larger range of output levels.

$$s = T(r) = c * log(r + 1)$$

Where s is the output image, r is the input image. Scaling constant is c which is provided by:

$$c = 255 / log(1 + Maximum\ pixel\ value\ from\ the\ input\ image)$$

When a digital image undergoes a logarithmic transformation, the darker intensity values are given brighter values, making the features existing in the image's darker or grayer portions more visible to the human eye. By using the logarithmic adjustment, the brighter intensity levels are also scaled down to lower values. Though not by as much, the darker intensity values are scaled up more than the lighter intensity ones. For a digital picture with intensity values ranging from 0 to 255, the transformation log(r+1) produces values in the range of 0 to 2.41.

Log transformation is based on the following steps: a) Convert the image to a grayscale or color space, b) Apply the log function to each pixel value in the image, c) Scale the resulting values to the appropriate dynamic range.

6.3.1.1 *Advantages of Log Transformation*

The capacity of log transformation to improve the visibility of features in dark areas of an image is one of its key advantages. This is accomplished by reducing the image's dynamic range, which can improve contrast in dark areas without overexposing bright ones. The capacity of log transformation to lessen the impact of noise in an image is another benefit. This is so that the amplitude of noise in the image can be reduced by the logarithmic function's compression of the intensity values. A straightforward and effective computational approach that may be used on photos of any size is the log transformation.

6.3.1.2 *Limitations of Log Transformation*

The fact that log transformation is a global adjustment i.e. it affects every pixel in the image uniformly one of its fundamental drawbacks. This may cause some sections of the image to be overcorrected while other regions may be undercorrected. An odd or unrealistic image could come from this. Additionally, the bright areas of an image may lose information as a result of log transformation. This is because the logarithmic function reduces the image's dynamic range, which might make features in bright areas harder to see.

6.3.2 Power Law Transformation or Gamma Correction

A common method for changing the brightness and color balance of digital photographs is gamma correction. An image's intensity values are elevated to a power *gamma ( )* that is normally between 0.5 and 2.5 as a result of this nonlinear adjustment. The following equation is used for power law transformation or gamma correction.

$$s = c * r^{\gamma}$$

Where s and r are the output and input pixel values respectively, c and γ are the positive constants. γ is the value of the exponent or the degree of adjustment called gamma (in digital image processing it is also referred as gamma encoding or gamma compression). Gamma is defined by

$$V_{out} = V_{in}^{gamma}$$

where, V_{out} is output luminance value, V_{in} is the input luminance/actual luminance value. Power-law curves with fractional values of γ (gamma) of $\gamma > 1$ for a small range of dark input values into a wider range of output values and $\gamma < 1$ *for* a high range of dark input values and a narrow range of output values or vice versa. The transformation is known as gamma correction and is used to adjust the brightness of an image.

Gamma correction is based on the following steps: a) Convert the image to a grayscale or color space, b) Apply the gamma function to each pixel value in the image, and c) Scale the resulting values to the appropriate dynamic range.

The value of gamma for the different display devices is different. For example, the Gamma of the Cathode Ray Tube lies between 1.8 and 2.5. When monitor calibration and adjustment of contrast setting happens that time, we control gamma. The industry has converged on a standard display gamma of 2.2.

6.3.2.1 Advantages of Gamma Correction

One of the key advantages of gamma correction is its capacity to modify the luminance and color balance of an image. It can be used to correct for the nonlinear response of display systems, which can cause images to look darker or brighter than intended. By adjusting the gamma value, the image can be made to appear more visually pleasing and accurate. This is achieved by increasing the contrast in these areas, which can make details that were previously difficult to see more visible. Gamma correction is also a straightforward and computationally effective approach that may be used to process photos of any size.

6.3.2.2 Limitations of Gamma Correction

The fact that gamma correction is a global adjustment i.e. it affects every pixel in the image equally—is one of its main drawbacks. This may cause some image sections to be overcorrected while other regions may be undercorrected. An odd or unrealistic image could come from this. Gamma correction can also make noise in an image appear louder, especially in areas of darkness where the signal-to-noise ratio is poor. This can lead to an increase in visible noise and a reduction in image quality.

6.3.3 Piecewise Linear Transformation or Contrast Stretching

An analyst may stretch specific histogram values for more augmentation in particular places when the histogram distribution in a picture is bi- or multimodal. A piecewise linear contrast stretch is the name of this contrast-enhancing technique. The selection of several linear enhancement steps that widen the brightness ranges in the modes of the histogram constitutes a piecewise linear contrast enhancement.

A method used in image processing to increase contrast in images is contrast stretching, often referred to as contrast enhancement. To use the full range of intensities, it includes extending the dynamic range of the pixel values in an image. Different methods, including linear stretching, piecewise linear stretching, and nonlinear stretching, can be used to achieve this. We will give an overview of contrast stretching, its benefits and drawbacks, and its applications in numerous fields in this review.

Contrast-Stretching transformation is an example of a linear method that uses piecewise linear functions, that are automatically derived from the input image's histogram. By changing the location of points (r_1, s_1) and (r_2, s_2), we can control the shape of the transformation function. For example,

$$\begin{aligned} s &= lr, & 0 \le r \le a \\ s &= m(r-a)+v, & a \le r < b \\ s &= n(r-b)+w, & b \le r < L-1 \end{aligned}$$

where l, m, n are slopes for transformation, w and v are constants.

The contrast stretching algorithm is based on the following steps:

1) Compute the minimum and maximum pixel values in the image. 2) Compute the desired minimum and maximum pixel values. 3) Compute the transformation function that maps the input intensity values to the output intensity values. 4) Apply the transformation function to each pixel in the image.

6.3.3.1 Advantages of Contrast Stretching

By extending the dynamic range of the pixel values to encompass the whole range of intensities available, contrast stretching can improve the contrast of an image. This can enhance the image's visual quality and make it simpler to draw out details and information from the image. The ease of use and computational efficiency of contrast stretching are further benefits.

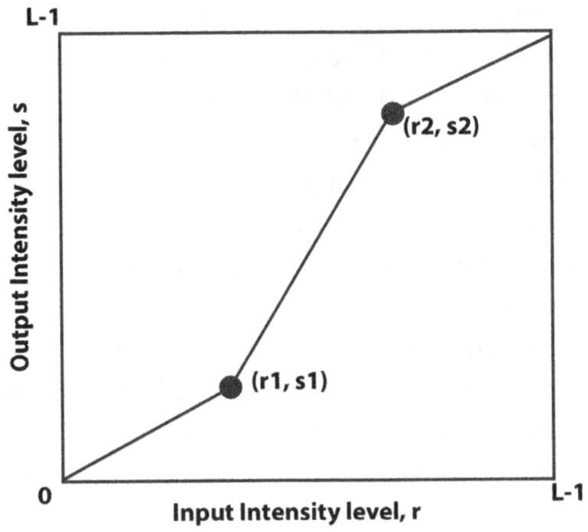

Figure 6.1 Output intensity level after contrast stretching.

It is a non-parametric technique that works with photos of any size and doesn't require any prior knowledge of image statistics. Figure 6.1 shows the output intensity level after the contrast stretching.

6.3.3.2 *Limitations of Contrast Stretching*

Contrast stretching has several drawbacks: it can amplify noise in an image because the algorithm stretches the intensity values of all pixels, including those that may be affected by noise; it may not be effective for images with low contrast; and it can cause overcorrection or under correction in some areas of the image, which can produce an image that appears unnatural or unrealistic.

6.3.4 Histogram Equalization

Similar to a bar chart, a histogram is a graphical representation of data values. Histogram equalization is a method used in image processing to describe the intensity of the white component of the image. We can increase the contrast in the image by doing this. A typical method for improving the contrast of photos with low dynamic range is histogram equalization. With the help of a non-linear transformation, the intensity values of an image are redistributed so that the histogram that results has a uniform distribution. We will give an overview of histogram equalization, its benefits and drawbacks, and its uses in a variety of sectors in this review.

One of the most beneficial methods of nonlinear contrast enhancement is histogram equalization. When the histogram of an image is equalized, all of the image's pixel values are redistributed so that about the same number of pixels are allocated to each of the user-specified output gray-scale classes (for example, 32, 64, and 256). At the brightest point in the histogram's most populated range (the "peaks"), contrast is raised. It automatically lessens contrast in areas of an image that are extremely light or dark and correspond to the tails of a normally distributed histogram. If there are few output values spread across a wide range, histogram equalization can also divide pixels into multiple groups.

The histogram equalization algorithm is based on the following steps: a) Compute the histogram of the input image, b) Compute the CDF of the histogram, c) Compute the transformation function that maps the input intensity values to the output intensity values, d) Apply the transformation function to each pixel in the image.

6.3.4.1 *Advantages of Histogram Equalization*

One of histogram equalization's key benefits is its capacity to improve an image's contrast by spreading the intensity values so that the resulting histogram has a uniform distribution. This can enhance the image's visual quality and make it simpler to draw out details and information from the image. The ease of use and computational efficiency of histogram equalization are further benefits. It is a non-parametric technique that does not require any prior knowledge of the picture statistics, and it may be used for photos of any size.

6.3.4.2 *Limitations of Histogram Equalization*

The fact that histogram equalization is a global adjustment i.e. it affects every pixel in the image equally—is one of its key drawbacks. This may cause some image sections to be overcorrected while other regions may be under corrected. An odd or unrealistic image could come from this. Histogram equalization can sometimes make noise in an image appear louder. This is so that all pixels' intensity levels, including those that might be impacted by noise, are equalized by the algorithm.

6.3.5 Contrast-Limited Adaptive Histogram Equalization (CLAHE)

We can generate many histograms for a single image during this histogram equalization process. Each histogram will be assigned to a certain area of the image. The procedure is used to a local set of pixels in the image during histogram equalization. The image is separated into equal-sized, discrete areas called tiles. Every area of the image is improved separately as a result.

Contrast Limited Adaptive Histogram Equalization (CLAHE) is a well-known image enhancement method that has been thoroughly researched and used in numerous industries, including remote sensing, surveillance, and medical imaging. By reducing the amplification of noise and maintaining local contrast, It is a variant of conventional histogram equalization (HE) that tackles HE's drawbacks. When using CLAHE, an image is divided into smaller areas, and histogram equalization is applied to each region separately.

The CLAHE algorithm is based on the following steps: a) Divide the image into non-overlapping tiles or regions, b) Compute the histogram of each tile, c) Apply the contrast limitation by clipping the histogram at a predetermined maximum value, d) Compute the cumulative distribution function (CDF) of each tile, e) Map the intensity values of each tile using the CDF of that tile. Combine the results of all tiles to form the final image.

The main distinction between CLAHE and conventional HE is the contrast restriction step. By clipping the histogram at a predefined maximum value, it prevents noise from being amplified further. Based on the size and quantity of pixels in the tile, the maximum value is calculated. This makes sure that noise amplification is constrained in tiny tiles with few pixels, while noise amplification is permitted to a greater extent in larger tiles with more pixels.

6.3.5.1 Advantages of CLAHE

Limiting the amplification of noise is one of CLAHE's key advantages over conventional HE. This is made possible by the contrast limiting step, which limits the amount to which small tiles with low pixel counts are amplified and, in turn, minimizes the impact of noise. The capacity of CLAHE to maintain local contrast is another benefit. To do this, the image is divided into smaller tiles, and then histogram equalization is applied to each tile separately. This makes it possible to maintain the contrast of high-contrast parts while enhancing low-contrast regions. CLAHE is also computationally effective and applicable to photographs of any size.

6.3.5.2 Limitation of CLAHE

However, this approach has had several downsides, which are Adaptive histogram equalization may over-amplify contrast in portions of the image that are nearly constant. Over amplification of the constant region is one of the AHE's shortcomings. Now, the amplified noise is lessened by the contrast limiter. Limits of contrast between their neighboring pixels are defined by the contrast limiter.

6.3.6 Adjustment Function

The 'adjustment function' in MATLAB's Image Processing Toolbox is used to improve or change the appearance of images. We can stretch or compress the intensity range by using it to translate the intensity values in the input image to new values in the output image. To regulate the adjustment, we can specify the input and output intensity ranges.

You can adjust visual characteristics including brightness, contrast, color harmony, and intensity with these functions. The following MATLAB adjustment functions are frequently used. *imadjust*: This feature allows you to change the brightness and contrast of a picture.

$$s = imadjust(r)$$

It maps the intensity values in grayscale image r to new values in s. By default, imadjust saturates the bottom 1% and the top 1% of all pixel values. This operation increases the contrast of the output image s. Histogram equalization, carried out by this function, improves the contrast of an image by redistributing pixel intensities. It makes details easier to see in both bright and dark areas.

6.4 Results and Discussion

Peak Signal-to-Noise Ratio (PSNR) and Mean Squared Error (MSE) values are used to measure the performance of the image enhancement technique. The PSNR indicates how well the enhanced image compares to the original, with higher values indicating higher quality, while the MSE measures the average squared difference between the original and enhanced images. The quality of the compressed or rebuilt image is better the higher the PSNR value and the more realistic the image, the closer our MSE value is to 0.

Here, we have considered five different biomedical images for analysis. The images (MRI and X-Ray images) are collected from the http://radiopaedia.org/cases. The details of the images are given below.

Image No. 1: Brain MRI
Image No. 2: Irregular phalanx foot X-RAY
Image No. 3: Intrathoracic stomach X-RAY
Image No. 4: Intramuscular elbow lipoma MRI
Image No. 5: Choroid plexus Papilloma MRI

6.4.1 Images and Histograms for Different Images Using Different Enhancement Methods

We have shown the original images, the images after enhancement, and the histogram of each enhanced image. In the first column, we have mentioned the name of the images, in the second column we have shown the images of the corresponding techniques and in the last column, there are the histograms of each image. Tables 6.1, 6.2, 6.3, 6.4 and 6.5 show the results corresponding to the different image enhancement techniques for Image 1, 2, 3, 4, and 5 respectively.

Table 6.1 Results for image 1 using different image enhancement techniques.

Method	Image (Brain MRI)	Histogram
Original Image	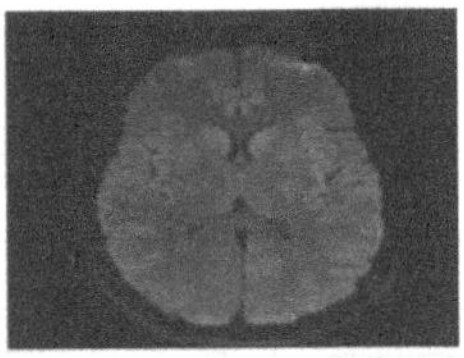	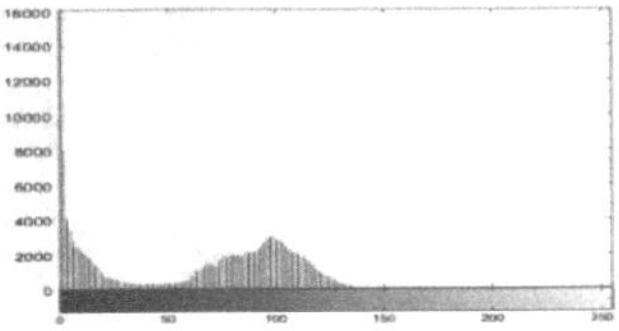
Histogram Equalization	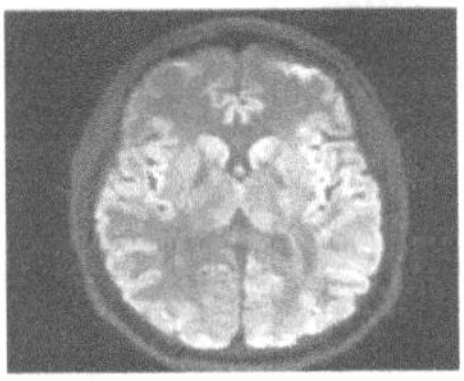	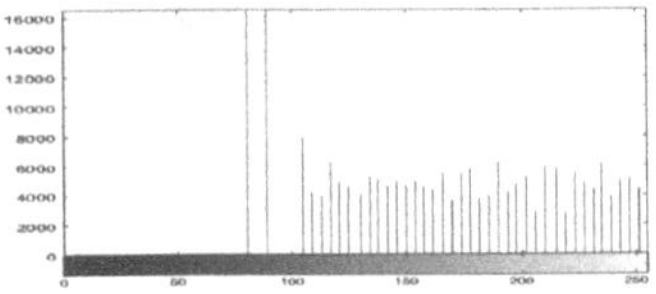
Adaptive Equalization	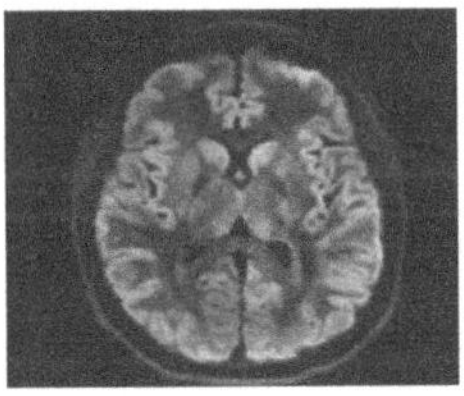	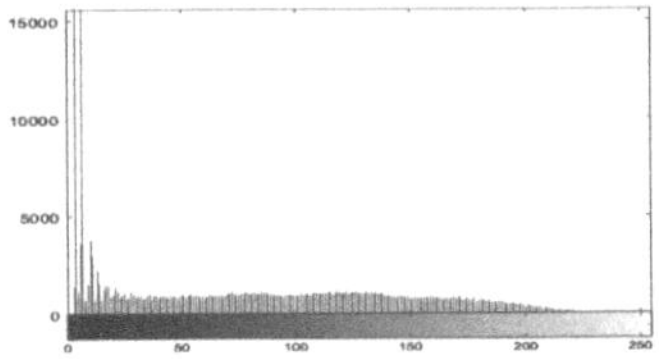

(*Continued*)

Table 6.1 Results for image 1 using different image enhancement techniques. (*Continued*)

Method	Image (Brain MRI)	Histogram
Image Adjustment	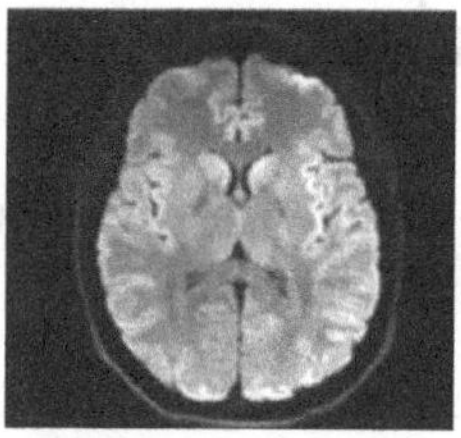	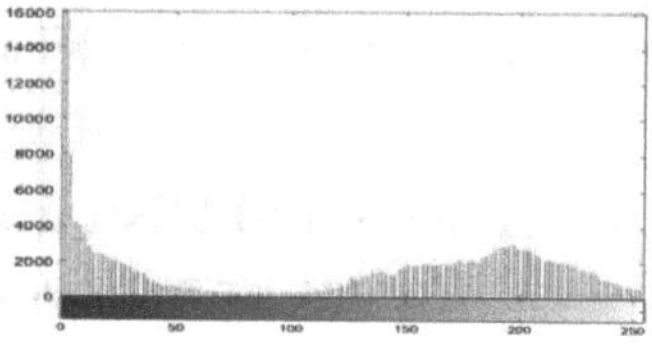
Log Transformation	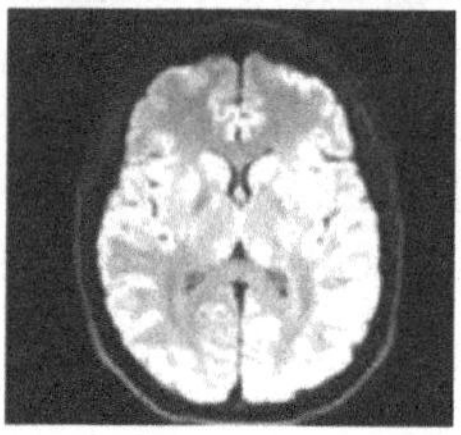	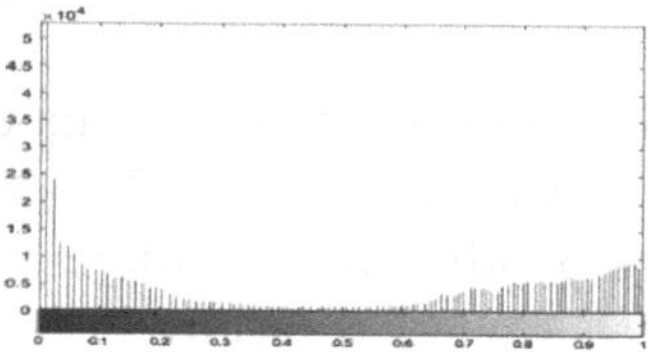
Contrast Stretching	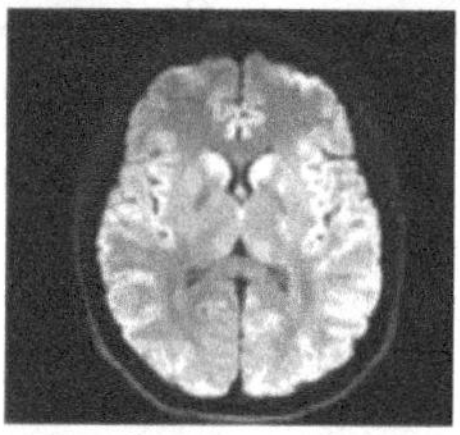	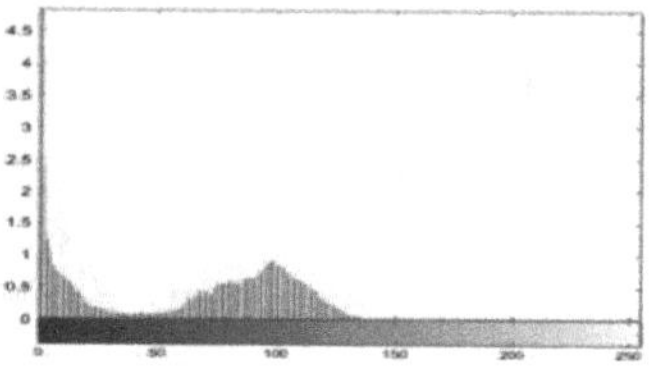
Gamma Correction	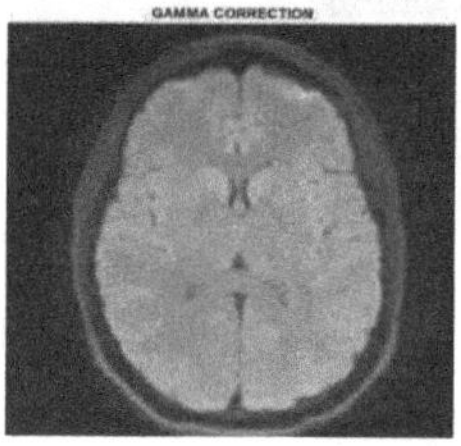	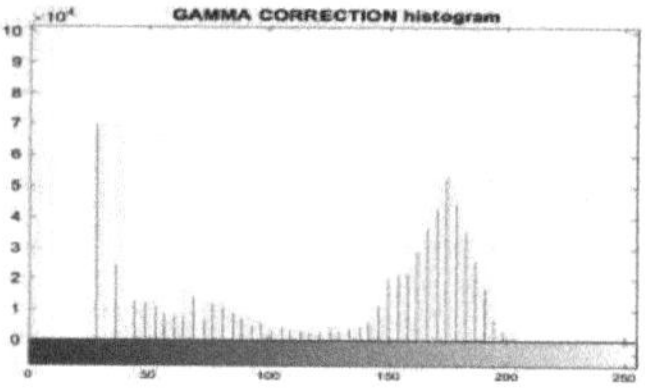

Table 6.2 Results for image 2 using different image enhancement techniques.

Method	Image (Foot X-ray)	Histogram
Original Image	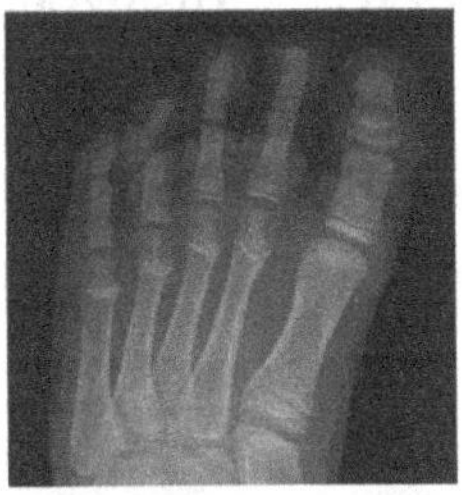	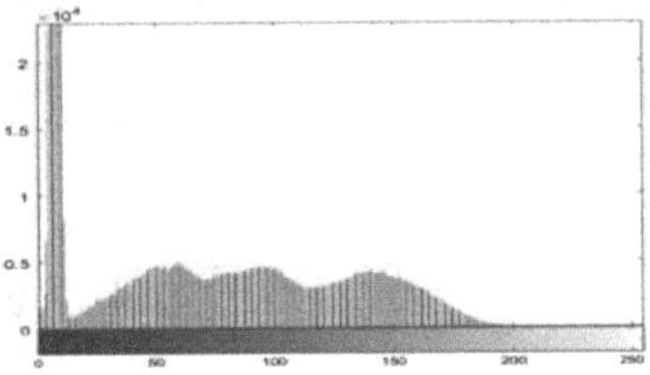
Histogram Equalization	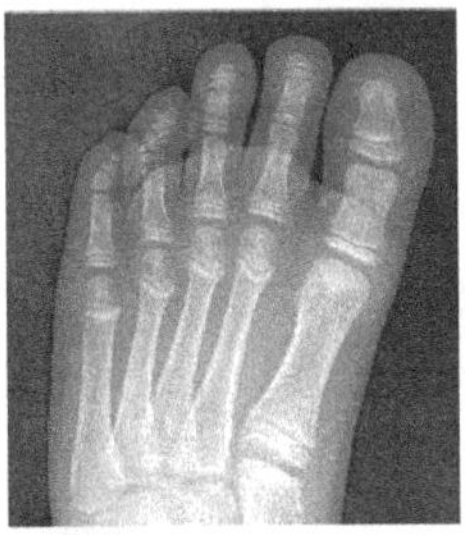	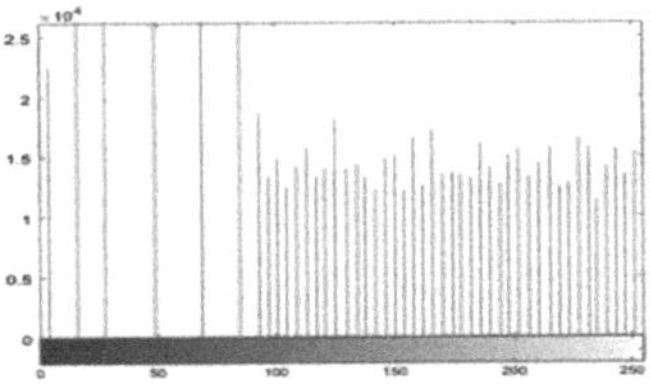
Adaptive Equalization	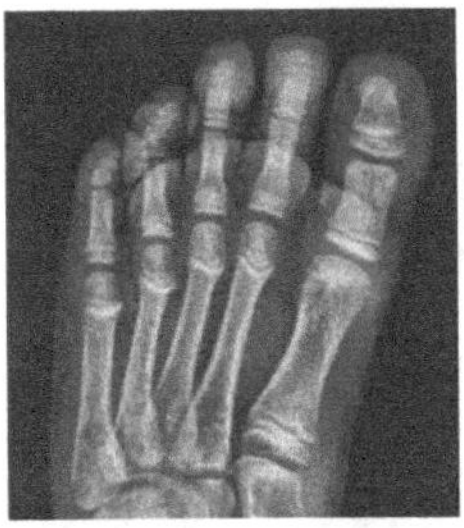	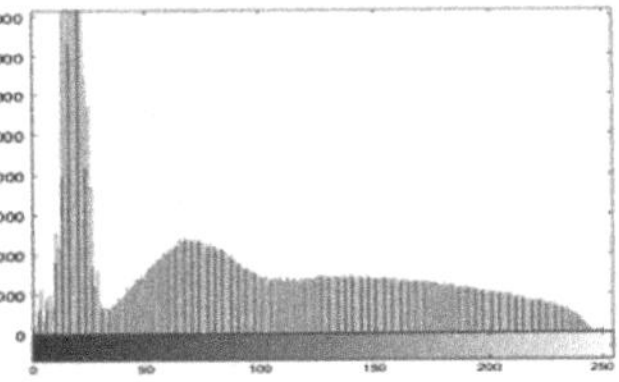
Image Adjustment	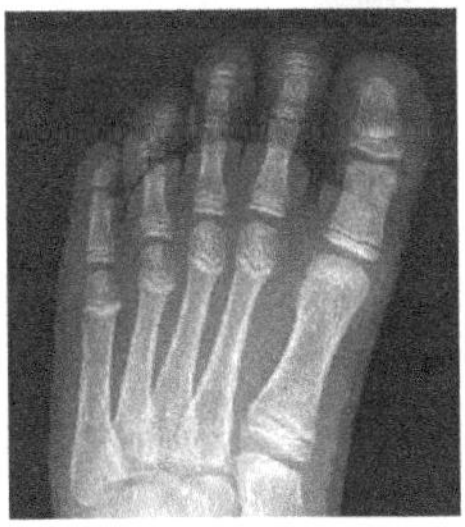	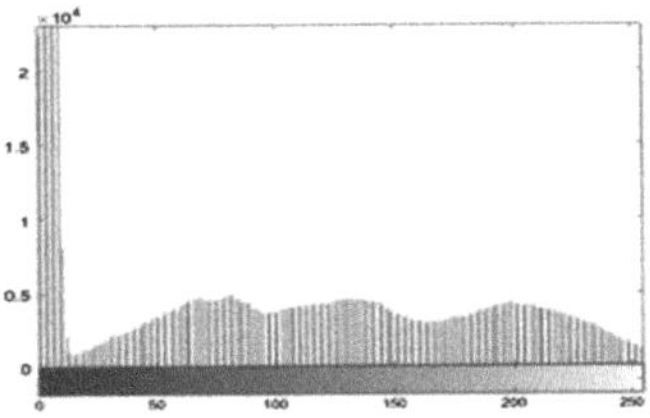

(Continued)

Table 6.2 Results for image 2 using different image enhancement techniques. (*Continued*)

Method	Image (Foot X-ray)	Histogram
Log Transformation	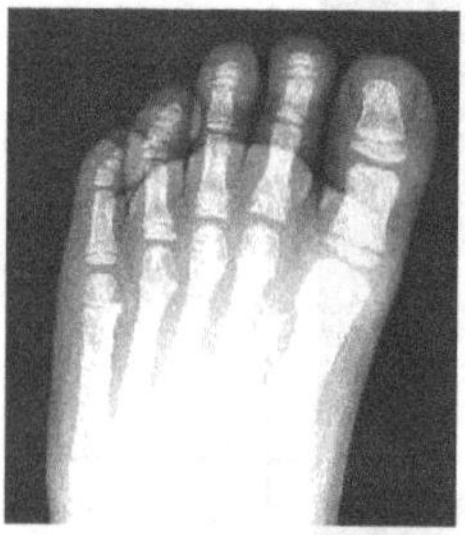	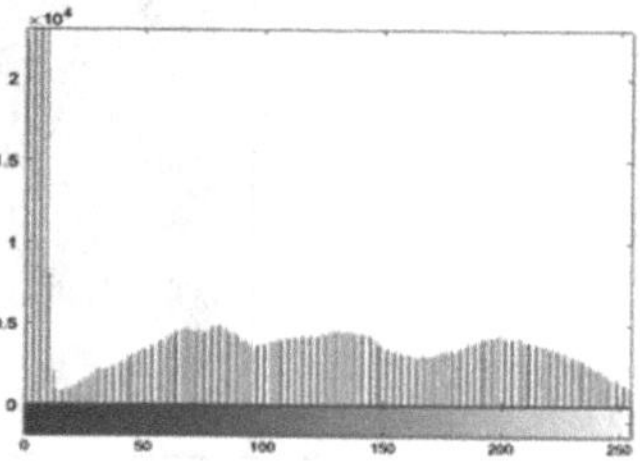
Contrast Stretching	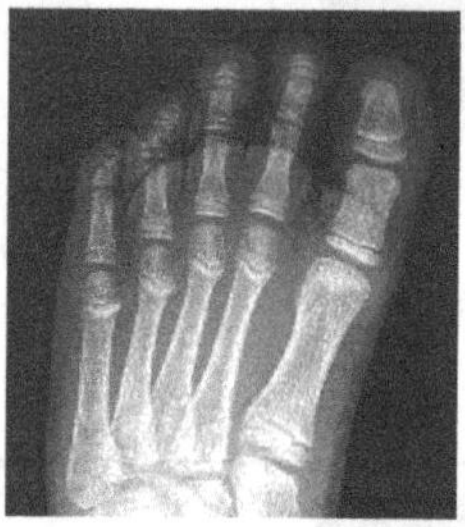	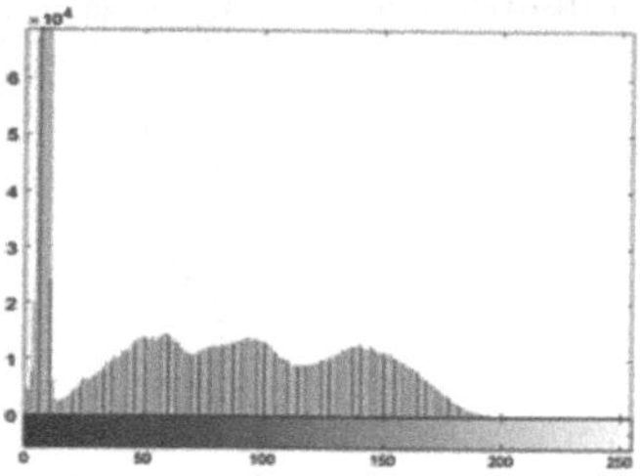
Gamma Correction	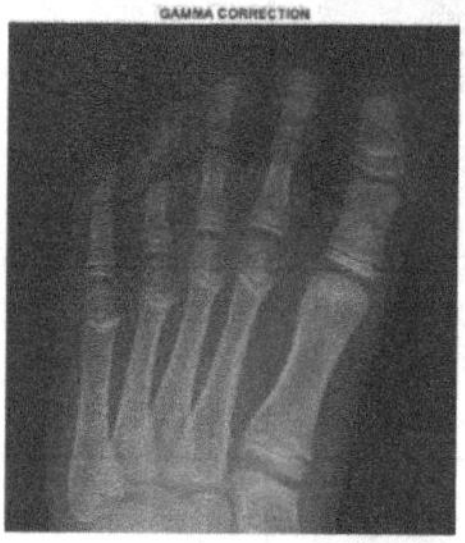	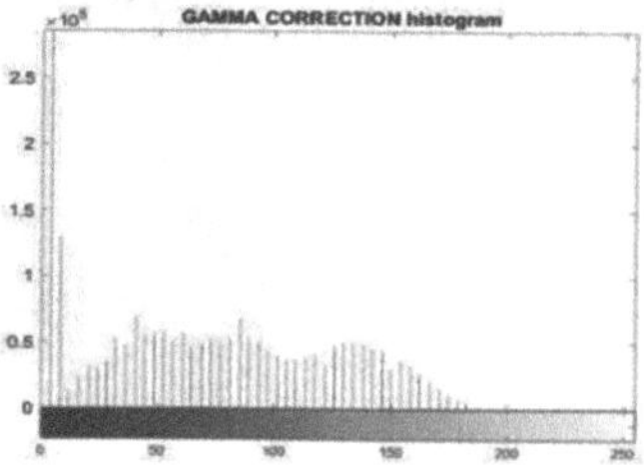

Table 6.3 Results for image 3 using different image enhancement techniques.

Method	Image (Stomach X-ray)	Histogram
Original Image	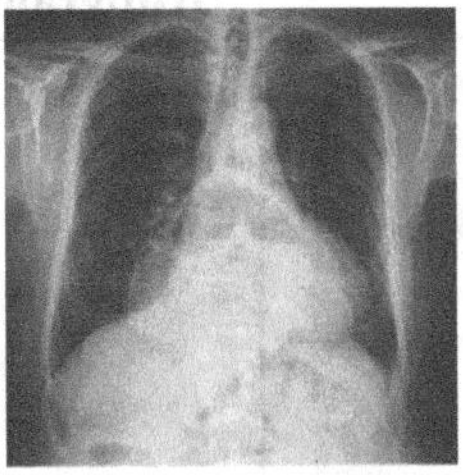	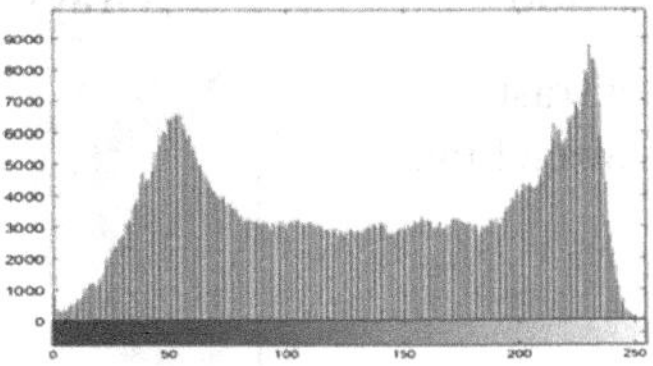
Histogram Equalization	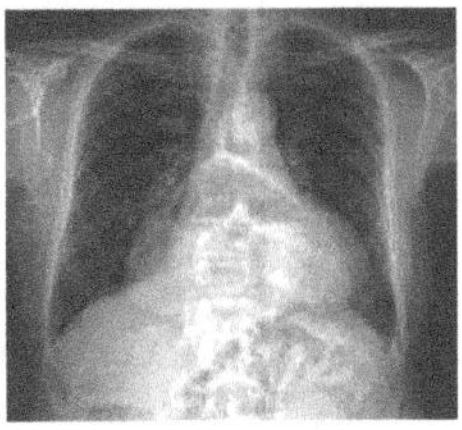	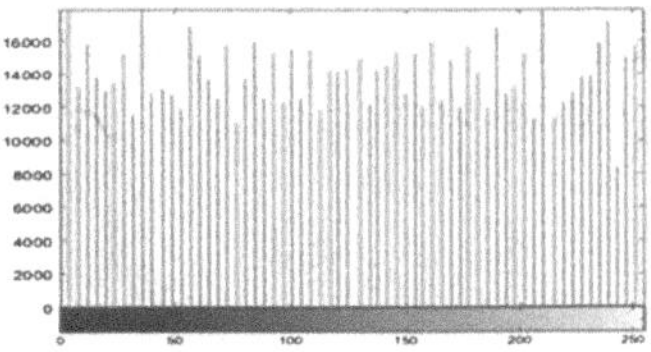
Adaptive Equalization	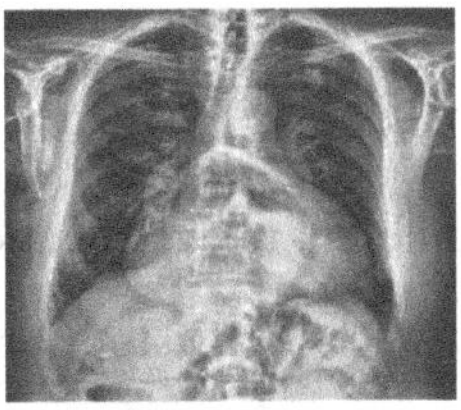	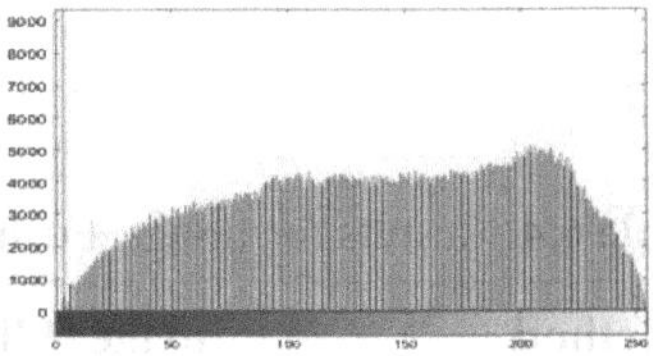
Image Adjustment	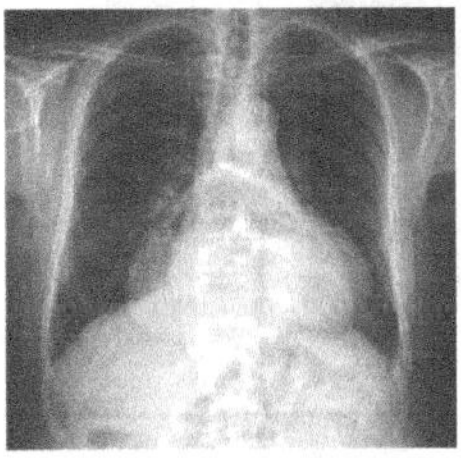	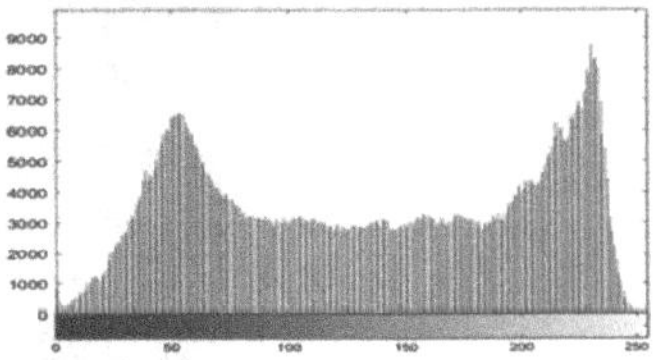
Log Transformation	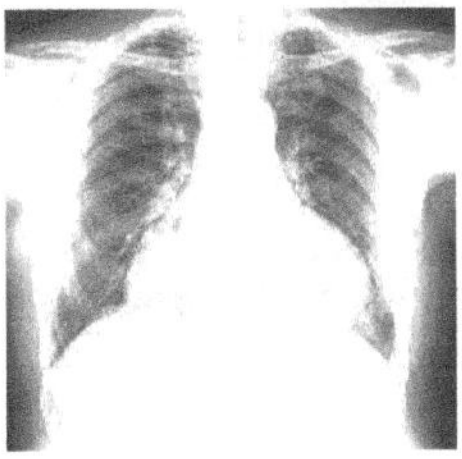	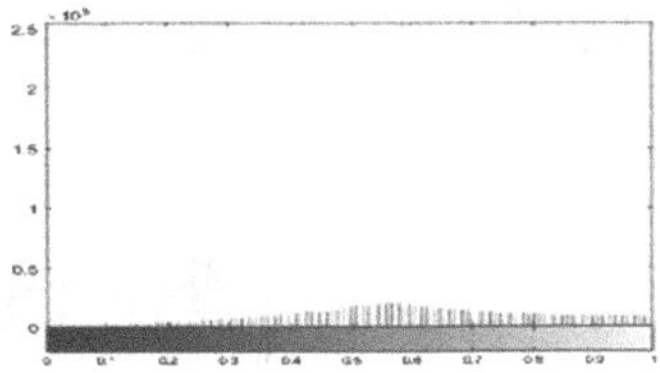

(*Continued*)

Table 6.3 Results for image 3 using different image enhancement techniques. (*Continued*)

Method	Image (Stomach X-ray)	Histogram
Contrast Stretching	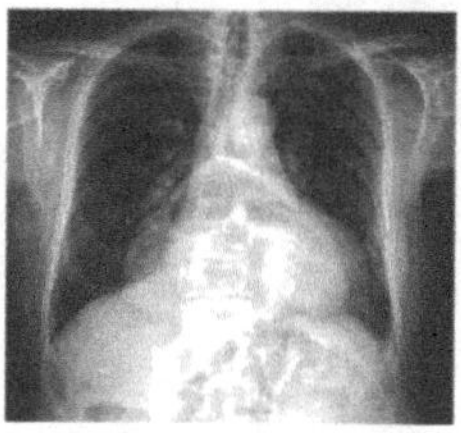	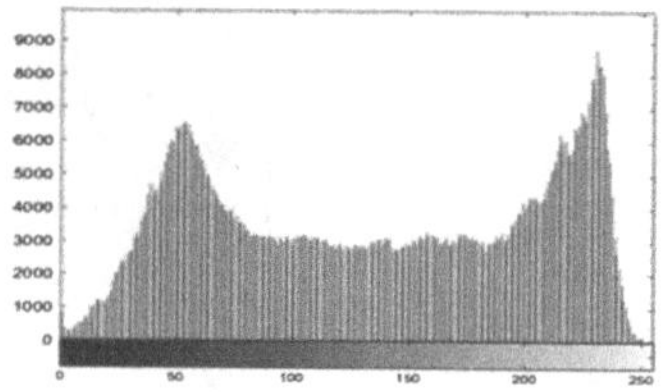
Gamma Correction	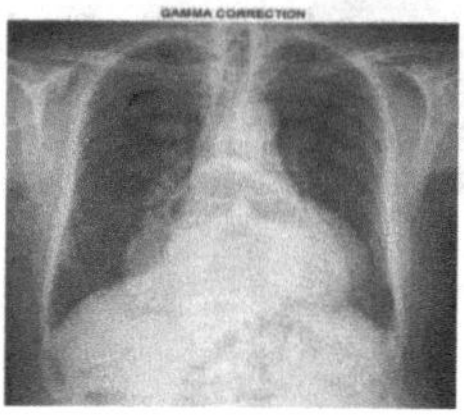	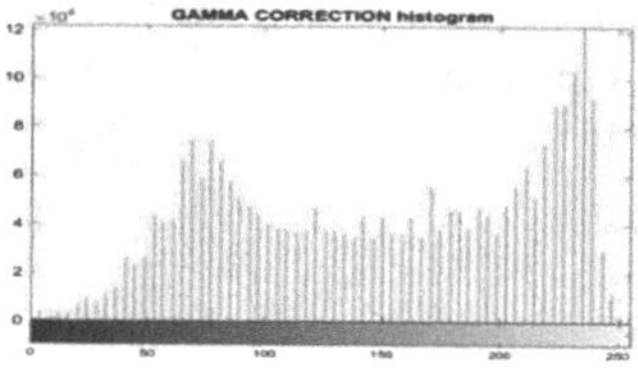

Table 6.4 Results for image 4 using different image enhancement techniques.

Method	Image (Elbow MRI)	Histogram
Original Image	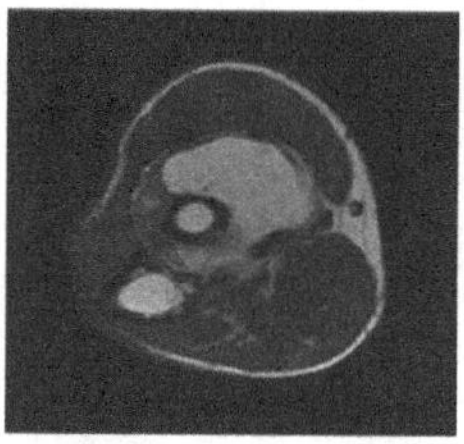	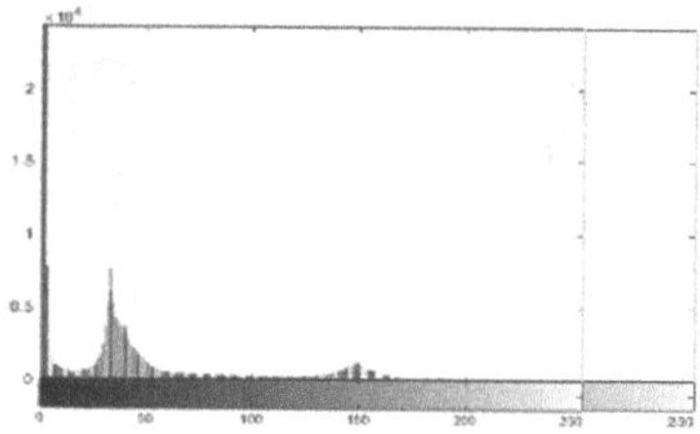
Histogram Equalization	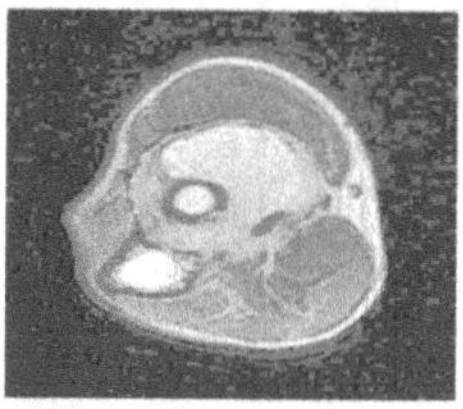	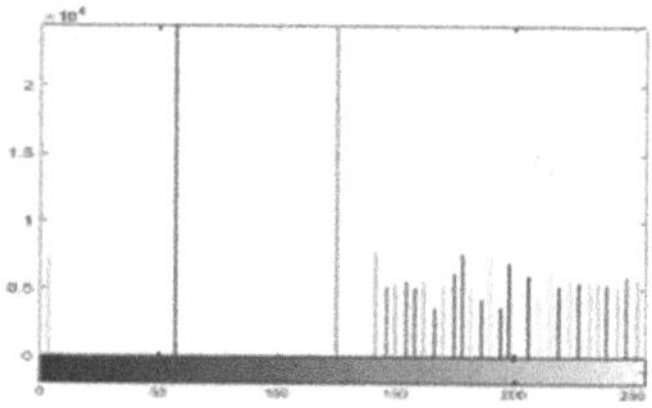

(*Continued*)

Table 6.4 Results for image 4 using different image enhancement techniques. (*Continued*)

Method	Image (Elbow MRI)	Histogram
Adaptive Equalization	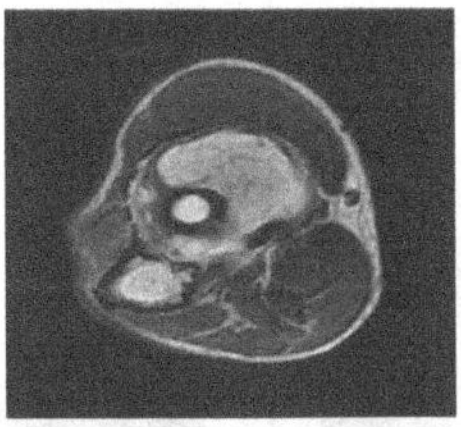	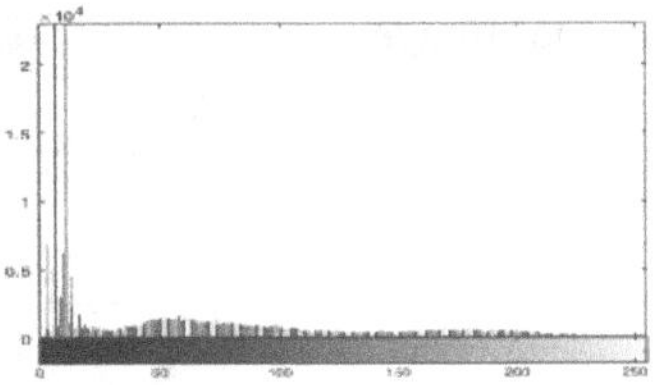
Image Adjustment	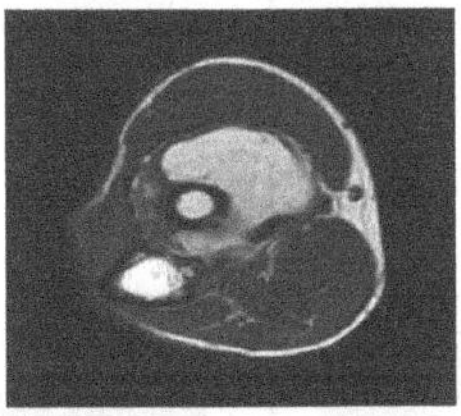	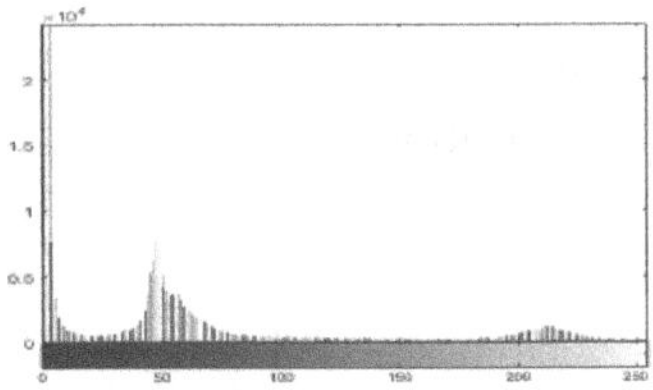
Log Transformation	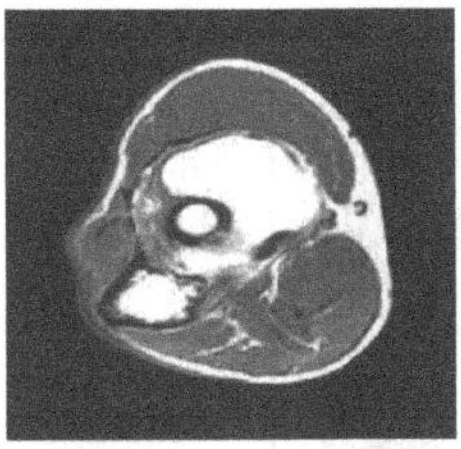	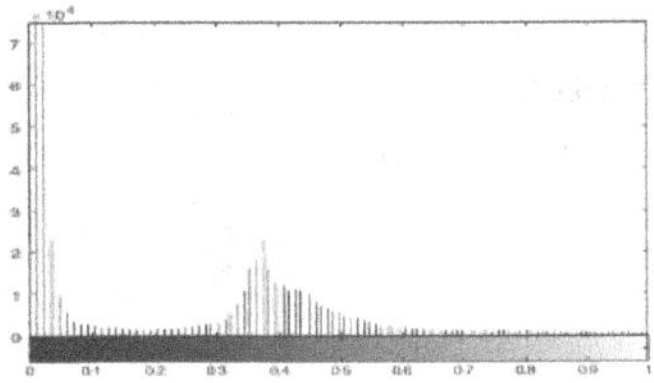
Contrast Stretching	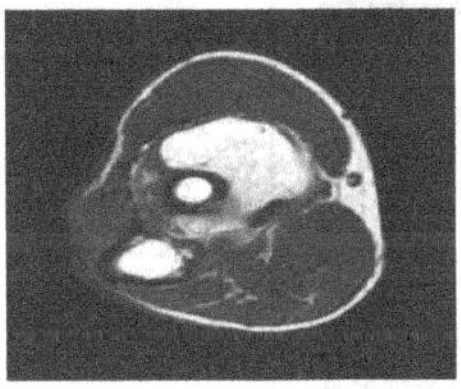	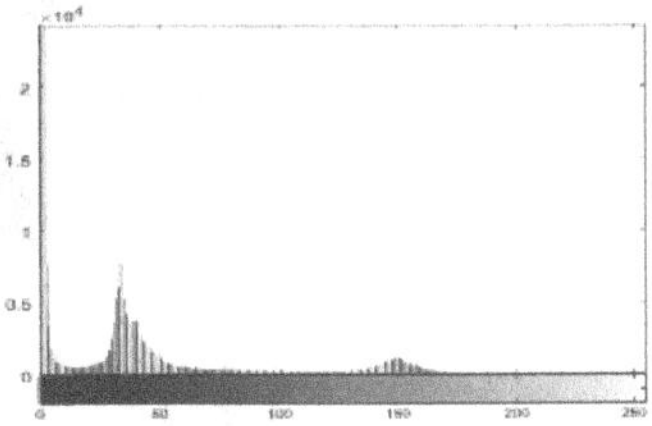
Gamma Correction	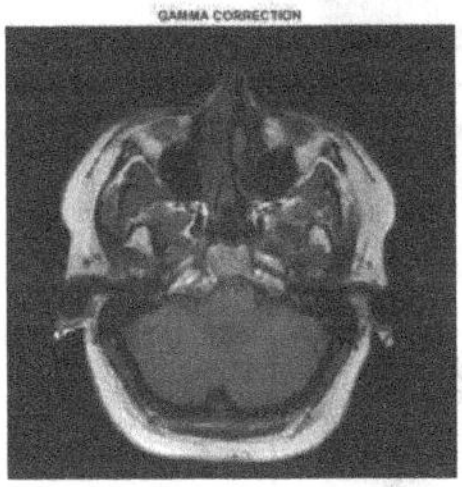	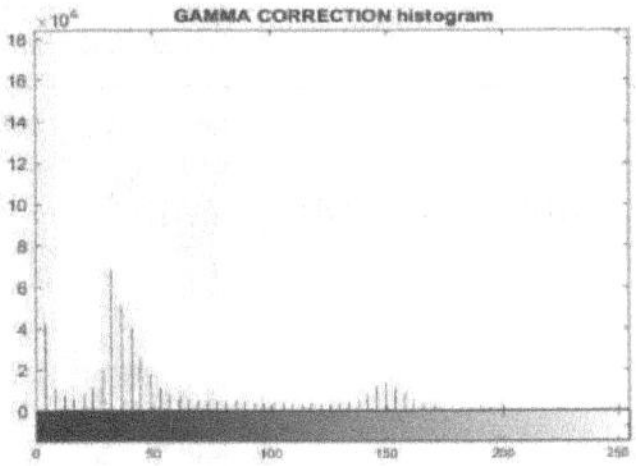

Table 6.5 Results for image 5 using different image enhancement techniques.

Method	Image (Papilloma MRI)	Histogram
Original Image	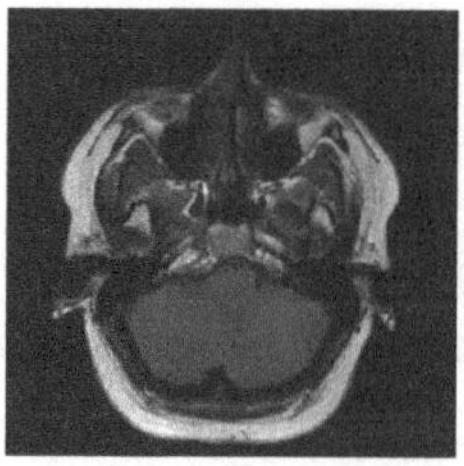	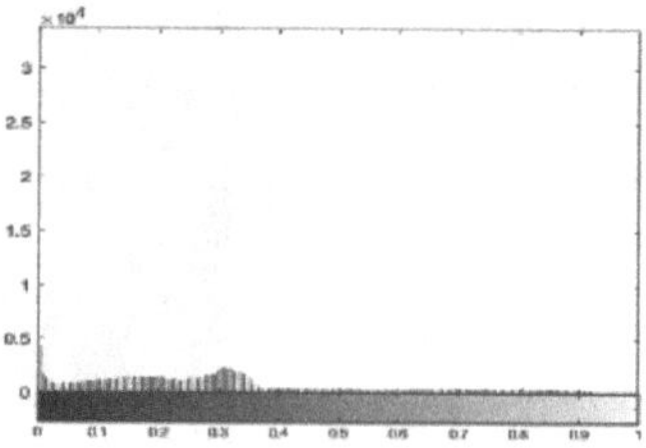
Histogram Equalization	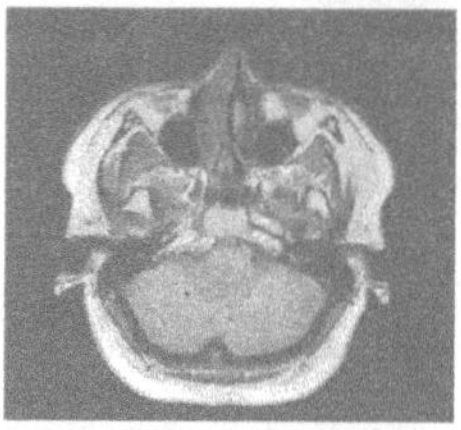	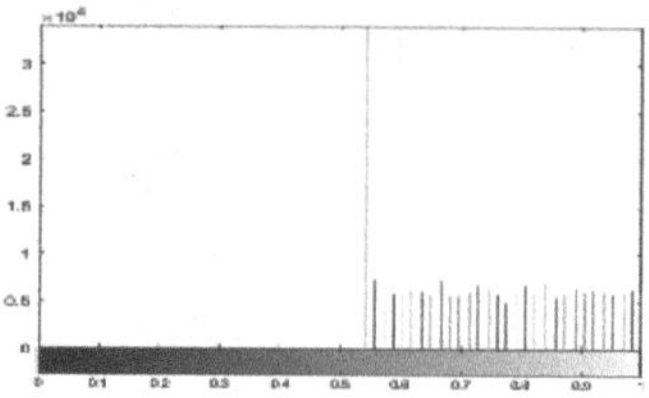
Adaptive Equalization	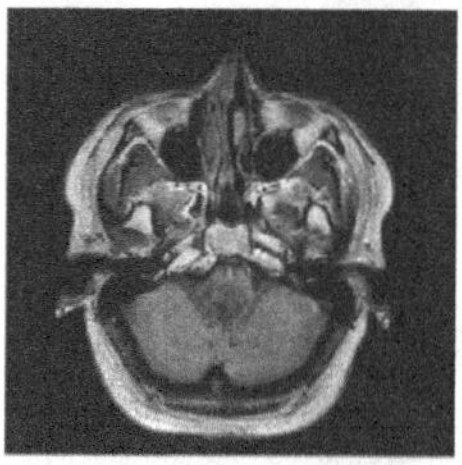	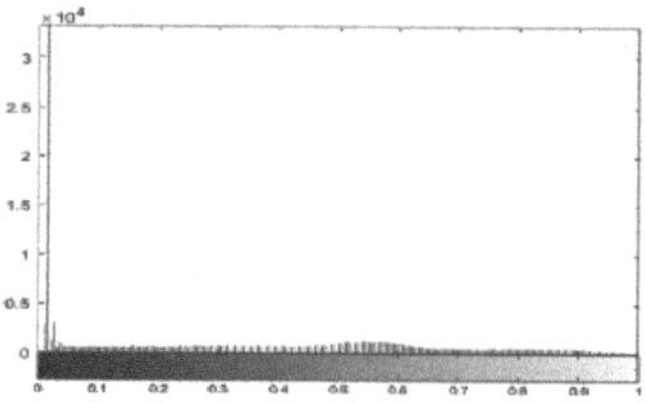
Image Adjustment	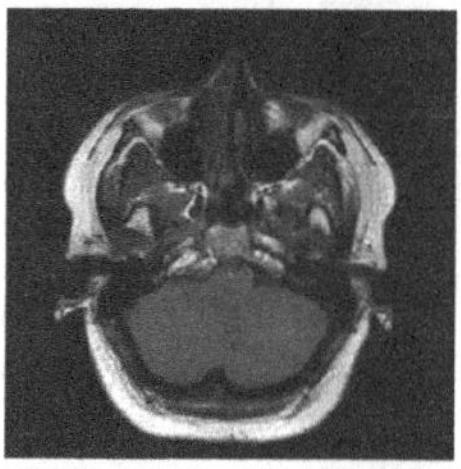	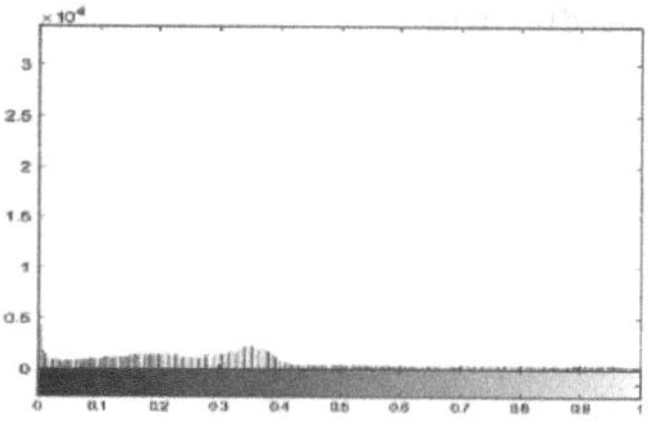
Log Transformation	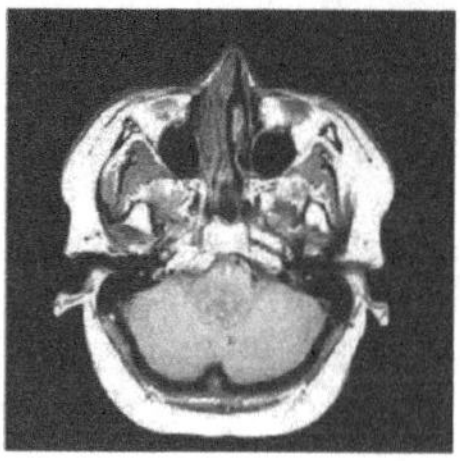	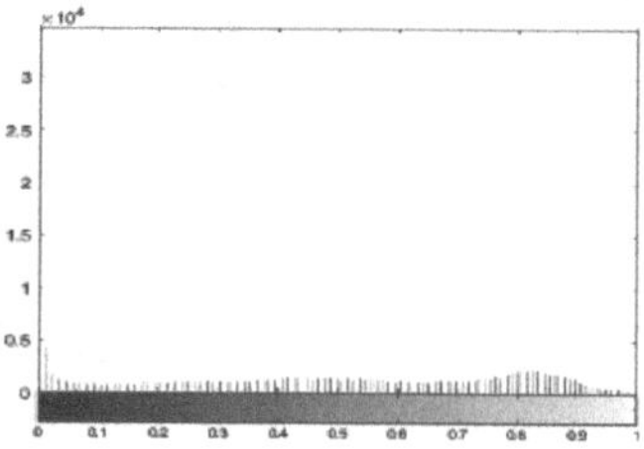

(Continued)

Table 6.5 Results for image 5 using different image enhancement techniques. (*Continued*)

Method	Image (Papilloma MRI)	Histogram
Contrast Stretching	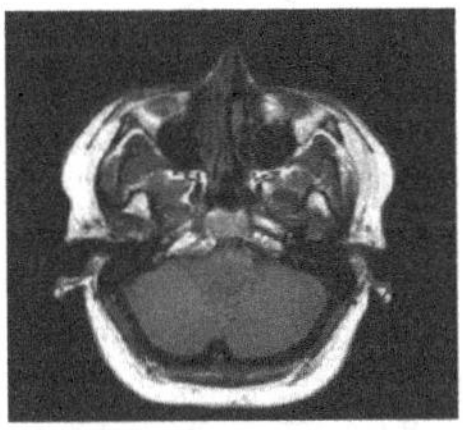	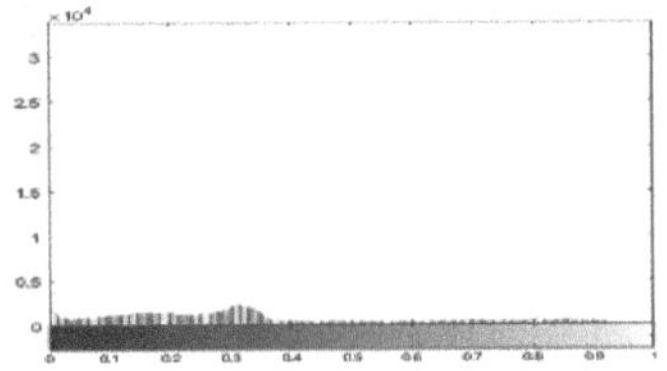
Gamma Correction	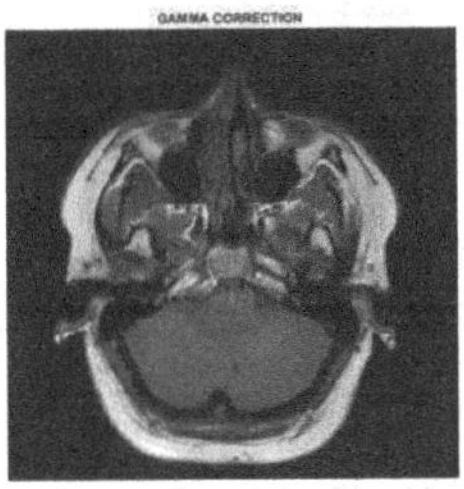	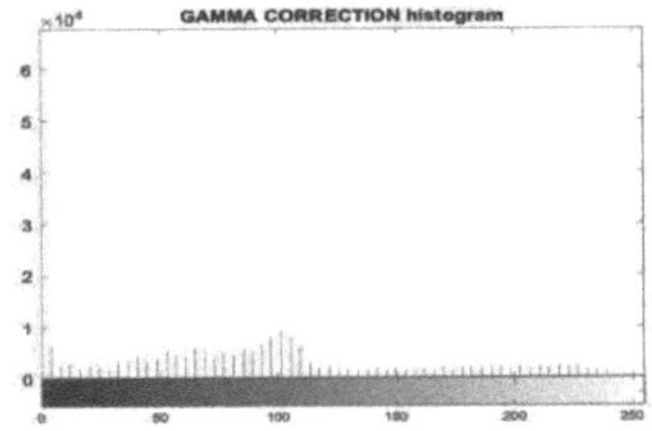

6.4.2 Comparison for Different Image Enhancement Techniques

Here, we have shown PSNR and MSE values for different images after applying Image Enhancement techniques in Tables 6.6 and 6.7 respectively.

We can see from these two tables of MSE and PSNR values that the adjustment technique has the best PSNR overall and the log transformation technique has the best number of MSE values. Here in case of MSE value the adjustment technique has the least number of MSE value and in another side in PSNR table the adaptive histogram technique and histogram equalization technique have least number of PSNR values. For the maximum values of each row in the two tables, we utilize bold italic font here.

The "Adjustment" technique consistently outperformed the others across all of the images among the techniques tested, including Histogram Equalization, Adaptive Histogram Equalization, Adjustment, Log Transformation, and Contrast Stretching. In comparison to the original images, it showed lower MSE values and higher PSNR values, indicating better image quality.

Table 6.6 Different MSE values of the above images.

	MSE values					
Methods	**Histogram equalization**	**Adaptive histogram equalization**	**Adjustment**	**Log transformation**	**Contrast stretching**	**Gamma correction**
Image-1	10208.00	930.85	3703.30	***0.1386***	5397.90	3153.90
Image-2	4635.40	1417.00	1224.60	***0.2286***	2111.00	42.86
Image-3	85.53	1227.10	100.73	***0.5497***	264.95	242.73
Image-4	11508.00	725.24	491.98	***0.0952***	1212.00	20.51
Image-5	0.2522	***0.0133***	0.00098	0.1470	0.0065	132.90

Table 6.7 Different PSNR values of the above images.

	PSNR values					
Methods	**Histogram equalization**	**Adaptive histogram equalization**	**Adjustment**	**Log transformation**	**Contrast stretching**	**Gamma correction**
Image-1	8.04	***18.44***	12.44	8.58	10.80	-34.98
Image-2	11.47	16.61	***17.25***	6.40	14.88	-16.32
Image-3	**28.80**	17.24	28.09	2.59	23.89	-23.85
Image-4	7.52	19.52	***21.21***	10.21	17.29	-13.12
Image-5	5.98	18.76	***30.07***	8.32	21.85	-21.23

More specifically, the PSNR values ranged from 12.4449 to 30.0755, while the MSE values for the Adjustment technique ranged from 3.7033e+03 to 9.8277e-04. These findings show that the Adjustment technique significantly decreased the mean squared differences between the original and enhanced images, resulting in higher fidelity and better image clarity.

When choosing an image enhancement technique, it is important to take the specific requirements and context into account. The Adjustment technique, however, stands out as the most promising option for enhancing biomedical images in this study based on the provided MSE and PSNR values. Their consistencies in performance across numerous images demonstrate how well it can enhance image quality and retain crucial details.

To confirm these results and investigate the applicability of other image enhancement techniques in various biomedical imaging scenarios, more investigation and testing may be required. Nevertheless, the analysis' findings show that the Adjustment technique has a lot of potential for improving the visual appeal and usefulness of biomedical images.

6.5 Conclusion

Techniques for enhancing photographs are essential for raising the caliber and readability of biomedical images. In this review article, we investigated and assessed six frequently applied image enhancement methods for biomedical images: log transformation, power law transformation or gamma correction, piecewise linear transformation or contrast stretching, histogram equalization, Contrast-Limited Adaptive Histogram Equalization (CLAHE), and adjustment method. It offers insightful information on the benefits and drawbacks of each technique, assisting academics, practitioners, and medical professionals in picking the best method for a particular biomedical imaging scenario. To identify the approach that yields the best results, we have also evaluated the performance of various strategies using Mean Square Error (MSE) and Peak Signal-to-Noise Ratio (PSNR) metrics.

These approaches' performance might be evaluated objectively by comparing them to the MSE and PSNR parameters. While PSNR evaluated the quality of the improved image in contrast to the original image, MSE calculated the average squared difference between the enhanced and original images. It is crucial to remember that these measures might not fully represent the perceived quality of the enhanced photos; therefore, a more thorough evaluation should also take into account expert subjective judgment. Our investigation showed that the optimum strategy depends on the particular image properties and the intended improvement objectives.

From our review work we get the PSNR values ranged from 12.4449 to 30.0755, while the MSE values for the Adjustment technique ranged from 3.7033e+03 to 9.8277e-04. From the above discussion of table values we can observe that the image adjustment technique has the best PSNR and log transformation technique has the best MSE. In image processing, if an image has greater PSNR and lesser MSE then it is considered as a good enhanced image.

Future studies should concentrate on investigating cutting-edge methods, like machine learning-based strategies and deep learning models, to further boost picture enhancement in the biomedical industry.

References

1. Abdallah, Y.M.Y. and Alqahtani, T., Research in medical imaging using image processing techniques. *Med. Imaging-Principles Appl.*, 10, 1–16, 2019.
2. Rana, S.B. and Rana, S.B., A review of medical image enhancement techniques for image processing. *Int. J. Curr. Eng. Technol.*, 5, 2, 1282–1286, 2015.
3. Amusa, K.A., Idowu, O.A., Adejumobi, I.A., Adebayo, G.A., Tri-modal technique for medical images enhancement. *Int. J. Adv. Appl. Sci.*, 11, 3, 199–210, 2022.
4. Shruthishree, S.H. and Tiwari, H., A review paper on medical image processing. *Int. J. Res. Granthaalayah*, 4, 4, 21–29, 2017.
5. Perumal Vels., S., Preprocessing by contrast enhancement techniques for medical images. *Int. J. Pure Appl. Math.*, 118, 18, 3681–3688, 2018.
6. Ikhsan, I.A.M., Hussain, A., Zulkifley, M.A., Tahir, N.M., Mustapha, A., An analysis of x-ray image enhancement methods for vertebral bone segmentation, in: *10th international colloquium on signal processing and its applications*, IEEE, pp. 208–211, 2014.
7. Jalab, H.A., Ibrahim, R.W., Hasan, A.M., Karim, F.K., Al-Shamasneh, A.A.R., Balcanu, D., *A new medical image enhancement algorithm based on fractional calculus*, vol. 68, no. 2, pp. 1467–1483, Computers, Materials & Continua, 2021.
8. Gopika, G.S., Prasad, P.H., Lekshmi, A.G., Lekshmypriya, S., Sreesaila, S., Arunima, C., Kumar, M.S., Anil, A., Sreekumar, A., Pillai, Z.S., Chemistry of cyanine dyes-A review. *Mater. Today Proc.*, 46, 3102–3108, 2021.
9. Naseera, S., Rajini, G.K., Venkateswarlu, B., Priyadarisini, M., A review on image processing applications in the medical field. *Res. J. Pharm. Technol.*, 10, 10, 3456–3460, 2017.
10. Ibrahim, M.S., Hamid, S.A., Muhammad, Z., Leh, N.A.M., Abdullah, S., Bakar, S.J.A., Osman, M.K., Fadhlullah, S.Y., CNN Comparative Study for Apple Quality Classification, in: *2022 IEEE 12th International Conference on Control System, Computing and Engineering (ICCSCE)*, IEEE, pp. 53–58, 2022, October.

7

Exploring Parkinson's Disease Progression and Patient Variability: Insights from Clinical and Molecular Data Analysis

Amit Kumar[1], Neha Sharma[1*] and Korhan Cengiz[2]

[1]Chitkara University Institute of Engineering & Technology, Chitkara University, Rajpura, Punjab, India
[2]Department of Information Technologies, Faculty of Informatics and Management, University of Hradec Kralove, Kralove, Czech Republic

Abstract
One common neurodegenerative disease with few treatment options is Parkinson's disease (PD). Disease progression prediction is essential for patient care and therapeutic development. Clinical, genetic, and molecular data related to Parkinson's disease (PD) are collectively gathered by the Accelerating Medicines Partnership (AMP). The goal of this study is to determine the parameters linked to Parkinson's disease (PD) progression and evaluate the appropriateness of the AMP PD dataset for predictive modeling through an Exploratory Data Analysis (EDA). Preprocessing of the data, feature selection, and visualization are all part of our EDA. We use correlation analysis to identify connections with illness progression scores by analyzing the distribution of clinical, demographic, genetic, and biomarker data. Initial results point to possible predictors, providing a framework for the creation of more predictive models in the future. The discovery of therapies and our understanding of Parkinson's disease are greatly aided by cooperative initiatives like AMP. Our EDA of the AMP PD dataset, in conclusion, provides information about the course of Parkinson's disease. The creation of predictive models in the future is informed by our research, which also advances our understanding of this intricate neurological illness.

***Keywords*:** Parkinson's disease, exploratory data analysis (EDA), accelerating medicines partnership (AMP), unified parkinson's disease rating scale (UPDRS)

**Corresponding author*: nehasharma0110@gmail.com

Pijush Dutta, Sudip Mandal, Korhan Cengiz, Arindam Sadhu and Gour Gopal Jana (eds.) Design and Forecasting Models for Disease Management (181–200) © 2025 Scrivener Publishing LLC

7.1 Introduction

Parkinson's disease (PD) is a neurological condition that progresses over time and causes a variety of motor and non-motor symptoms. Since there is currently no recognized treatment for Parkinson's disease (PD), the main goals of care are to manage and reduce the symptoms. Our objective in this research paper is to estimate patient scores on the Movement Disorder Society Unified Parkinson's Disease Rating Scale (MDS-UPDRS) [1] to forecast the course of Parkinson's disease. An essential evaluation tool for determining the severity of Parkinson's disease (PD) and monitoring its evolution over time is the MDS-UPDRS. It consists of fifty thorough questions covering both motor and non-motor symptoms shown in Figure 7.1 related to the illness. There is a maximum total score of 272 points, and each item is graded on a scale from 0 (normal) to 4 (severe impairment). The evaluation is broken down into four sections: Non-motor aspects of daily life experiences are evaluated in Part I. These symptoms include depression, motivation, initiative, sleep, and more. There are thirteen questions on it, and the answers can range from 0 to 52.

Section II: Motor Aspects of Everyday Experiences. This section assesses motor symptoms, including speech, salivation, swallowing, eating, dressing, and tremors. There are 13 questions in all, and the scores go

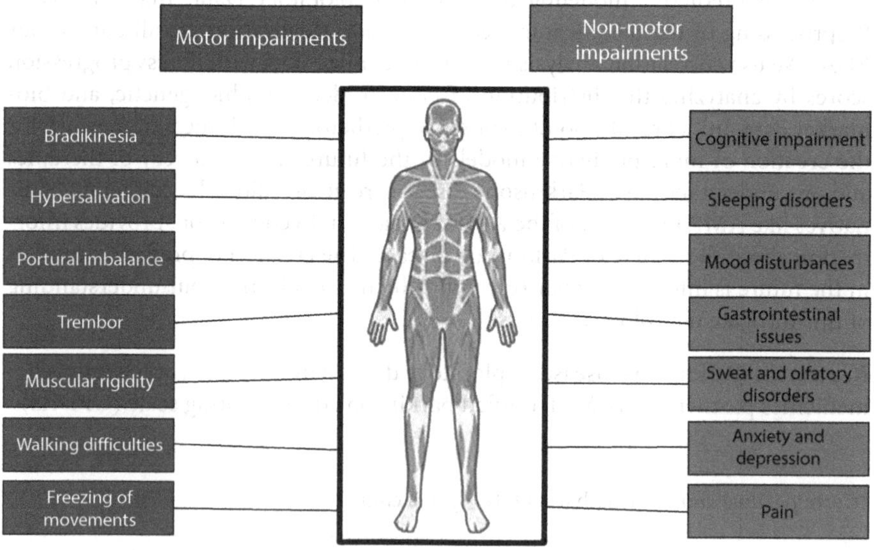

Figure 7.1 Motor and non-motor impairment.

from 0 to 52. Section III: Motor Examination: A doctor performs a physical examination in this segment, looking at 34 motor-related factors. The possible scores are 0 through 136. Section IV: Motor Complications: This section addresses motor issues associated with Parkinson's disease, including off-duration. There are six questions on it, and the answers range from 0 to 24. Predicting MDS-UPDRS scores for each clinic visit (e.g., updrs_1 for Part I, updrs_2 for Part II, etc.) and scores for prospective visits at 6, 12, and 24 months in the future is our job in this research. We anticipate that our forecasts will be based on two essential components: Previous Score Values (Time Series Viewpoint): We will use patient history MDS-UPDRS scores to evaluate disease progression. We will be able to forecast future scores and comprehend how the condition has changed over time with the use of this time series data. CSF Proteins and Peptides (Parkinson's Disease Biomarkers): We will include information regarding cerebrospinal fluid (CSF) proteins and peptides in addition to the patient's history of scores [2, 3]. These biomarkers will support our forecasts and serve as pertinent indications of the disease's progression. It's crucial to remember that the research organizers don't offer any other patient data, such as gender or age. As a result, we will only make predictions based on the CSF data that is now available and the patient's prior MDS-UPDRS scores [1]. In conclusion, forecasting the course of Parkinson's disease using a combination of biomarker data and history score data is a difficult but important problem that this research offers [6]. The effective creation of precise prediction models has the potential to greatly advance our knowledge of the course of Parkinson's disease (PD) and assist in both management and treatment [4].

7.2 Literature Review

One of the neurodegenerative disorders known as Parkinson's disease (PD) involves the progressive loss of dopaminergic neurons in the brain's substantia nigra [7]. Patients with Parkinson's disease (PD) can use the Unified Parkinson's Disease Rating Scale (UPDRS) to gauge how severe their motor and non-motor symptoms are [8]. Studies have indicated that UPDRS scores typically rise with time, indicating the disease's progressive character.

Finding PD biomarkers has attracted more attention in recent years. Proteins and peptides have become attractive options for biomarker discovery. Analysis of proteins and peptides in PD patients has been done using mass spectrometry-based methods [9, 12], which are comparable to the data presented in this work. The objective of these investigations is to identify distinct molecular indicators linked to the ailment, potentially facilitating prompt diagnosis [10] and illness surveillance [11].

Furthermore, interest in the idea of patient-specific variability in Parkinson's disease has grown. Since the course of a patient's illness might vary greatly, customized treatment plans are essential. Research is being done to better understand the causes of variability, including hereditary and environmental variables.

The literature concludes by emphasizing the significance of monitoring UPDRS scores over time, investigating molecular indicators including proteins and peptides, and identifying patient-specific variability in Parkinson's disease [13]. These lines of inquiry advance our knowledge of the course of Parkinson's disease (PD) and aid in the creation of individualized treatment plans.

7.3 Data Review

This multi-source dataset taken from Kaggle appears to be comprehensive and suitable for conducting a detailed analysis [5]. Here's a breakdown of the data files: Train Clinical Data: Contains information on clinic visits. Includes the target variable, UPDRS scores, which are crucial for assessing Parkinson's Disease progression. It likely includes demographic, diagnostic, and treatment-related information. Train Peptides Data: Provides mass spectrometry data at the peptide level. Mass spectrometry data can offer insights into the molecular composition and changes associated with Parkinson's Disease. Train Proteins Data: Offers aggregated mass spectrometry data at the protein expression frequency level. Aggregated data can provide a broader view of molecular changes, potentially identifying key protein biomarkers associated with disease progression. Supplemental Clinical Data: Contains clinical records but lacks associated peptide and protein data. Useful for understanding the overall dynamics of Parkinson's Disease progression from a clinical perspective. Example Test Files: Provides test files that can be used for debugging and submitting predictions. Ensures compatibility and accuracy in the submission process. To gain insights into Parkinson's Disease progression and potentially develop predictive models, you may need to perform data preprocessing, integration, and exploratory data analysis (EDA). Integrating clinical, peptide, and protein data can uncover valuable relationships and biomarkers related to disease progression. The UPDRS scores in the trained clinical data can serve as your target variable for predictive modeling. Ensure that you follow best practices in data analysis, including data cleaning, feature engineering, and model evaluation, to maximize the utility of this dataset for Parkinson's Disease research.

7.3.1 Clinical Data

For an analysis to be meaningful, it is crucial to comprehend the variables present in the clinical dataset. The following is a quick summary of the major clinical data fields: visit_id: A unique identifier for every visit, created by combining the patient's ID with the month of the visit. This distinct number enables long-term patient progress monitoring. The month of the visit about the patient's initial visit is called visit_month. comprehending the course of a disease across time requires a comprehension of this temporal information. patient_id: A special number that only each patient has. For longitudinal analysis, this enables you to group visits made by the same patient. updrs_[1–4]: The scores of the patient for the various sections (1–4) of the Unified Parkinson's Disease Rating Scale (UPDRS) are indicated in these fields. Aspects of the disease such as behavior and mood (Part 1) and motor functions (Part 3) are evaluated in each section separately. More severe symptoms are indicated by higher ratings. clinical_state_of_upd23b_medication: This binary variable indicates whether the patient took medicine at the time of the UPDRS assessment, such as Levodopa. The particular relevance of this information can be seen in Part 3 of the UPDRS, which evaluates motor function. The medication status must be taken into account because it has a big influence on the evaluation's outcome. 'train_clinical_data.csv' and 'supplemental_clinical_data.csv,' two datasets, indicate that both training and supplementary clinical data are available. Combining these datasets and managing outliers and missing information correctly can help create predictive models and offer a thorough understanding of how Parkinson's disease progresses. Throughout the analytic process, you can investigate the connections among UPDRS scores, visit months, medication status, and patient demographics to learn more about the course of the disease and how medication affects the intensity of symptoms. Research on the dynamics of Parkinson's disease and the assessment of possible therapies depend heavily on this data.

In Figure 7.2, the clinical information you gave is structured as follows: 2615 rows (samples) in total, Eight columns (features) in total. This indicates that you have a dataset with 2615 rows, each of which represents a distinct patient visit, and 8 columns that contain diverse clinical and demographic data, such as medication status and UPDRS scores. This dataset should offer a significant amount of data for thorough studies and the creation of models that predict the course of Parkinson's disease.

```
Clinical data shape: (2615, 8)
```

Out[2]:

	visit_id	patient_id	visit_month	updrs_1	updrs_2	updrs_3	updrs_4	upd23b_clinical_state_on_medication
0	55_0	55	0	10.0	6.0	15.0	NaN	NaN
1	55_3	55	3	10.0	7.0	25.0	NaN	NaN
2	55_6	55	6	8.0	10.0	34.0	NaN	NaN
3	55_9	55	9	8.0	9.0	30.0	0.0	On
4	55_12	55	12	10.0	10.0	41.0	0.0	On

Figure 7.2 Clinical data shape (2615, 8).

In Figure 7.3, the additional clinical information that you provide is structured as follows: 2223 is the number of rows (samples). Eight columns (features) in total. There are 2223 rows in this extra clinical dataset, and each row corresponds to a distinct clinical record or visit. Like the core clinical dataset, it has eight columns that include different clinical and demographic variables. Despite the absence of correlated peptide and protein information, this dataset can nonetheless offer insightful information about the general dynamics of Parkinson's disease progression from a clinical standpoint. In conjunction with the primary clinical dataset, it facilitates a more thorough examination and comprehension of the illness.

```
Supplemental clinical data shape: (2223, 8)
```

Out[3]:

	visit_id	patient_id	visit_month	updrs_1	updrs_2	updrs_3	updrs_4	upd23b_clinical_state_on_medication
0	35_0	35	0	5.0	3.0	16.0	0.0	NaN
1	35_36	35	36	6.0	4.0	20.0	0.0	NaN
2	75_0	75	0	4.0	6.0	26.0	0.0	NaN
3	75_36	75	36	1.0	8.0	38.0	0.0	On
4	155_0	155	0	NaN	NaN	0.0	NaN	NaN

Figure 7.3 Supplemental clinical data shape (2223, 8).

Key indicators can be summarized using aggregated statistics for the combined clinical data (train and supplementary). Here is a condensed version of what Figure 7.4 depicts:

```
Clinical data aggregated statistics (train):

Out[6]:
```

	patient_id	visit_month	updrs_1	updrs_2	updrs_3	updrs_4
count	2,615.00	2,615.00	2,614.00	2,613.00	2,590.00	1,577.00
mean	32,651.74	31.19	7.11	6.74	19.42	1.86
std	18,535.76	25.20	5.53	6.32	15.00	3.02
min	55.00	0.00	0.00	0.00	0.00	0.00
25%	16,574.00	10.50	3.00	1.00	6.00	0.00
50%	29,417.00	24.00	6.00	5.00	19.00	0.00
75%	50,611.00	48.00	10.00	10.00	29.00	3.00
max	65,043.00	108.00	33.00	40.00	86.00	20.00

Figure 7.4 Clinical data aggregated statistics (train and supp combined).

Key indicators can be summarized using aggregated statistics for the training clinical data (train). This is a condensed version of what Figure 7.5 depicts:

```
Clinical data aggregated statistics (train and supp combined):

Out[5]:
```

	patient_id	visit_month	updrs_1	updrs_2	updrs_3	updrs_4
count	4,838.00	4,838.00	4,624.00	4,622.00	4,808.00	2,872.00
mean	32,571.92	22.79	6.49	6.64	21.03	1.40
std	18,580.88	22.46	5.10	5.77	13.95	2.61
min	35.00	0.00	0.00	0.00	0.00	0.00
25%	16,687.25	5.00	3.00	2.00	11.00	0.00
50%	31,154.00	18.00	5.00	5.00	20.00	0.00
75%	49,239.00	36.00	9.00	10.00	30.00	2.00
max	65,530.00	108.00	33.00	40.00	86.00	20.00

Figure 7.5 Clinical data aggregated statistics (train).

Have a look at the missing data in Figure 7.6. The missing numbers for updrs_1, updrs_2, updrs_3, and updrs_4 show that the assessment was not carried out. The missing value for upd23b_clinical_state_on_medication indicates that the medication status is unclear.

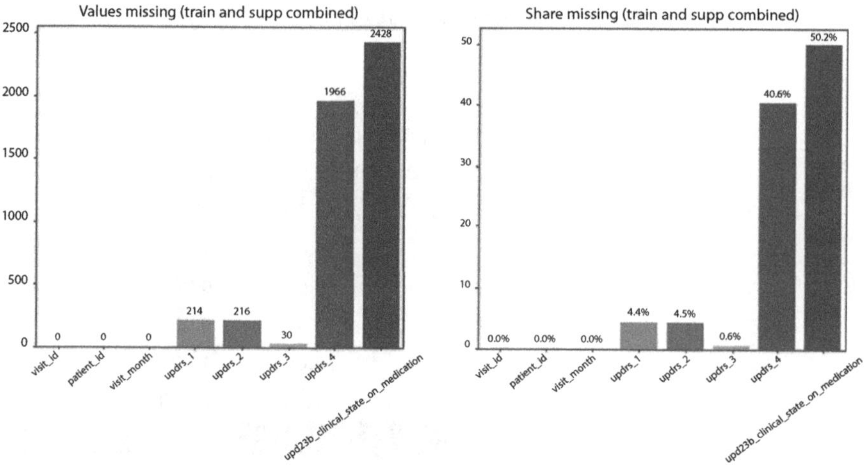

Figure 7.6 Value and share missing (train and support combined).

About the supplemental dataset, there are 771 distinct patients, compared to 248 in the train. The unique values for each column in the dataset are computed to find the total number of unique values in the combined clinical data (Train and Supplemental). The following is an example of how to present the data in Figure 7.7.

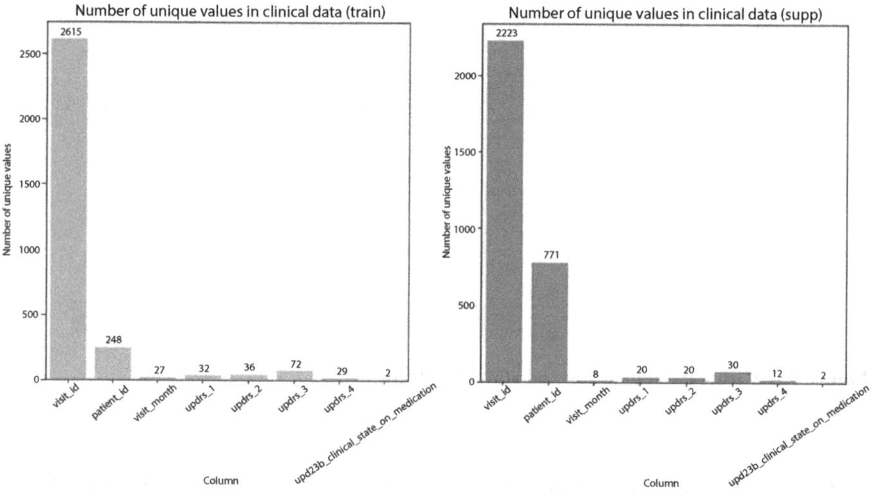

Figure 7.7 Number of unique values in clinical data (train and support).

The data is categorized by patient ID and then count the number of visits for each patient to determine the number of visits per patient in the combined clinical data (Train and Supplemental) shown in Figure 7.8.

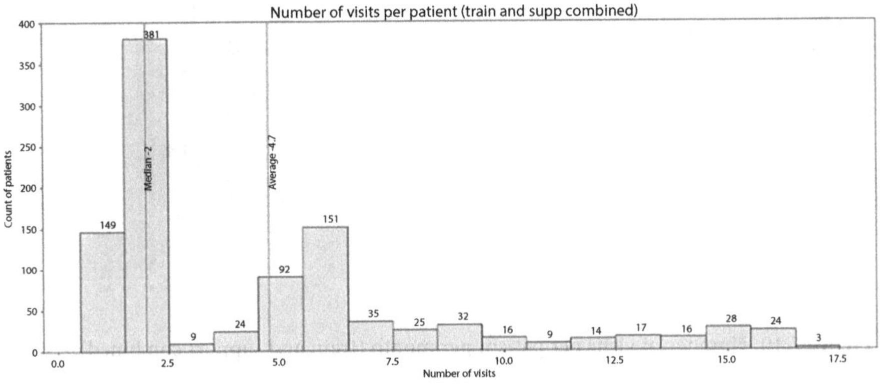

Figure 7.8 Number of visits per patient (train and support combined).

The visits for each patient are counted after grouping the data by patient ID to determine the number of visits per patient in the training clinical data depicted in Figure 7.9.

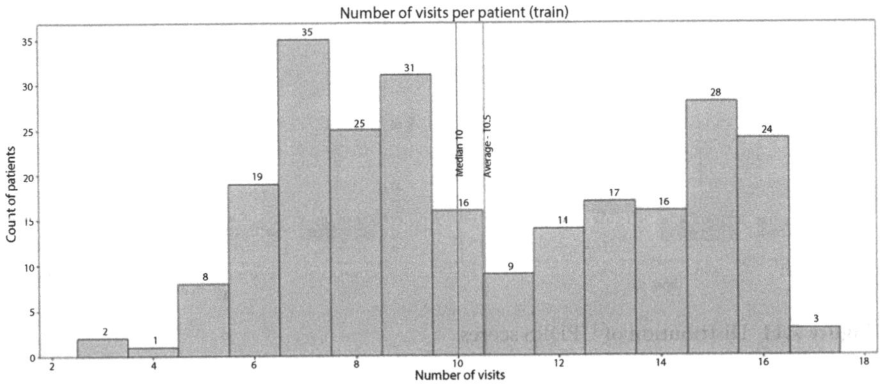

Figure 7.9 Number of visits per patient (train).

The grouping of data is done by patient ID and visit month, then count the number of visits for each combination to determine the number of visits per patient per month in the combined clinical data (Train and Supplemental) shown in Figure 7.10.

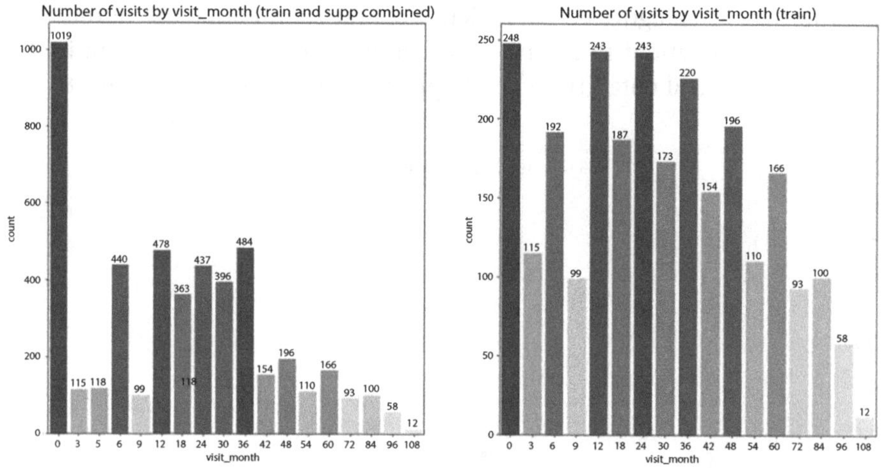

Figure 7.10 Number of visits per patient per month (train and support combined).

Histograms and box plots can be used to show the distribution of UPDRS scores in your clinical data. The dispersion and central tendency of the scores for the various UPDRS scale components are shown in Figure 7.11.

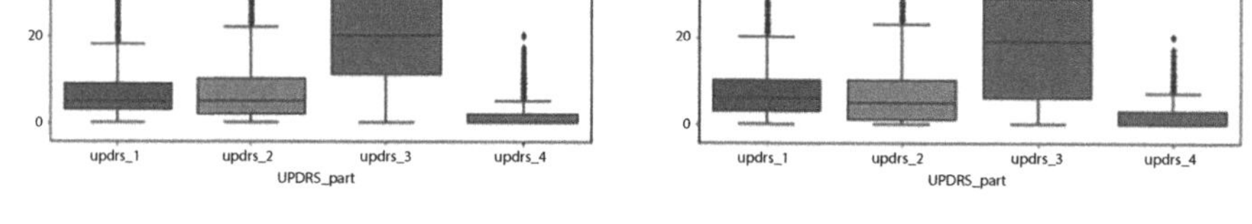

Figure 7.11 Distribution of UPDRS scores.

It is not unexpected that UPDRS scores in many Parkinson's disease cases tend to rise over time shown in Figure 7.12. This finding is in line with the disease's progressive nature, which frequently results in symptoms getting worse with time. The purpose of the UPDRS scale is to evaluate both non-motor symptoms and motor performance of Parkinson's disease. A decline in the patient's state may be indicated by an increase in UPDRS scores.

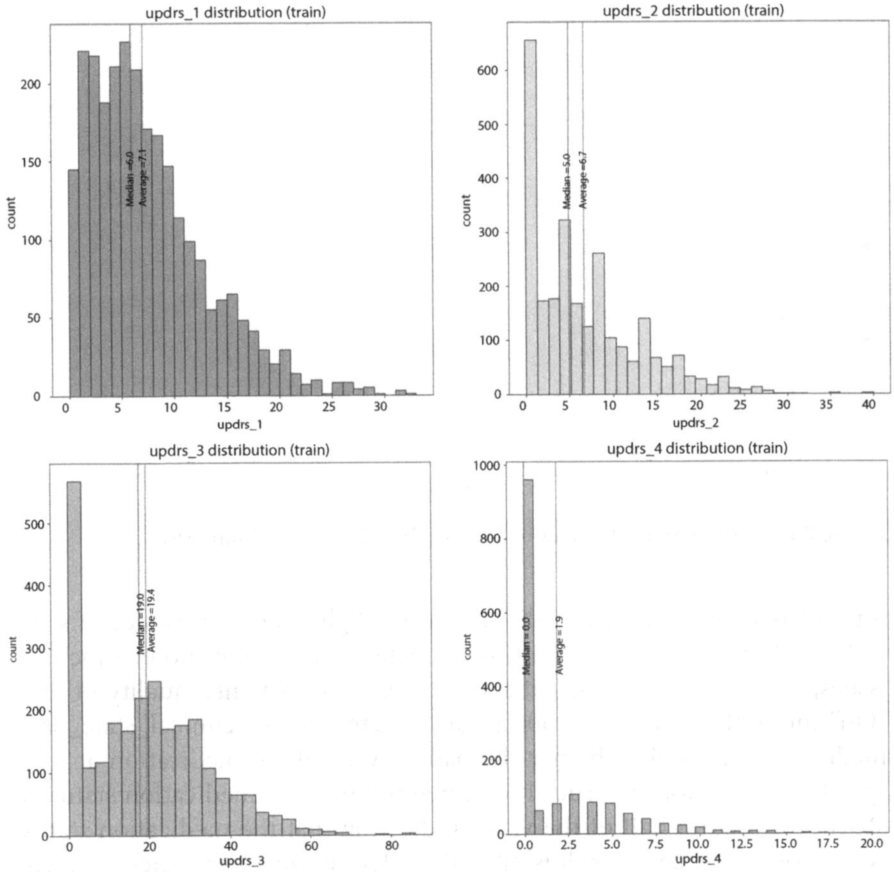

Figure 7.12 updrs_1, updrs_2, updrs_3, and updrs_4 distribution (train).

After doing an analysis of the UPDRS scores for each visit month for the combined clinical dataset (train and support), I discovered patterns that shed light on the evolution of Parkinson's disease. The UPDRS score distribution across the UPDRS scale's different portions (updrs_1, updrs_2, updrs_3, and updrs_4) is notable for continuously displaying the highest scores by visit month for updrs_3. This research is consistent with the clinical notion that, in comparison to other components of the scale, UPDRS Part 3 largely examines motor function and tends to show more significant variations and deterioration with time shown in Figure 7.13. This result emphasizes the significance of routinely monitoring UPDRS Part 3 scores and emphasizes the crucial function of motor symptom assessment, particularly in the context of treatment evaluation and disease management

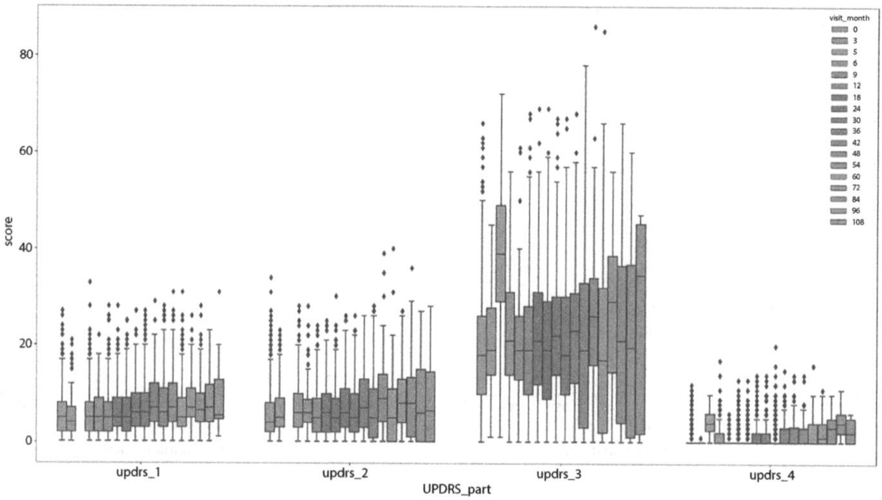

Figure 7.13 UPDRS Score distribution by Visit_Month (train and support).

for patients with Parkinson's disease. It highlights the importance of specialized therapies and interventions targeted at resolving motor function issues, considering their significant influence on patients' quality of life. "On" and "Off," the two appropriate answers to upd23b_clinical_state_on_medication), indicate whether the patient was taking medication at that specific visit or not. It's crucial to remember that the medication status is absent for around half of the visits, indicating that the medication status is unknown in such cases. It is difficult to determine how medicine affects UPDRS scores for those particular visits because of the missing data. It's also critical to note that the test set does not have access to this medication status data, therefore predictions about medication status in the test set can be dependent on other features or models. In data analysis, dealing with missing data is a typical difficulty. Several tactics can be used to handle it efficiently, such as imputation techniques or, if appropriate, treating the missing data as a different category. Careful examination of this missing data is essential because, in the context of Parkinson's Disease research, understanding the impact of medication on symptom severity and disease progression is crucial.

7.3.2 Peptides Data

Essential components of cells, proteins and peptides perform vital biological tasks. The term "peptides" refers to molecules made up of two to fifty amino acids. Important biomarkers for Parkinson's disease include certain peptides. Data on peptides comprises the following:

- visit_id is the visit's ID code.
- visit month: The month of the visit in relation to the patient's initial visit.
- patient_id - The patient's ID code.
- UniProt: The linked protein's UniProt ID code. A protein frequently consists of several peptides.
- Peptide: The arrangement of amino acids that make up a peptide. For the applicable codes, see this table.

There may be rare annotations that are not listed in the table. Peptides absent from the train set might be present in the test set. The frequency of an amino acid in a sample is known as its peptide abundance. This is peptide-level mass spectrometry data. In other words, visit_id and peptide can be combined to uniquely identify each row.

981,834 rows (samples) in total,6 columns (features) in total. Furthermore, the dataset contains no duplicate peptide entries, suggesting that every visit_id and peptide combination is distinct. This dataset is useful for in-depth research and investigation of peptide-level information relevant to Parkinson's disease because it looks to be rather substantial, with a large number of samples. In Figure 7.14, data can be used by researchers to find possible biomarkers and investigate the function of peptides in the processes and course of disease.

The fact that the data is only accessible for a portion of visitors should be noted. Considering the data you have given:

- A total of 1,068 visits containing peptide data
- 40.8% of visits included peptide data

Out[23]:

	visit_id	visit_month	patient_id	UniProt	Peptide	PeptideAbundance
0	55_0	0	55	O00391	NEQEQPLGQWHLS	11254.3
1	55_0	0	55	O00533	GNPEPTFSWTK	102060.0
2	55_0	0	55	O00533	IEIPSSVQQVPTIIK	174185.0
3	55_0	0	55	O00533	KPQSAVYSTGSNGILLC(UniMod_4)EAEGEPQPTIK	27278.9
4	55_0	0	55	O00533	SMEQNGPGLEYR	30838.7

Figure 7.14 Peptide abundance.

Out[25]:

	count	mean	std	min	25%	50%	75%	max
visit_month	981,834.00	26.11	22.91	0.00	6.00	24.00	48.00	108.00
patient_id	981,834.00	32,603.47	18,605.93	55.00	16,566.00	29,313.00	49,995.00	65,043.00
PeptideAbundance	981,834.00	642,890.25	3,377,989.09	11.00	28,174.25	74,308.30	221,338.75	178,752,000.00

Figure 7.15 Peptide data aggregated statistics.

This indicates that 1,068 visits, or roughly 40.8% of all visits in the dataset, have peptide data available for them. When performing studies, researchers and analysts should be aware of this subset and take into account the implications of missing peptide data when making inferences or creating models.

Peptide data aggregated statistics shown in Figure 7.15, can provide light on the features of the Peptide dataset. The precise statistics shown in Figure 7.15 may be computed, however, will depend on the goals of your study. The peptide data can be better understood by researchers, and they can find potential biomarkers and patterns associated with Parkinson's disease with the aid of this aggregated information. Your analysis's relevant statistics might be determined by considering your particular research objectives.

7.3.3 Protein Data

An outline of the protein data is essential for comprehending Parkinson's disease's molecular components. Proteins are essential for the proper operation of cells, and studying them can provide important information on the underlying causes and biomarkers of disease. Below is a synopsis of the most important details on the protein data:Visit_id: A unique reference for every dataset data point, serving as an identifier for every visit. Visit_month: To record and analyze protein data over time, the month of the visit concerning the patient's initial visit.patient_id: A special number that serves as a patient's identity, allowing data to be grouped according to the patient for longitudinal analysis.UniProt: The linked protein's UniProt ID code. Protein-specific variants can be studied because a single protein might be linked to several peptides. Normalized protein expression, or NPX, is a measure of how frequently or abundantly a protein occurs in a sample. Since some proteins contain many copies of a single peptide, it might not have a one-to-one relationship with component peptides.

A thorough understanding of the proteins connected to Parkinson's disease can be obtained from the mass spectrometry data at the protein level. Precise tracking and analysis of protein alterations over time is ensured by the combination of visit_id and UniProt, which uniquely identify each row. Furthermore, the incorporation of UniProt IDs facilitates the investigation of protein-level alterations linked to the illness by connecting proteins to the peptides that make them up. It is noteworthy that proteins absent from the train set might be present in the test set, which presents an opportunity to uncover novel molecular insights related to Parkinson's disease. By using this data, researchers can find biomarkers and learn more about the underlying molecular causes of the disease.

The following features are included in the protein data shown in Figure 7.16: 232,741 rows (samples) in total, Five columns (features) in total. Furthermore, the dataset contains no duplicate protein entries, indicating that every visit_id and UniProt combination is distinct. This dataset looks to be vast and sample-rich, making it a useful tool for analyzing protein-level data associated with Parkinson's disease. For the objectives of research and analysis, data dependability and integrity are ensured by the lack of duplicate entries. Using this data, researchers can investigate how particular proteins contribute to the development of diseases and the identification of biomarkers.

It is significant to remember that only a portion of visits have access to protein data. Considering the data you have given: Visits containing Protein Data: 1,068, Visits Including Protein Data as a Share: 40.8%. this indicates that 1,068 visits out of the total, or about 40.8% of the visits in the dataset, have protein data accessible. When performing studies, researchers and analysts should be aware of this subset and take into account the implications of missing protein data when making inferences or creating models.

Out[35]:

	visit_id	visit_month	patient_id	UniProt	NPX
0	55_0	0	55	O00391	11254.3
1	55_0	0	55	O00533	732430.0
2	55_0	0	55	O00584	39585.8
3	55_0	0	55	O14498	41526.9
4	55_0	0	55	O14773	31238.0

Figure 7.16 Proteins data shape (232741, 5).

Out[37]:

	count	mean	std	min	25%	50%	75%	max
visit_month	232,741.00	26.10	22.87	0.00	6.00	24.00	48.00	108.00
patient_id	232,741.00	32,593.88	18,608.48	55.00	16,566.00	29,313.00	49,995.00	65,043.00
NPX	232,741.00	2,712,076.94	22,241,547.32	84.61	29,464.40	113,556.00	563,894.00	613,851,000.00

Figure 7.17 Protein data aggregated statistics.

With 232,741 samples in the protein collection, data integrity is guaranteed by the absence of duplicate entries. For 40.8% of the 1,068 visits, protein data is available, providing a significant amount of coverage. Researchers can compute important statistics, such as the number of distinct proteins, mean protein expression, distribution of expression levels, and temporal patterns, to efficiently analyze this data to find putative biomarkers and learn more about the mechanisms behind Parkinson's disease.

For protein data, aggregated statistics can offer insightful information on the features of the protein dataset. The precise statistics shown in Figure 7.17, may like to compute will depend on the goals of your study. When analyzing protein data, take into account the following typical aggregated statistics.

7.4 Parkinson's Dynamic for Patients in Train

UPDRS scores for 3 patients in the train data shown in Figure 7.18 can help reveal important information about how Parkinson's disease progresses differently from patient to patient.

To better understand how individual patient scores vary over time, Figure 7.18 shows separate line plots or subplots for each of the various sections (updrs_1, updrs_2, updrs_3, and updrs_4). The general method for visualizing each patient's UPDRS score is as follows: Utilize patient_id to group the train data. Figure 7.18 shows the separate plot for each patient with the visit_month (time) on the x-axis and the UPDRS scores for the relevant portion (updrs_1, updrs_2, updrs_3, or updrs_4) on the y-axis. Plot the UPDRS scores of every patient on a single plot, being sure to identify each patient with a separate color or marker. For every component of the UPDRS scale, repeat these steps. UPDRS scores vary from patient to patient, spot trends or patterns, and learn more about how each patient's disease develops at a time thanks to this visualization. In the context of Parkinson's disease research and clinical analysis, it can be a potent tool.

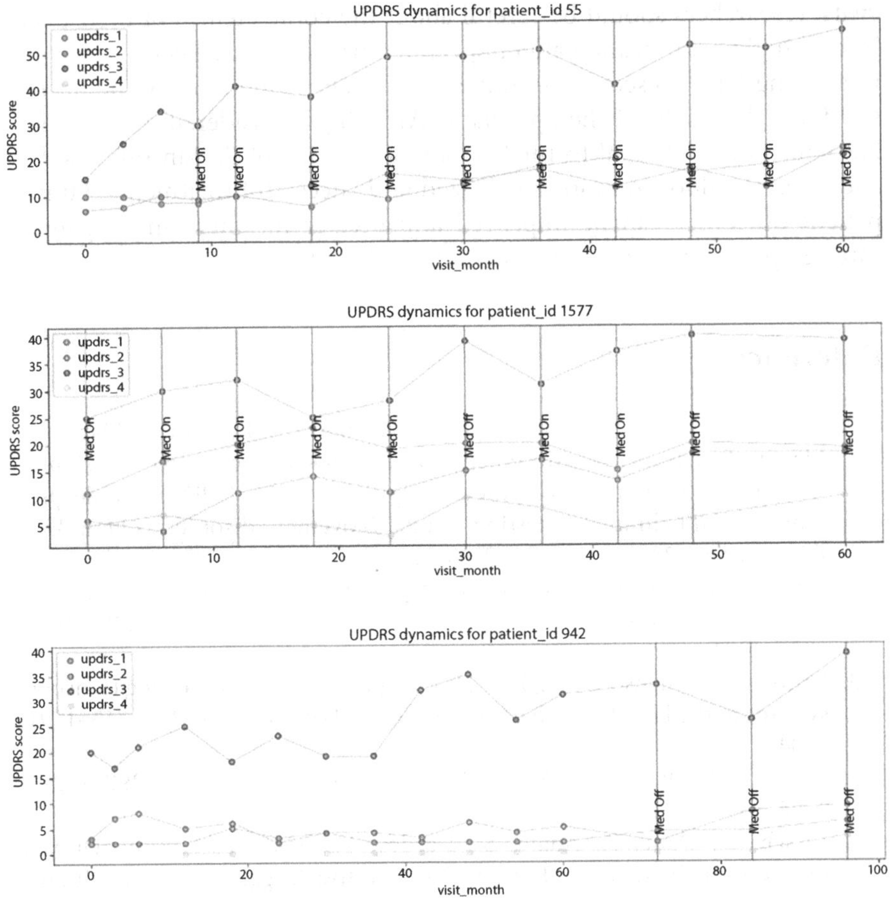

Figure 7.18 UPDRS scores for 3 patients in the train data.

7.5 Conclusion

To sum up, the examination of the Parkinson's Disease dataset has yielded noteworthy revelations. Crucial indicators of the severity of the condition, UPDRS scores, usually exhibit a progressive trend over time, reflecting the course of the illness. The difficulty in obtaining missing drug status information highlights how difficult it is to determine how therapies affect the severity of symptoms. A potential direction for the investigation of molecular biomarkers and protein-level changes associated with Parkinson's disease is presented by the incorporation of peptides and proteins data. However, since peptides and proteins are only available for a portion of visits, researchers should be careful about the completeness of their data. Interestingly, significant patient-specific variability has been shown by visualizing UPDRS scores for individual patients, emphasizing the necessity for individualized therapy plans. All things considered, this dataset has enormous potential to further our knowledge of Parkinson's disease and open the door to more specialized and successful treatments. Future patient outcomes can be improved and greater insights can be gained through predictive modeling and additional study.

References

1. Goetz, C.G., Tilley, B.C., Shaftman, S.R., Stebbins, G.T., Fahn, S., Martinez-Martin, P., LaPelle, N., Movement Disorder Society-sponsored revision of the Unified Parkinson's Disease Rating Scale (MDS-UPDRS): scale presentation and clinimetric testing results. Movement disorders. *Off. J. Mov. Disorder Soc.*, 23, 15, 2129–2170, 2008.
2. Seyfert, S., Kunzmann, V., Schwertfeger, N., Koch, H.C., Faulstich, A., Determinants of lumbar CSF protein concentration. *J. Neurol.*, 249, 1021–1026, 2002.
3. Smith, D.E., Johanson, C.E., Keep, R.F., Peptide and peptide analog transport systems at the blood–CSF barrier. *Adv. Drug Deliv. Rev.*, 56, 12, 1765–1791, 2004.
4. Bloem, B.R., Okun, M.S., Klein, C., Parkinson's disease. *Lancet*, 397, 10291, 2284–2303, 2021.
5. Kirsch, L., Dane, S., Adam, S., Dardov, V.. AMP®-Parkinson's Disease Progression Prediction. Kaggle, 2023, https://kaggle.com/competitions/amp-parkinsons-disease-progression-prediction.
6. Looney, S.W. and Hagan, J.L., *Analysis of biomarker data: a practical guide*, John Wiley & Sons, 2015.
7. Sakharkar, M.K., Kashmir Singh, S.K., Rajamanickam, K., Mohamed Essa, M., Yang, J., Chidambaram, S.B., A systems biology approach towards the identification of candidate therapeutic genes and potential biomarkers for Parkinson's disease. *PLoS One*, 14, 9, e0220995, 2019.
8. Pappala, K., Garuda, B., Seepana, G., Thalabaktula, S., Uppaturi, A., Non-motor symptoms of Parkinson's disease: its prevalence across various stages and its correlation with the severity of the disease and quality of life. *Ann. Mov. Disord.*, 2, 3, 102–102, 2019.
9. Letourneau, D.R. and Volmer, D.A., Mass spectrometry-based methods for the advanced characterization and structural analysis of lignin: A review. *Mass Spectrom. Rev.*, 42, 1, 144–188, 2023.

10. Kaur, S., Gupta, S., Singh, S., Koundal, D., Zaguia, A., Convolutional neural network based hurricane damage detection using satellite images. *Soft Comput.*, 26, 16, 7831–7845, 2022.
11. Panwar, A., Yadav, R., Mishra, K., Gupta, S., Deep learning techniques for the real time detection of Covid19 and pneumonia using chest radiographs, in: *EUROCON 2021 - 19th IEEE International Conference on Smart Technologies, Proceedings*, pp. 250–253, 2021, doi: 10.1109/EUROCON52738.2021.9535604.
12. Kumar, A. and Sharma, N., An Investigation of Advancements in Blockchain Consensus Algorithms & Leading Protocols, in: *2023 4th IEEE Global Conference for Advancement in Technology (GCAT)*, pp. 1–6, IEEE, 2023.
13. Goyal, J., Sharma, N., Kumar, A., Anand, A., Prabha, C., Classifying Diabetes Using Artificially Intelligent Techniques: A Comparative Analysis, in: *2023 IEEE 2nd International Conference on Industrial Electronics: Developments & Applications (ICIDeA)*, pp. 353–358, IEEE, 2023.

8

A Survey-Based Comparative Study on Machine Learning Techniques for Early Detection of Mental Illness

Prachi Majumder[1], Sompadma Mukherjee[2], Shreyashi Saha[3], Tamasree Biswas[4], Mousumi Saha[3], Deepanwita Das[5] and Suchismita Maiti[3]*

[1]Accenture, Kolkata, India
[2]Capgemini, Mumbai, India
[3]Narula Institute of Technology, Kolkata, India
[4]Aston University, Birmingham, UK
[5]National Institute of Technology, Durgapur, India

Abstract

The onset of the COVID-19 pandemic and the practice of social isolation have led to a rise in instances of mental health issues. Advances in machine learning in the present decade have created additional opportunities for detection and the identification of mental health issues. Within this document, we perform a comparative study about the earlier studies that have been performed for automatic prediction and diagnosis of mental illness using machine learning algorithms. Mainly concentrating on various data collection methods, we categorize them to determine the most widely used algorithm. A comparative study of the quality metrics is also carried out simultaneously. For any particular machine learning algorithm, all the quality metrics are compared and analyzed to deduce the one with a better outcome.

***Keywords*:** Mental illness, random forest, Naïve Bayes, decision tree, support vector machine, k-nearest neighbor

8.1 Introduction

As the COVID-19 pandemic has been sweeping across the globe for almost 2 years, it started showing its effect on the mental health of most individuals.

**Corresponding author*: suchismita2006@gmail.com

Pijush Dutta, Sudip Mandal, Korhan Cengiz, Arindam Sadhu and Gour Gopal Jana (eds.) Design and Forecasting Models for Disease Management (201–214) © 2025 Scrivener Publishing LLC

The root cause of increased mental illness may be financial pressure, change in daily routine, social isolation, etc. As the pandemic persists, feelings of depression, stress, anxiety, and insomnia have become ingrained in nearly every individual within society. Currently, it is crucial to offer mental health services to the entire population. To achieve this, mental illness detection is the most preliminary task. Due to technological advancements, Machine Learning (ML) can be extensively applied to identifying mental illness [1]. There are a large number of machine learning methods employed in predicting this outcome. After reviewing 26 papers we compare and analyze the 5 most prominent algorithms like Random Forest, Naïve Bayes, Decision Tree, Support Vector Machine, and k-nearest neighbor are among the machine learning techniques utilized. Despite having numerous machine learning algorithms, the above-mentioned 5 approaches are considered for study because a majority of the past works were carried out using the chosen algorithms. The purpose of this research is to review the Earlier studies or prior research works that focus on the machine learning approach for mental illness. Nevertheless, apart from the machine learning techniques, the statistical analysis method is also used for making such forecasts. Alongside the recent advances in machine learning, in this research, we shall be focusing only on the machine learning approaches for the analysis. Moreover, the quality metrics that shall be considered in our work are f1-score, Precision, recall, and accuracy metrics factor. The objective of this research is to offer a comprehensive insight into the recent advancement in the said field. Further, it would also examine the major scope of future research. The paper is organized as follows: Section 8.2 presents the context of this research, delving into the origins and commencement of the study. Section 8.3 describes the related research endeavors that comprise the previous pieces that have been done in the past. Section 8.4 is the comparative result which centers on the comparison of the utilization of different algorithms. Section 8.5 consists of the discussion where the analysis of our work is depicted and Section 8.6 is the conclusion which sums up the entire work.

8.2 Background

Mental illness has a significant effect on the well-being of an individual irrespective of age, socio-economic status, demography, etc. To better understand the mental state of an individual, machine learning plays a very prominent function in early identification. Machine learning (ML) applications utilize data obtained from standard questionnaires or collected social media content for the early identification of mental health issues. This early detection process helps mental health providers obtain better diagnoses in a clinical setting and subsequent steps treatment [2]. Advances in technology have enabled us to gather a diverse set of data about mental health researchers at a very fast pace and ML is a robust technique effectively utilized for the analysis of this data. In this respect, it is rightly said that "ML involves the use of advanced statistical and probabilistic techniques to construct systems with an ability to automatically learn from data" [3]. This alteration feature enables accurate identification of patterns in data and consequently the predictions from data sources are accurate [4]. Thus, we can deduce that the amalgamation of ML with mental health is very pertinent taking into consideration the technological advancement and the pandemic era.

8.3 Review of Previous Works

In the present work, we have divided the research works into two groups depending on the data collection method, namely, approaches involving questionnaires and those centered on social media content. Within the questionnaire-centric approach, the data is collected through a set of predefined questions for the research subjects. The latter approach gathers the social-media content of each research subject from different social media platforms through their user handlers. In both methodologies, data collection adheres to ethical standards.

8.3.1 Standard Questionnaire

A standard questionnaire for mental health is written under the supervision of a psychologist. It is used to measure mental health scale. Participants are systematically presented with identical questions in a consistent format, and their responses are then uniformly documented [5].

- Anxiety, Elderly individuals can experience depression at any given moment. Most of the It often goes unnoticed for a period. This could be Because of the absence of clear symptoms in the elderly or their reluctance to undergo clinical diagnosis, it often remains undetected. So, Sau, A., and Bhakta, I. developed an appropriate predictive model among elderly patients using machine learning algorithms. Ten classifiers (Bayesian Network, Logistic Regression, Multi-Layer

Perceptron, Naive Bayes, Random Forest, Decision Tree, Support Vector Machine, Random Subspace, and k-nearest Neighbors) were assessed using a dataset comprising 510 elderly patients. The evaluation included testing with a ten-fold cross-validation methodology where RF is the appropriate classifier for the proposed model based on the highest F1 measure 89% [6].

- M. Srividya & S. Mohanavalli1 & N. Bhalaji have proposed a research work, where they have developed a machine learning model to ascertain the mental health status of a specific group by employing clustering algorithms such as K Means, Hierarchical, Density-based, K-medoids, and their variations, along with classification algorithms. - namely Utilizing Support Vector Machines (SVM), Decision Trees (DT), Naïve Bayes (NB), K-Nearest Neighbor (KNN), and Logistic Regression (LR) classifiers were employed. 656 samples were gathered from two target groups, one consisting of 300 people and another group consisting of 356 people. They were given a questionnaire that consisted of 20 questions. Then these samples were subjected to various clustering algorithms to recognize potential groups within two specified target groups. After clustering the data, training set data is then fed into multiple classifiers. Among 5 algorithms, according to accuracy score support Vector Machine and K-Nearest Neighbors exhibited the most effective performance when considered as individual classifiers, both showing the same accuracy score of 0.89 [7].
- In 2019 D. Leightley *et al.* proposed a model based on identifying probable PTSD involving a total of 13,690 individuals who are currently serving or have previously served in the UK Armed Forces. The mental health of any of the soldiers is a significant element as it affects their field performance retention, health, and well-being. Early detection of PTSD will lead to early treatment. Most of the time PTSD occurs due to any traumatic incident in war or could occur after retirement. Due to the large and complex nature of data, machine learning methods have been used in this paper. Data collection started from 2004-2009 [8, 9]. Data has been Gathered in two stages, with the first phase spanning from 2004 to 2006 and phase 2 collected from 2007-2009. The employed machine learning methods include Support

Vector Machines (SVM), Random Forests (RF), Artificial Neural Networks (ANN), and Bagging. Based on the accuracy score, Random Forests achieved the highest accuracy (0.97), while Support Vector Machines exhibited the highest sensitivity rates (0.70) [10].

- Another study conducted by A. Sau and I. Bhakta in 2019 on Applying machine learning to discern anxiety and depression in seafarers. Many studies have shown that seafarers suffer from depression and anxiety due to loneliness, fear of losing jobs, and separation from family. So A. Sau and I. Bhakta proposed an automated machine-learning model that has been created to detect anxiety and depression in their early stages. The model was trained using data gathered from 470 seafarers who underwent interviews at Haldia Dock Complex., India. It showed another algorithm, named Catboost, to be one of the best algorithms for the detection of depression among 4 other algorithms (LR, NB, RF, SVM) with an accuracy of 82.6% and precision of 84.1% [11].
- Anu Priya *et al.* Suggested a model for the identification of anxiety, depression, and stress utilizing the Depression Anxiety Stress Scale Questionnaire (DASS 21). This questionnaire consists of questionnaire consists of 21 questions, with seven questions assigned to each stress, anxiety, and depression scale. The research involved 348 participants spanning ages 20 to 60, encompassing both genders and individuals employed or unemployed via Google Forms Subsequently, the data was categorized employing machine learning algorithms, including Decision Tree, Random Forest Tree, Naïve Bayes, Support Vector Machine, and KNN. Comparing the algorithms in terms of F1 score, Random Forest Tree outperforms [12].
- In a study by Tuka Alhanai *et al.*, we came across different types of sequence models used by researchers to identify depression using automated depression detection algorithms. The dataset included texts and audio collected from 142 individuals undergoing depression screening which was further separated into three groups: training, development, and testing sets. Since the experiment was divided into three parts, the following mentioned models were used namely: Context-free Modeling, Weighted Modeling, Sequence-Modeling. The sequence-modeling is again divided into

two models: Long Short-Term Memory (LSTM) model and multi-modal Long Short-Term Memory (LSTM) model. The algorithms and attributes that were utilized in different parts of the experiments are Logistic Regression, Word2Vec, Covarep, LSTM, Support vector machine (SVM), etc. [13].

8.3.2 Social Media Content

Social media plays a significant role in depression detection using machine learning as it provides a huge amount of data to analyze for researchers. Many people consider Social Network Sites (SNS) and social media as individuals who use their diaries to express their emotional state. Twitter is one of the well-known social media sites used by people of almost all age groups. Twitter can collect extensive data through the Twitter API, enabling researchers to identify potential signs of depression by analyzing publicly available tweet domains, using machine learning algorithms. Other SNS such as Facebook, and Instagram are also popular and supply a substantial amount of data for analysis, by keeping ethical practices in check.

- Depression and distress play a significant role in the number of deaths by suicide. In a different research investigation, we have seen the usage of Examining linguistic patterns and word frequency software (LIWC, version 2015), which is a computer-based text analysis program. Here, the researchers analyzed the textual content of the data collected from Twitter. The data for this study has been collected from participants selected through Amazon's Mechanical Turk (MTurk, www.mturk.com). The algorithm which is used in this study is the Decision Tree, which clearly showed the results into two sub-group categories, that is, suicidal and non-suicidal. With a total of 85 females and 50 males, the study concluded that 17 individuals were significantly clinically suicidal and the remaining 118 were non-suicidal (suicidal and non-suicidal as defined by DSI-SS cut-off criteria) [14].
- Islam *et al.* proposed a model that aims to involve the analysis of Facebook data to identify potential factors indicative of depression among relevant users. The research explores linguistic cues to detect events that evoke emotions. The machine learning algorithms employed for depression detection include a Support Vector Machine (SVM), Decision Tree, Ensemble, and k-Nearest Neighbor (KNN).

Comparing each classifier using the F1 measure, the Decision tree outperforms with highest the F1 score - 0.73 [15].

- In 2018, Rafiqul Islam *et al.* introduced a novel approach to identify depression in Facebook users. Despite the various algorithms employed by researchers for detecting depression on social media, each paper reports differing accuracy scores or F1 scores. Rafiqul Islam *et al.* utilized only the KNN classifier, including Fine KNN, Medium KNN, Coarse KNN, Cosine KNN, Cubic KNN, and Weighted KNN. Data was collected from Facebook using NCapture and then analyzed with LIWC2015 to assess emotion levels in user comments. The KNN classifier was applied to classify various feature types. Results indicated that Coarse KNN emerged as the optimal model with the highest F1 scores in emotional process (0.71), linguistic style (0.70), and temporal process (0.70) [16].
- Hatoon S. ALSAGIRI *et al.* demonstrated the identification of depression levels by employing a dataset that includes 111 Twitter user profiles and over 300,000 tweets. Various machine learning algorithms, including Support Vector Machine (SVM) with different kernels, Naive Bayes (NB), and Decision Tree (DT), were utilized for the analysis. Results showed that SVM-linear has been the most favorable one with an F1- score of 0.79 and an accuracy of 82.5% [17].
- Kuhaneswaran A/L Govindasamy and Naveen Palanichamy introduced an alternative model designed to identify user depression through tweet analysis. The classification was carried out using two machine learning algorithms, specifically Naïve Bayes and a hybrid model named NBTree. Upon evaluating the accuracy scores of both classifiers, it was observed that both algorithms demonstrated comparable performance, achieving the same level of accuracy, 92.34 accuracy for 1000 tweets and 97.31 for 3000 tweets [18].
- In a different study conducted by Alina Trifan *et al.*, an approach utilizing bag-of-words features and a rule-based estimator was employed to detect depressed users on the Reddit social media platform, analyzing psycholinguistic patterns. The dataset used in this research was initially introduced by Yates *et al.* The classifiers implemented in the study include Multimodal Naive Bayes (MNB), Support Vector Machine (SVM), and Passive Aggressive (PA). The most

favorable outcomes were obtained with the Tf-idf weighting method, particularly with the passive-aggressive (PA) classifier [19].

8.4 Comparative Result

In this study, the research papers are classified according to the type of datasets used in the individual papers. Each study has used unique datasets and produced significant results.

We can see that in most of cases the supervised algorithms such as SVM (with different kernels), DT, and NB produce results with greater precision than the unsupervised ones. As illustrated in a research paper by A Sau *et al.*, which was conducted on 510 geriatric patients, showed that the Random Forest classifier was the best among the other 10 classifiers used in this particular experiment [17]. Again in another study, by Hatoon S.ALSAGRI *et al.*, that was conducted on Twitter users, with data collected from 111 user profiles and 300,000 plus tweets, the results achieved in this particular study showed that SVM-l showed the optimum results among the other two algorithms namely Decision Tree and Naive Bayes [20]. The more we improve the dataset and feed data to the machine learning model, the more accurate and precise results we can achieve. We have seen that the studies conducted in each of the papers are of different datasets but the goal is more or less the same; that is to differentiate or discern between individuals experiencing depression and those who are not individuals. We also noticed that data collected through social media platforms is larger in comparison to data collected through questionnaires. The researchers mainly used four types of quality metrics to draw a comparison among the algorithms in their respective papers, namely Accuracy, Precision, Recall, and F1-measure. Maximum studies declared the best algorithm according to their respective work concerning F1 measure since the F1 score provides the optimum score in measuring complex data. The F1-Score, also known as the F-score, serves as a metric to assess a model's accuracy concerning a dataset. It is particularly applicable in the evaluation of binary classification systems that categorize examples into 'positive' or 'negative' outcomes. The formula for calculating the F1-Score is depicted in Equation 8.1.

$$F1-SCORE = 2 * \frac{(precision * recall)}{(precision + recall)} \tag{8.1}$$

Table 8.1 Algorithms used by author their papers.

Author	DT	SVM	RFT	NB	KNN	BN	MLP	RS	KS	SMO	Hybrid NB tree	Ensemble	LR	Catboost	ANN
Anu Priya *et al.*	√	√	√	√	√										
Sau A. *et al.*			√	√		√	√	√	√	√					
Govindasamy *et al.*				√							√				
Islam *et al.*	√	√			√							√			
M. Srividya *et al.*	√	√		√	√								√		
A. Sau *et al.*		√	√	√									√	√	
D. Leightley *et al.*		√	√									√			√
R. Islam *et al.*					√										
Hatoon S. *et al.*	√	√		√											
Scott R. *et al.*	√														

In Table 8.1 we see what algorithms the individual authors of their respective papers use in their datasets to identify depressed and non-depressed users. We see that SVM was used by the maximum number of authors in their research work followed by DT, NB and the rest.

Based on data collection techniques i.e. Standard Questionnaire approach and Social media content approach, we have calculated the maximum algorithm used in a paper. Figure 8.1 shows that 25% of research work is evaluated using SVM, RFT, and KNN classifiers. Figure 8.2 shows that 30% of research work is evaluated using a Decision Tree classified for evaluating the performance of their model every researcher used accuracy, precision, recall, and F1 score. We are considering only the F1 score as it gives the most accurate result than the others. After studying 26 papers, we are considering 3 paper to judge their performance using the F1 score.

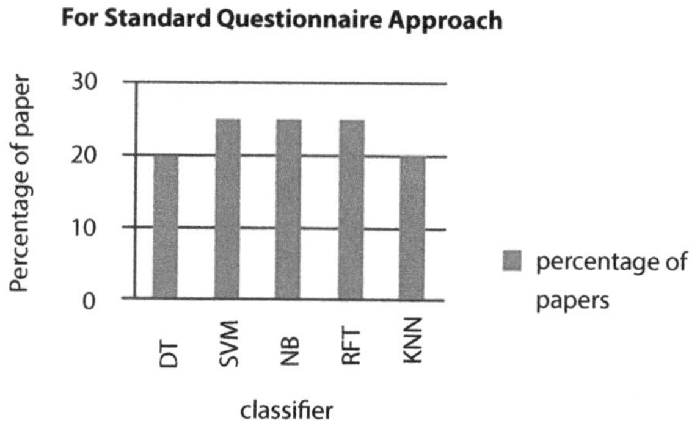

Figure 8.1 Percentage of papers using a classifier for SQA.

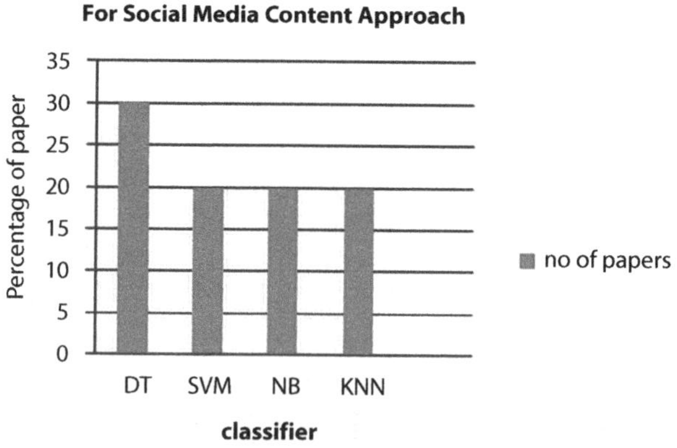

Figure 8.2 Percentage of papers using a classifier for SMCA.

Anu Priya *et al.* used DT, RF, SVM, NB, and KNN classifiers. From Figure 8.3 it is clear that the F1 score of Naïve Bayes is highest for depression and RF for stress. Based on the F1 score RF has been considered as the best model for predicting stress, depression, and anxiety. Hatoon S. Alsagiri *et al.* used SVM, NB, and DT classifiers in their research work. According to Figure 8.4 F1 score of SVM-linear is highest.

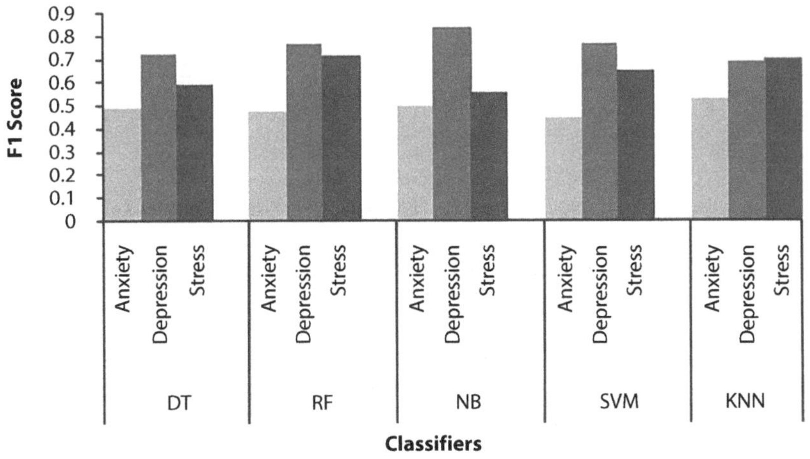

Figure 8.3 Predicting anxiety, depression & stress using ML.

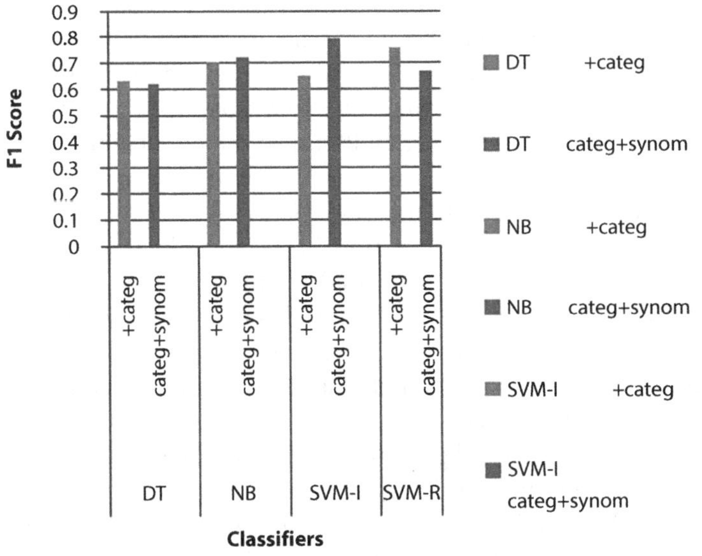

Figure 8.4 Depression detection in twitter.

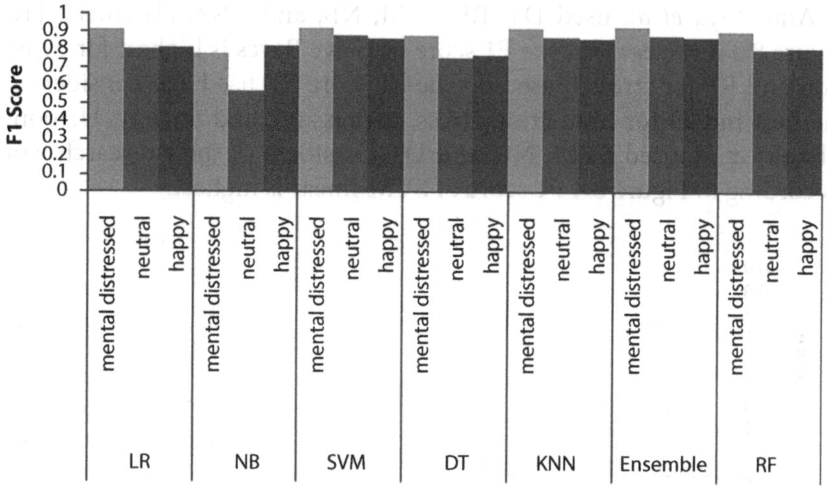

Figure 8.5 Behavioral modeling for mental health using ML algorithms.

M. Srividya & S. Mohanavalli1 & N. Bhalaji used LR, NB, SVM, DT, KNN, Ensemble, and RF classifier in his paper. According to the F1 score in Figure 8.5 SVM and KNN are the highest.

8.5 Discussion

This paper aims to figure out prevalent machine learning approaches applied for identifying mental health issues and the usage of the metrics assessing the quality of particular algorithms. Due to the pandemic, the last two years have seen a swift increase in mental illness in the population due to several reasons. To ensure quality mental health, early detection of mental illness and subsequently, appropriate treatment is to be carried out. In achieving the target, the use of ML is unquestionable. ML can be broadly classified as i) supervised learning, ii) unsupervised learning, and iii) semi-supervised learning. In the course of our research in this area paper, we have seen that the majority of the previous works were done using supervised learning algorithms. So it opens up a potential area for future research using unsupervised or semi-supervised learning. There is a significant scope to explore how efficiently unsupervised or semi-supervised learning can be applied.

Furthermore, certain considerations are to be made while performing data collection. As ML requires the access to training data, the outcome is dependent on the response of the research subject. If collaboration is established between the researchers and the clinicians then the accurate outcome can be obtained. Moreover, greater collaboration between data science and mental health experts is required to implement the settings in the real world.

8.6 Conclusion

This paper explored various categories of machine learning algorithms used in depression detection by researchers. We explored different types of datasets by categorizing them into two broad parts: Questionnaire-based and social media based. A comparative study is done among the papers in which it is observed that supervised machine learning algorithms performed better in comparison to the unsupervised ones. This study may be extended in the future by exploring the different aspects of not only unsupervised machine learning algorithms but also any new method for identifying depression through machine learning.

References

1. Thieme, A., Danielle, B., Gavin, D., Machine learning in mental health: A systematic review of the HCI literature to support the development of effective and implementable ML systems. *ACM Trans. Comput.-Hum. Interact. (TOCHI)*, 27, 1–53, 2020.
2. Su, C., Zhenxing, X., Jyotishman, P., Fei, W., Deep learning in mental health outcome research: a scoping review. *Transl. Psychiatry*, 10, 1, 1–26, 2020.
3. Shatte, A.BR., Hutchinson, D.M., Teague, S.J., Machine learning in mental health: a scoping review of methods and applications. *Psychol. Med.*, 49.9, 1426–1448, 2019.
4. Jordan, M., II and Tom, M.M., Machine learning: Trends, perspectives, and prospects. *Science*, 349.6245, 255–260, 2015.
5. Dutta, P., Pal, S., Kumar, A., Cengiz, K., *Artificial Intelligence for Cognitive Modeling: Theory and Practice*, CRC Press, Florida, USA, 2023.
6. Sau, A. and Bhakta, I., Predicting anxiety and depression in elderly patients using machine learning technology. *Healthcare Technol. Lett.*, 4, 238–243, 2017. 10.1049/htl.2016.0096.
7. Srividya, M., Mohanavalli, S., Bhalaji, N., Behavioral Modeling for Mental Health using Machine Learning Algorithms. *J. Med. Syst.*, 3;42, 5, 88, 2018. doi:10.1007/s10916-018-0934-5, PMID: 29610979.
8. Fear, N.T., Jones, M., Murphy, D., Hull, L., Iversen, A.C., Coker, B., Wessely, S., What are the consequences of deployment to Iraq and Afghanistan on the

mental health of the UK armed forces? A cohort study. *Lancet*, 375, 1783–1797, 2010. doi:10.1016/S0140-6736(10)60672-1.
9. Hotopf, M., Hull, L., Fear, N.T., Browne, T., Horn, O., Iversen, A., Wessely, S., The health of UK military personnel who deployed to the 2003 Iraq war: a cohort study. *Lancet*, 367, 1731–1741, 2006. doi:10.1016/S0140-6736(06)68662-5.
10. Leightley, D., Williamson, V., Darby, J., Fear, N.T., Identifying probable post-traumatic stress disorder: applying supervised machine learning to data from a UK military cohort. *J. Ment. Health*, 28, 1, 34–41, 2019. doi:10.1080/09638237.2018.1521946. Epub, PMID: 30445899.
11. Arkaprabha, S. and Ishita, B., Screening of anxiety and depression among the seafarers using machine learning technology. *Inf. Med. Unlocked*, 16, 100149, 2019. ISSN 2352-9148, doi.org.
12. Anu, P., Shruti, G., Neha, P.T., Predicting Anxiety, Depression and Stress in Modern Life using Machine Learning Algorithms. *Procedia Comput. Sci.*, 167, 1258–1267, 2020. ISSN 1877-050.
13. Hanai, T., Ghassemi, M., Glass, J., *Detecting Depression with Audio/Text Sequence Modeling of Interviews*, pp. 1716–1720, ISCA - International Speech Communication Association, France, 2018, 10.21437/Interspeech.2018-2522.
14. Braithwaite, S.R., Giraud, C.C., West, J., Barnes, M.D., Hanson, C.L., Validating Machine Learning Algorithms for Twitter Data Against Established Measures of Suicidality. *JMIR Ment. Health*, 3, 2, e21, 2016. Published 2016 May 16. doi:10.2196/mental.4822.
15. Islam, M.R., Kabir, M.A., Ahmed., A., Kamal, A.R.M., Wang, H., Ulhaq, A., Depression detection from social network data using machine learning techniques. *Health Inf. Sci. Syst.*, 27;6, 1, 8, 2018. doi:10.1007/s13755-018-0046-0. PMID: 30186594; PMCID: PMC6111060.
16. Kumar, M.R., Pooja, K., Udathu, M., Prasanna, J.L., Santhosh, C., Detection of depression using machine learning algorithms. *Int. J. Online Biomed. Eng. (iJOE), 18*, 04, 155–163, 2022.
17. AlSagri, H.S. and Ykhlef, M., Machine learning-based approach for depression detection in twitter using content and activity features. *IEICE Trans. Inf. Syst., 103*, 8, 1825–1832, 2020.
18. Jain, K., Mohapatra, S., Bohra, R., Varun, V.V.S.S., Evaluation of Depression Detection in Sentiment Analysis Through Machine Learning Model, in: *International Conference on Communication and Intelligent Systems*, Springer Nature Singapore, Singapore, pp. 567–575, 2022.
19. Trifan, A., Understanding Depression from Psycholinguistic Patterns in Social Media Texts. *Adv. Inf. Retr.*, 12036, 402–409, 2020.
20. Dutta, P., Paul, S., Kumar, A., Comparative analysis of various supervised machine learning techniques for diagnosis of COVID-19, in: *Electronic devices, circuits, and systems for biomedical applications*, Academic Press, pp. 521–540, 2021.

Part 2

CLINICAL DECISION SUPPORT SYSTEM FOR EARLY DISEASE DETECTION AND MANAGEMENT

9

Diagnostics and Classification of Alzheimer's Diseases Using Improved Deep Learning Architectures

Mainak Dey[1]*, Pijush Dutta[2] and Gour Gopal Jana[2]

[1]*Department of Electronics & Communication Engineering, Narula Institute of Technology, Kolkata, India*
[2]*Department of Electronics & Communication Engineering, Greater Kolkata College Engineering and Management, Baruipur, Phultala, West Bengal, India*

Abstract
One of the most common neurodegenerative illnesses that affect older adults and mostly impair memory in the brain is Alzheimer's disease. Alzheimer's disease is a global condition that primarily affects those over 65. Accurate identification of this condition requires early diagnosis. Because there are so many people who come with the ailment in such cases manual diagnosis is labor-intensive and prone to inaccuracy. Earlier numerous methods have been used to diagnose and categorize Alzheimer's disease; however, they do not provide up-to-the-mark precision. The model in this study proposes an efficient net B2, efficient net B3, and efficient net B4 architecture-based deep learning solution for detecting and categorizing Alzheimer's disease. The efficient Net B3 outperformed than other two algorithms based on training and testing accuracy, AUC and Cohen's Kappa (k) value values for Efficient B3 architecture are 0.972, 0.882, 0.92.5, and 0.93 respectively, while the accuracy values for Efficient B2 and Efficient B4 were 0.819 and 0.7192 & 0.873 and 0.8125. Furthermore, it has been also observed that this model improved the accuracy of other state-of-the-art techniques.

Keywords: Deep learning, Alzheimer's disease, cohen's kappa, area under curve (AUC)

**Corresponding author*: mainakdey2001@gmail.com

Pijush Dutta, Sudip Mandal, Korhan Cengiz, Arindam Sadhu and Gour Gopal Jana (eds.) Design and Forecasting Models for Disease Management (217–236) © 2025 Scrivener Publishing LLC

9.1 Introduction

AD, the most prevalent type of dementia, needs intensive medical attention. For effective AD patient therapy, a timely study of AD prognosis is required [1]. AD is a long-term brain disease that impairs thinking, a loss of the ability to do fundamental tasks, and destroys brain tissue [2]. Physicians categorize AD using CAD techniques and neuroimaging in its early phases. From the World Alzheimer's Association review, it has been observed that a very significant number of patients survived over 4.7 million Alzheimer's disease [3]. They projected that 60 million people might be impacted by AD in the next 50 years. One person develops dementia every three seconds, with AD accounting for 60% of cases [4]. Alzheimer's dementia is classified into Mild dementia, moderate dementia, and Severe Dementia. During mild Dementia: Individuals with mild dementia experience cognitive deficits that can occasionally interfere with their day-to-day activities. Memory loss, ambiguity, personality changes, disorientation, and trouble carrying out daily duties are some symptoms. Moderate Dementia: The patient needs greater care and help as their daily life gets more complicated. The symptoms are similar to high but mild dementia. Even comb-overs may become more difficult for certain people. They may also show notable personality changes, such as sudden increases in paranoia or irritability. During severe Dementia, the symptoms might get worse. Some patients might not be able to communicate, in which case they would need full-time care. A person may lose control of their bladder, making it difficult for them to carry out simple tasks like sitting in a chair and holding their head up in a normal position. Research is being done to identify this condition early to prevent aberrant brain degeneration, lower the cost of medical care, and guarantee better outcomes. There are several research has been performed for early detection and classification of Alzheimer's disease [5]. Many new diagnostic criteria reflect the growing dependence of a wide range of neuroimaging techniques on dementia diagnosis. By utilizing machine learning, neuroimaging improves the accuracy of diagnosis for different kinds of dementia. It takes specific preprocessing measures to put machine learning algorithms into practice. The steps in the ML-based classification process include feature extraction and selection, feature dimensionality reduction, and classifier algorithm [6]. These methods need a high level of expertise and several optimization phases, which can be laborious. Large-scale multimodal techniques have recently been investigated for early detection and automated AD categorization neuroimaging datasets. MRI, Positron Emission Tomography (PET), and genetic sequencing data are among the several techniques for studying AD. Making a decision requires a lot of time spent evaluating many modalities [7]. We think that the MRI modality's superior tissue contrast, less ionizing radiation, increased imaging flexibility, and capacity to offer insightful data on human brain architecture are all beneficial to this research project [8]. It is thought to be crucial to create an improved computer-aided diagnosis system that can decipher MRI images and identify if a patient has Alzheimer's disease. Advanced deep learning algorithms (DL) approaches were adopted for AD classification on raw MRI images by feeding the CNN with information from the cortical surface. This research presents a model that extracts discriminative features using a convolutional neural network. Class imbalance is handled utilizing the SMOTE approach [9]. The model is implemented in such a manner that it lowers the computational cost and parameters cost along with improving the accuracy of AD classification. By training the models using the MRI dataset from Kaggle, the models are assessed [10]. The following are some noteworthy contributions made by this study. The advanced CNN algorithms EfficientNetB2, EfficientNetB3, and EfficientNetB4 were suggested. To identify AD phases, we additionally evaluated the suggested model's accuracy, the area under the curve, and Cohen's kappa score with deep features and manually created features. Section 9.2 highlights the relevant work, and Section 9.3 discusses the suggested course of action. Section 9.5 provides a summary of the paper's conclusions and future directions, whereas Section 9.4 examines the comprehensive experimental results.

9.2 Related Works

Recently, several deep learning techniques have been presented as Alzheimer's disease diagnosis tools, helping medical professionals make well-informed judgments. In this part, as indicated in Table 9.1, we highlight a few of the research that is closely linked to this study.

However, implementing the classification model for parametric optimization and reducing the class imbalance of the multiclass AD dataset still have a scope. In this research, a hybrid SMOTE-based Efficient Net CNN model is proposed to address this issue.

Table 9.1 Literature survey of Alzheimer's diseases.

Authors	Methodology	Outcomes from research
Lu *et al.* [11]	Multistage multimodal Deep neural network (DNN)	The classification model achieved the maximum accuracy for mild, moderated, and non-demented AD classes are about 82.4%, 94.23 %, and 86.3% respectively.
Gupta *et al.* [12]	Two different data sources ADNI and NRCD (National Research Centre for Dementia) are used for classifying AD	Maximum accuracy obtained from the ADNI dataset was about 96.42%.
Ahmed *et al.* [13]	An ensemble CNN model with a feature extraction method	This model obtains an accuracy of 90.05%.
Basher *et al.* [14]	Localization of target regions from MRI Dataset of AD patient.	The maximum accuracy obtained for the left and right hippocampal were 94.82% and 94.02% respectively.
Nawaz *et al.* [15]	pre-trained Alex net model with three classifier models SVM, KNN, and RF was used	Maximum accuracy obtained by Random Forest (RF) was used at about 99.21%
Ieracitano *et al.* [16]	A class & multiclass of 2D images applied by the CNN model for the AD dataset.	The accuracy of class and multiclass for AD, MCI, and HC were 89.8% and 83.3%, respectively.
Jain *et al.* [17]	A hybrid pre-trained feature extraction VGG16 integrated transfer learning (PFSECTL) method was proposed	Maximum accuracy in classifying three categories of ADNI was 95.73%.

(*Continued*)

Table 9.1 Literature survey of Alzheimer's diseases. (*Continued*)

Authors	Methodology	Outcomes from research
Mehmood *et al.* [18]	The proposed model was applied to the AD dataset using a VGG pre-trained & layer-wise transfer learning model.	classification accuracy obtained about 83.72% and 98.73% respectively.
Shi *et al.* [19]	For small and large ADNI datasets Deep polynomial network is Proposed.	The model achieved a maximum accuracy of about 55.34% for multi and binary classification
Liu *et al.* [20]	Deep Siamese neural networks (SNN) were proposed to investigate the discriminative capacity of brain asymmetry associated with AD and MCI.	The maximum accuracy was 92.72%.
Wang *et al.* [21]	A probability-based 3D-Dense Nets CNN fusion approach applied to the ADNI dataset	The classification model accuracy achieved 97.52%.
Shankar *et al.* [22]	A hybrid Grey wolf optimization (GWO) integrated ML classification model was proposed to diagnose AD	The maximum accuracy obtained by the hybrid GWO-CNN model is about 96.23 %.
Janghel [23]	A classification model proposed by SVM, Linear discrimination (LD), K means clustering, and decision tree (DT) algorithm followed by pre-trained VGG16 on the ADNI database.	Maximum accuracy achieved for MRI images and PET images was 99.95% and 73.46 % respectively.

(*Continued*)

Table 9.1 Literature survey of Alzheimer's diseases. (*Continued*)

Authors	Methodology	Outcomes from research
Ge *et al.* [24]	3D multiscale deep learning (DL) approach adopted	Maximum test accuracy and average accuracy obtained for the proposed algorithms are 93.53% and 87.24 % respectively.
Bi and Wang [25]	Spike Convolutional Deep Boltzmann Machine learning model (CDB-ML) investigated	The proposed methodology shows fruitful results using high-level extraction and demonstrates more efficiency than state-of-the-art methods.
Sarraf *et al.* [26]	A deep-learning-based CNN model is investigated on ADNI MRI images	The functional MRI and MRI images achieved maximum accuracy of about 94.32% and 97.88% respectively.
Afzal *et al.* [27]	A data augmentation framework was proposed for the OASIS dataset	Classification accuracy of the single view and 3D view of the OASIS dataset was 98.41% and 95.11% respectively.

9.3 Method

The most common neurodegenerative disease is AD which affects people in different ways in the world. Individuals may experience brain damage, memory loss, and behavioral changes, and even they forget their past, likes, and dislikes, and even become unable to recognize family members. In this section, we introduce our proposed research method. We used the EfficientNetB3 model Paper on the prediction of Alzheimer's disease [28, 29]. The EfficientNetB3 architecture shown in Figure 9.1 offers uniform scales for all dimensions of a network including depth, width, and resolution. This CNN architecture can find the proper scaling coefficient by applying a grid search strategy under a fixed resource constraint [29–31]. The efficient Net model introduces a parameter size reduction algorithm

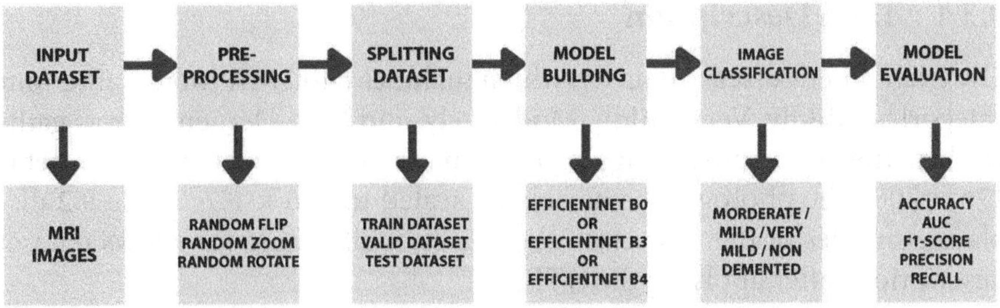

Figure 9.1 Flow diagram of the proposed method.

and floating-point operations per second (FLOPS), which help to improve the accuracy of existing CNNs [32].

Step 1: Input a dataset of Alzheimer's disease class 4 MRI images [33]. The record is retrieved from Kaggle.
Step 2: In this step, extraneous information is removed and preprocessing is performed. Used to expand data to make better predictions. The Image Data Generator class is responsible for this process required. Geometric transformations, random flips, random rotations, and random zooms were performed.
Step 3: In this step, we use 4098 images to train the dataset, and 1023 images to validate the dataset, we also test the dataset using 1279 images.
Step 4: In this section, we used the EfficientNetB3 model to complete the task. categorical cross-entropy in the form of cross-entropy loss or logarithmic loss was used to measure the performance of multi-classification problems with outputs greater than 2 (SoftMax), and 20 epochs were run.
Step 5: The EfficientNetB3 model is built into this study and used to classify AD on MRI images. Depending on the developmental stage, data is classified into four classes: mild, moderate, non-dementia, and very mild dementia. The dropout value is determined by dividing the dataset into training sets, test sets, and valid sentences. Also, the learning rate, activation function, number of hidden layers, epoch, cross-entropy, and All batch sizes were adjusted to optimize the neural network so that the model could be trained perfectly. Classification of datasets.
Step 6: The dataset used in this work was approximately balanced. Conventional Accuracy and Evaluation Determining the performance of the classification model in this study is subjective. performance. In this study, we used metrics and various evaluation performance metrics to evaluate the model. It was used to explore the limitations and capabilities of the EfficientNetB3 model by means of Accuracy, AUC, and F1 score [34–40].

9.3.1 Data Description

The Alzheimer's disease dataset, which includes 6400 MR Images into four categories: Mildly, Very Mildly Moderately, and Non-Demented was gathered via the open-source Kaggle platform. The picture size of the dataset is 176 × 208. The photographs have been scaled to 176 × 176. Table 9.2 displays examples of the four classes' pictures, while Table 9.3 shows dataset descriptions after applying SMOTE.

Data distribution is shown in Table 9.2, where a number of the acquired pictures indicate an obvious class imbalance. The SMOTE is a popular step in the preprocessing stage to fix the class imbalance problem to match the majority classes [35]. Reducing knowledge loss and minimizing overfitting are two advantages of employing SMOTE [41, 42]. With each class including 3200 photos, Table 9.3 displays the dataset distribution of each category shifted to 3200 followed by the SMOTE technique's resulting overall dataset to 12800 images.

It is evident from Figure 9.2's dataset distribution and a few photographs in the acquired dataset that there is a class imbalance in the dataset.

Table 9.2 Data set description before SMOTE.

Class of dataset	No. of samples
Mild Demented	896
Moderate Demented	3200
Non-Demented	2240
Very Mild Demented	64

Table 9.3 Dataset description after SMOTE.

Class of dataset	No. of samples
Mild Demented	3200
Moderate Demented	3200
Non-Demented	3200
Very Mild Demented	3200

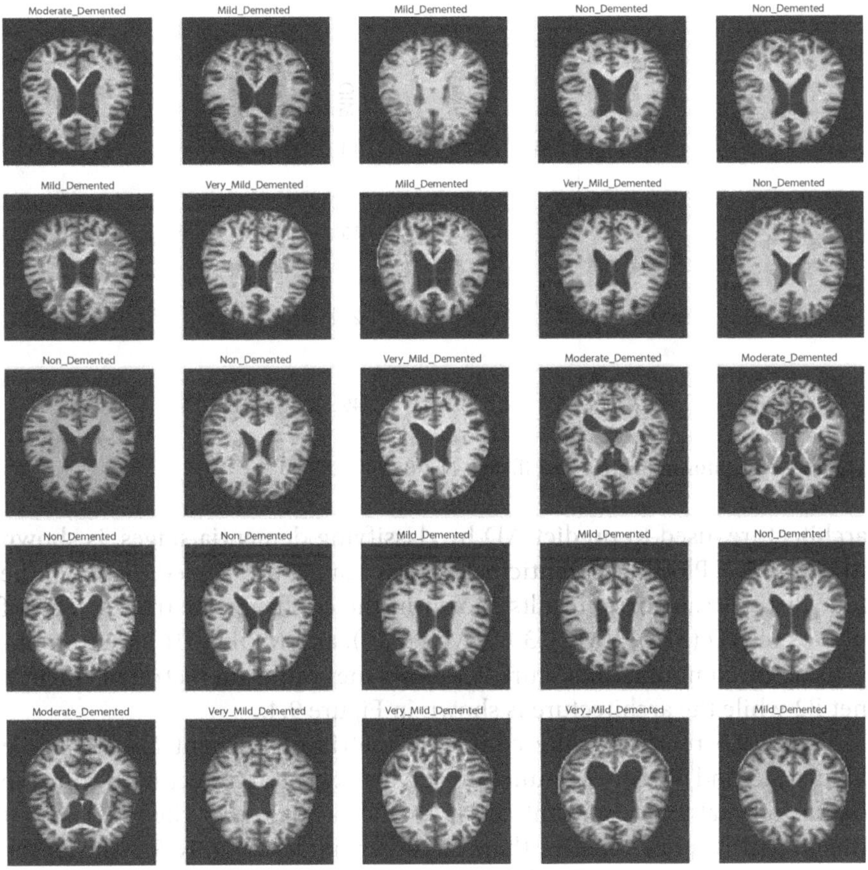

Figure 9.2 Description of a dataset of AD.

9.4 Result Analysis

Two scenarios, without the SMOTE and one with it, are used to train the present research. AUC is a property that categorizes positive and negative classes effectively for every epoch. The training and validation accuracy, AUC, and loss are displayed in Figure 9.4. Due to the overfitting without the SMOTE approach has a validation and training accuracy of about 62% and 76% respectively. The SMOTE-assisted model attains an approximate 97.2% training accuracy and an 85.4% validation accuracy. Following the training model, the confusion matrix is evaluated using a test set in which the pictures are available but were not given to the model during training. The confusion matrix obtained by improved SMOTE-Efficient Net CNN

	ND	VMD	MID	MOD
ND	294	10	14	8
VMD	7	281	15	6
MID	12	10	292	15
MOD	8	8	45	251

Figure 9.3 Confusion matrix for Efficient Net B2.

architecture, used to predict AD by classifying dementia stages, is shown in Figure 9.2. Plotting the anticipated class against the labeled classes of the four distinct categories results shown in the confusion matrix. A total of photos of ND (326), VMD (309), MD (329), and MOD (316) are the subjects of the computation. Figure 9.3 shows the confusion matrix of efficient net B2 while the architecture is shown in Figure 9.4.

Figure 9.5 represents the confusion matrix of Efficient Net B3 while Figure 9.6 and Figure 9.7 function for block and architecture. From the confusion matrix of Efficient B3, It has been noted that the precision of this model is more effective than the other models. Figure 9.8 shows the reflection of Training and validation loss and accuracy for efficient NetB3. Figure 9.9 represents the confusion matrix of Efficient Net B4 and Figure 9.10 represents its architecture of it.

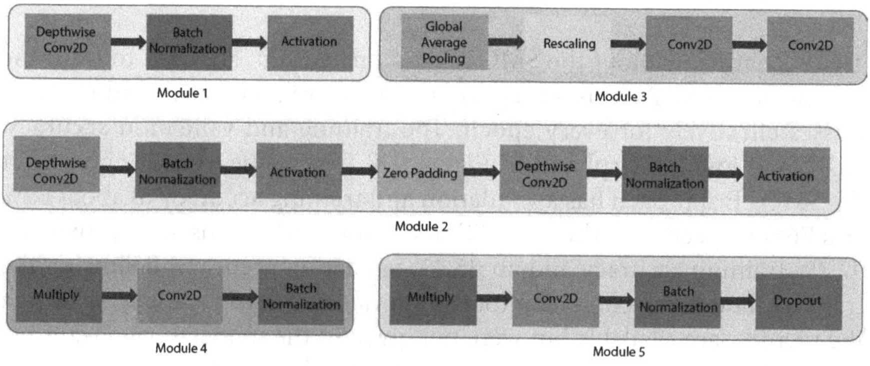

Figure 9.4 Efficient Net B2 architecture.

	ND	VMD	MID	MOD
ND	314	2	8	2
VMD	0	309	0	0
MID	2	0	322	5
MOD	5	0	37	274

Figure 9.5 Confusion matrix for Efficient Net B3.

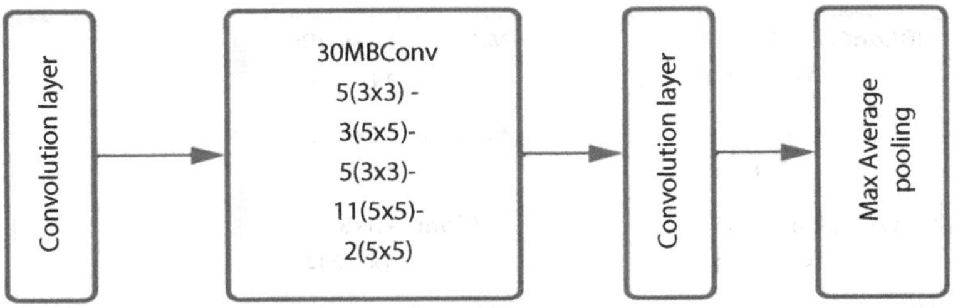

Figure 9.6 Efficient Net B3 block.

9.4.1 Performance Metrics

Several assessment indicators were employed in this experiment to evaluate the performance of the Alzheimer's disease model. Accuracy, F score, precision, recall, and confusion matrix were among these measurements.

Accuracy: Ratio of correct prediction to overall predictions mathematically represented by $\frac{TP+TN}{TP+TN+FP+FN}$

Precision: Ratio of correct predictions to the total predicted represented by $\frac{TP}{TP+FP}$

Recall or Sensitivity: Ratio of correct sample class to the actual sample class represented by $\frac{TP}{TP+FN}$

Specificity: Ratio of the actual 0 to the correctly predicted 0. It can be defined as $\frac{TN}{TP+TN}$

F1 score / F measure: It is a geometric mean of precision and recall if the data is imbalanced represented as $\frac{2*(recall*precision)}{recall+precision}$

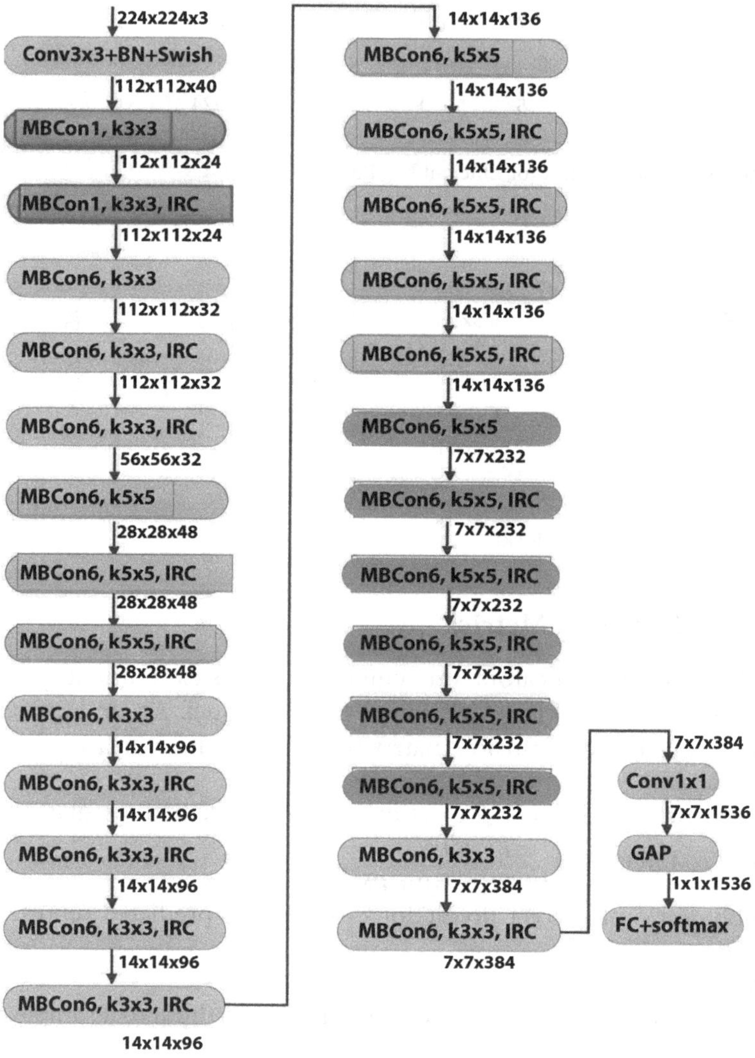

Figure 9.7 Efficient Net B3 architecture.

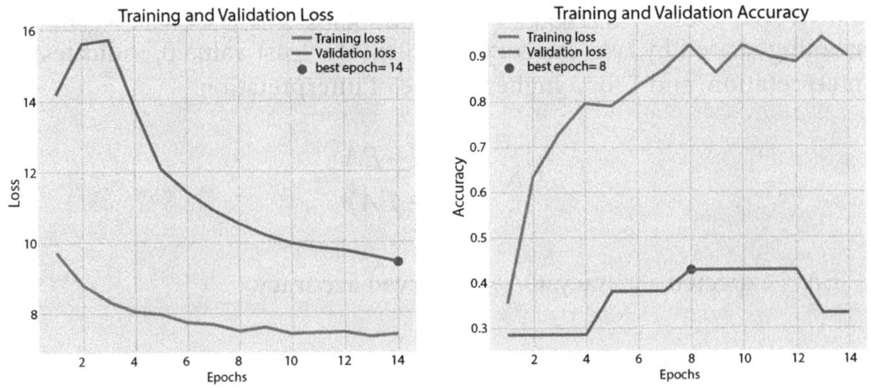

Figure 9.8 Training and validation loss and accuracy for Efficient NetB3.

	ND	VMD	MID	MOD
ND	304	6	12	4
VMD	4	289	12	4
MID	8	6	304	11
MOD	2	5	43	261

Figure 9.9 Confusion matrix for Efficient Net B4.

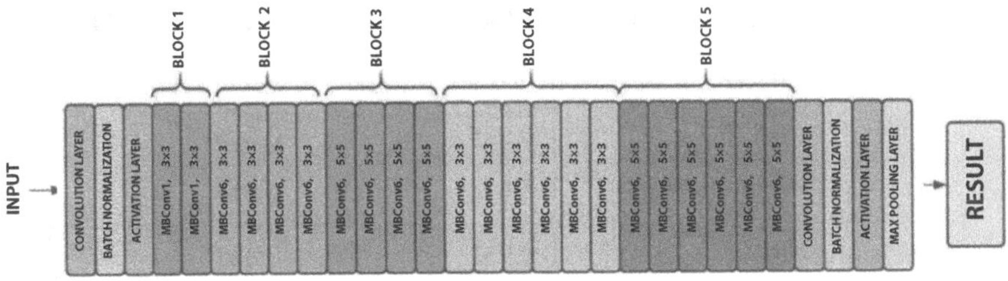

Figure 9.10 Efficient Net B4 architecture.

Cohen's Kappa: Cohen's Kappa (CK) ranking must be used to ascertain the meaning shared by two category datasets. The least value 0, indicates no interpretation, and 1 for a higher degree of interpretation.

$$CK = \frac{(OA - EA)}{(1 - EA)}$$

EA = expected accuracy, OA = observed accuracy,

9.4.2 Experimental Setup

The experiment was conducted utilizing the Google Collab Pro+ platform on an NVIDIA GPU V 100. Details of the experimental setup up are shown in Table 9.4. After the simulation, six metrics are obtained for all the algorithms namely: accuracy, AUC, F1 score, Precision, Recall, and k. Table 9.5 shows the comparative study of all these algorithms.

Table 9.4 Experimental setup.

Platform	**Google collab**
GPU	NVIDIA GPU V 100
RAM	16Gb
Tensor Flow	2.12.0
Keras Version	2.12.0

Table 9.5 Comparative study of the proposed algorithm.

	EfficientNetB2	**EfficientNetB3**	**EfficientNetB4**
Accuracy	0.819	0.972	0.873
AUC	0.825	0.925	0.896
F1_Score	0.766	0.956	0.872
Precision	0.696	0.952	0.823
Recall	0.759	0.8965	0.815
k	0.58	0.92	0.71

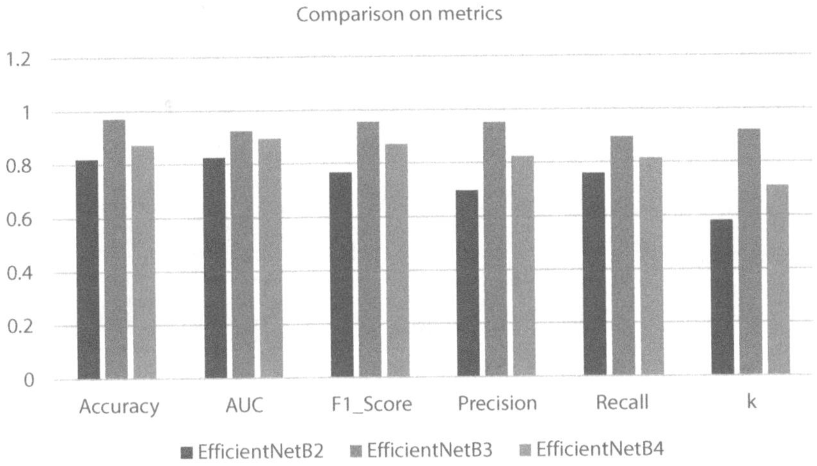

Figure 9.11 Comparative study on metrics.

Table 9.6 Comparative study based on Efficient Net B3 SMOTE and without SMOTE algorithm.

Parameters	Efficient NetB3 without SMOTE	Efficient NetB3 with SMOTE
Accuracy	0.85	0.972
AUC	0.812	0.925
F1_Score	0.806	0.956
Precision	0.709	0.952
Recall	0.815	0.8965

Table 9.7 Comparative analysis using state of art and suggested techniques.

References	Dataset	Accuracy	Precision	Recall	F1_score
DEMET +SMOTE [43]	Kaggle	95.23	95	95	95.27
VGG16 [17]	ADNI	95.73	96.23	88.85	89.85
MLP [44]	ADNI	89	85	87	89
CBLSTM+SMOTE [44]	ADNI	82	78	88	82
Present Research	Kaggle	97.2	95.2	89.65	95.6

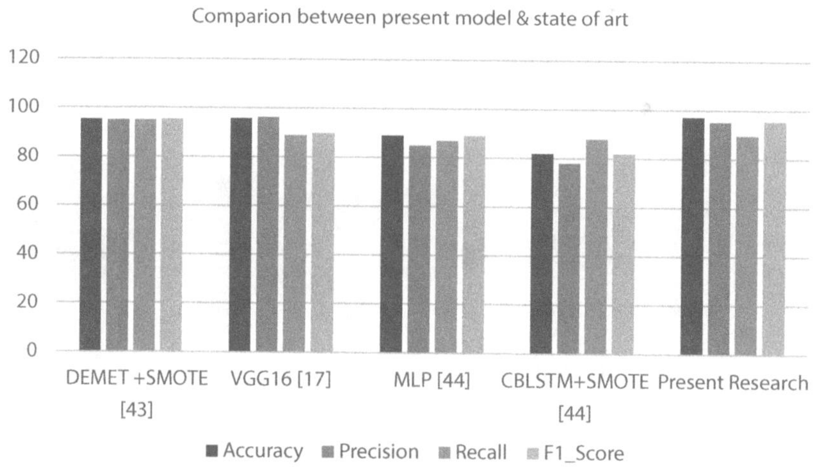

Figure 9.12 Comparative study between state-of-the-art and suggested methods.

The graphical measurement of six metrics applied on Efficient Net B2, Net B3, and Net B4 are shown in Figure 9.11. From the result analysis, it has been observed that from all the metrics perspective Efficient B3 outperformed the other two algorithms.

The accuracy of the model is constantly decreased by irregular data patterns, so to improve the data patterns or data imbalance SMOTE is applied in this research. Table 9.6 represents how SMOTE helps to improve the model accuracy for Efficient Net B3 algorithms.

In last section of this research, we analysis proposed model with other state of art techniques shown in Table 9.7 and Figure 9.12. From result it has been observed that proposed Efficient Net B3 CNN out performed than other state of art.

9.5 Conclusion

AD is a most common neurodegenerative disease that causes impaired memory, permanent brain damage, and other critical brain functions. As the progression of the disease is slow it is difficult to find it in the early stage. In this research, it has been observed that the EfficientNetB3 model of CNN on the AD dataset is best suited for the analysis of Brain Dementia. Deep learning (DL) techniques can extract potential features of AD and predict brain anatomical changes. CNN using MRI is the most appropriate one to know the presence of Alzheimer's disease. Overall research was performed in three stages: in the first stage data imbalance was overcome by SMOTE, second stage, three different improved CNN algorithms (Efficient NetB2, Efficient NetB3, and Efficient B4) were proposed for the classification purpose and the final stage statistical analysis explained in the result analysis section. However proposed model also validates other state-of-the-art techniques used in the same types of datasets. As per this paper, CNN-based models have been extensively used for AD presenting its strength and the reason is that techniques like ANN suffer from local minima. CNN is better as this has ended up being excellent for computer vision tasks in the field of medicine within a current couple of years. Deep learning gives promising benefits by more accurately modeling complicated data.

Data Availability

The following details on the availability of the data were provided: Kaggle offers Magnetic Resonance Imaging (MRI) scans.: https://www. kaggle. com/datasets/tourist55/alzheimers-dataset-4-class-of-images.

References

1. Liss, J.L., Seleri Assunção, S., Cummings, J., Atri, A., Geldmacher, D.S., Candela, S.F., Sabbagh, M.N., Practical recommendations for timely, accurate diagnosis of symptomatic Alzheimer's disease (MCI and dementia) in primary care: a review and synthesis. *J. Internal Med.*, 290, 2, 310–334, 2021.
2. Dindar, N. and Yücel, Y.Y., Alzheimer's, Its Diagnosis, Treatment And Future Approaches. *Curr. Res. Health Sci.*, 16, 237–268, 2022.
3. Lazli, L., Boukadoum, M., Mohamed, O.A., A survey on computer-aided diagnosis of brain disorders through MRI based on machine learning and data mining methodologies with an emphasis on Alzheimer disease diagnosis and the contribution of the multimodal fusion. *Appl. Sci.*, 10, 5, 1894, 2020.
4. Arora, S., Santiago, J.A., Bernstein, M., Potashkin, J.A., Diet and lifestyle impact the development and progression of Alzheimer's dementia. *Front. Nutr.*, 10, 1213223, 2023.
5. Rasmussen, J. and Langerman, H., Alzheimer's disease–why we need early diagnosis, in: *Degenerative neurological and neuromuscular disease*, pp. 123–130, 2019.
6. Abdulhammed, R., Musafer, H., Alessa, A., Faezipour, M., Abuzneid, A., Features dimensionality reduction approaches for machine learning based network intrusion detection. *Electronics*, 8, 3, 322, 2019.

7. Kas, A., Rozenblum, L., Pyatigorskaya, N., Clinical Value of Hybrid PET/MR Imaging: Brain Imaging Using PET/MR Imaging. *Magn. Reson. Imaging Clin.*, *31*, 4, 591–604, 2023.
8. Das, I.J., McGee, K.P., Tyagi, N., Wang, H., Role and future of MRI in radiation oncology. *Br. J. Radiol.*, *92*, 1094, 20180505, 2019.
9. Özdemir, A., Polat, K., Alhudhaif, A., Classification of imbalanced hyperspectral images using SMOTE-based deep learning methods. *Expert Syst. Appl.*, *178*, 1–44, 114986, 2021.
10. Chaihtra, D.H. and Shetty, S.V., Alzheimer's disease detection from brain MRI data using deep learning techniques, in: *2021 2nd Global Conference for Advancement in Technology (GCAT)*, pp. 1–5, IEEE, 2021.
11. Lu, D., Popuri, K., Ding, G.W., Balachandar, R., Beg, M.F., Multimodal and multiscale deep neural networks for the early diagnosis of Alzheimer's disease using structural MR and FDG-PET images. *Sci. Rep.*, *8*, 1, 5697, 2018.
12. Gupta, Y., Lee, K.H., Choi, K.Y., Lee, J.J., Kim, B.C., Kwon, G.R., Alzheimer's Disease Neuroimaging Initiative., Early diagnosis of Alzheimer's disease using combined features from voxel-based morphometry and cortical, subcortical, and hippocampus regions of MRI T1 brain images. *PLoS One*, *14*, 10, e0222446, 2019.
13. Ahmed, S., Choi, K.Y., Lee, J.J., Kim, B.C., Kwon, G.R., Lee, K.H., Jung, H.Y., Ensembles of patch-based classifiers for diagnosis of Alzheimer diseases. *IEEE Access*, *7*, 73373–73383, 2019.
14. Basher, A., Kim, B.C., Lee, K.H., Jung, H.Y., Volumetric feature-based Alzheimer's disease diagnosis from sMRI data using a convolutional neural network and a deep neural network. *IEEE Access*, *9*, 29870–29882, 2021.
15. Nawaz, H., Maqsood, M., Afzal, S., Aadil, F., Mehmood, I., Rho, S., A deep feature-based real-time system for Alzheimer's disease stage detection. *Multimed. Tools Appl.*, *80*, 35789–35807, 2021.
16. Ieracitano, C., Mammone, N., Bramanti, A., Hussain, A., Morabito, F.C., A convolutional neural network approach for classification of dementia stages based on 2D-spectral representation of EEG recordings. *Neurocomputing*, *323*, 96–107, 2019.
17. Jain, R., Jain, N., Aggarwal, A., Hemanth, D.J., Convolutional neural network based Alzheimer's disease classification from magnetic resonance brain images. *Cognit. Syst. Res.*, *57*, 147–159, 2019.
18. Mehmood, A., Yang, S., Feng, Z., Wang, M., Ahmad, A.S., Khan, R., Yaqub, M., A transfer learning approach for early diagnosis of Alzheimer's disease on MRI images. *Neuroscience*, *460*, 43–52, 2021.
19. Shi, J., Zheng, X., Li, Y., Zhang, Q., Ying, S., Multimodal neuroimaging feature learning with multimodal stacked deep polynomial networks for diagnosis of Alzheimer's disease. *IEEE J. Biomed. Health. Inf.*, *22*, 1, 173–183, 2017.
20. Liu, C.F., Padhy, S., Ramachandran, S., Wang, V.X., Efimov, A., Bernal, A., Alzheimer's Disease Neuroimaging Initiative., Using deep Siamese neural

networks for detection of brain asymmetries associated with Alzheimer's disease and mild cognitive impairment. *Magn. Reson. Imaging, 64*, 190–199, 2019.

21. Wang, H., Shen, Y., Wang, S., Xiao, T., Deng, L., Wang, X., Zhao, X., Ensemble of 3D densely connected convolutional network for diagnosis of mild cognitive impairment and Alzheimer's disease. *Neurocomputing, 333*, 145–156, 2019.
22. S. K, L.S.K., Khanna, A., Tanwar, S., Rodrigues, J.J.P.C., Roy, N.R., Alzheimer detection using group grey wolf optimization based features with convolutional classifier. *Comput. Electr. Eng.*, 77, 230–243, 2019. doi:10.1016/j.compeleceng.2019.06.001.
23. Janghel, R.R. and Rathore, Y.K., Deep convolution neural network based system for early diagnosis of Alzheimer's disease. *Irbm, 42*, 4, 258–267, 2021.
24. Ge, C., Qu, Q., Gu, I.Y.H., Jakola, A.S., Multiscale deep convolutional networks for characterization and detection of Alzheimer's disease using MR images, in: *2019 IEEE International Conference on Image Processing (ICIP)*, pp. 789–793, IEEE, 2019.
25. Bi, X. and Wang, H., Early Alzheimer's disease diagnosis based on EEG spectral images using deep learning. *Neural Netw., 114*, 119–135, 2019.
26. Sarraf, S., DeSouza, D.D., Anderson, J., Tofighi, G., Alzheimer's Disease Neuroimaging Initiativ, DeepAD: Alzheimer's disease classification via deep convolutional neural networks using MRI and fMRI. *BioRxiv*, 070441, 1–32, 2016.
27. Afzal, S., Maqsood, M., Nazir, F., Khan, U., Aadil, F., Awan, K.M., Song, O.Y., A data augmentation-based framework to handle class imbalance problem for Alzheimer's stage detection. *IEEE Access, 7*, 115528–115539, 2019.
28. Chandrasekar, M., Ganesh, M., Saleena, B., Balasubramanian, P., Breast Cancer Histopathological Image Classification using EfficientNet Architecture, in: *2020 IEEE International Conference on Technology, Engineering, Management for Societal impact using Marketing, Entrepreneurship and Talent (TEMSMET)*, pp. 1–5, IEEE, 2020.
29. Elmoufidi, A. and Ammoun, H., Diabetic retinopathy prevention using efficientNetB3 architecture and fundus photography. *SN Comput. Sci.*, 4, 1, 78, 2022.
30. Elmoufidi, A. and Amoun, H., *EfficientNetB3 Architecture for Diabetic Retinopathy Assessment using Fundus Images*, Research Square, USA, 2021.
31. Ganesh, M., Dulam, S., Venkatasubbu, P., Diabetic retinopathy diagnosis with InceptionResNetV2, Xception, and EfficientNetB3, in: *Artificial Intelligence and Technologies: Select Proceedings of ICRTAC-AIT 2020*, pp. 405–413, Springer Singapore, Singapore, 2021.
32. Salian, S.R. and Sawarkar, S.D., Melanoma skin lesion classification using improved efficientnetb3. *Jordanian J. Comput. Inf. Technol.*, 8, 1, 45–56, 2022.
33. https://www.kaggle.com/datasets/tourist55/alzheimers-dataset-4-class-of-images.

34. Dutta, P., Paul, S., Kumar, A., Comparative analysis of various supervised machine learning techniques for diagnosis of COVID-19, in: *Electronic devices, circuits, and systems for biomedical applications*, Academic Press, pp. 521–540, 2021.
35. Dutta, P., Paul, S., Majumder, M., *An efficient SMOTE based machine learning classification for prediction & detection of PCOS*, Research Square, USA, 2021.
36. Dutta, P., Paul, S., Obaid, A.J., Pal, S., Mukhopadhyay, K., Feature selection based artificial intelligence techniques for the prediction of COVID like diseases, *J. Phys.: Conf. Ser.*, 1963, 012167, IOP Publishing, 2021.
37. Dutta, P., Paul, S., Cengiz, K., Anand, R., Kumar, A., A predictive method for emotional sentiment analysis by deep learning from EEG of brainwave dataset, in: *Artificial Intelligence for Neurological Disorders*, Academic Press, pp. 25–48, 2023.
38. Dutta, P., Paul, S., Shaw, N., Sen, S., Majumder, M., Heart disease prediction: A comparative study based on a machine-learning approach, in: *Artificial Intelligence and Cybersecurity*, CRC Press, pp. 1–18, 2022.
39. Dutta, P., Pal, S., Kumar, A., Cengiz, K., *Artificial Intelligence for Cognitive Modeling: Theory and Practice*, CRC Press, USA, 2023.
40. Dutta, P., Paul, S., Cengiz, K., Anand, R., Majumder, M., A predictive method for emotional sentiment analysis by machine learning from electroencephalography of brainwave data, in: *Implementation of Smart Healthcare Systems using AI, IoT, and Blockchain*, Academic Press, pp. 109–130, 2023.
41. Dutta, P., Paul, S., Majumder, M., *Intelligent SMOTE Based Machine learning Classification for Fetal State on Cardiotocography Dataset*, Research Square, USA, 2021.
42. Dutta, P., Paul, S., Jana, G.G., Sadhu, A., Hybrid Genetic Algorithm Random Forest algorithm (HGARF) for improving the missing value Imputation in Hepatitis Medical Dataset, in: *2023 International Symposium on Devices, Circuits and Systems (ISDCS)*, vol. 1, pp. 01–05, IEEE, 2023.
43. Murugan, S., Venkatesan, C., Sumithra, M.G., Gao, X.Z., Elakkiya, B., Akila, M., Manoharan, S., DEMNET: a deep learning model for early diagnosis of Alzheimer diseases and dementia from MR images. *IEEE Access*, *9*, 90319–90329, 2021.
44. Stamate, D., Smith, R., Tsygancov, R., Vorobev, R., Langham, J., Stahl, D., Reeves, D., Applying deep learning to predicting dementia and mild cognitive impairment, in: *Artificial Intelligence Applications and Innovations: 16th IFIP WG 12.5 International Conference, AIAI 2020*, Neos Marmaras, Greece, Proceedings, Part II 16. June 5–7, 2020, pp. 308–319, Springer International Publishing, 2020.

10

Perform a Comparative Study Based on Conventional Machine Learning Approaches for Human Stress Level Detection

Pratham Sharma[1]*, Prerana Singh[1], Mahe Parah[1], Shyamapriya Chatterjee[1], Anirban Bhar[1], Soumya Bhattacharyya[1] and Pijush Dutta[2]

[1]*Department of Information Technology, Narula Institute of Technology, Kolkata, India*

[2]*Department of Electronics & Communication Engineering, Greater Kolkata College Engineering and Management, Baruipur, Phultala, West Bengal, India*

Abstract

One of the biggest issues in today's world is mental stress. Now that age is not taken into account when calculating stress levels, it makes no difference. Depression, heart attacks, and even suicide can result from extreme stress. Stress may have an impact on many facets of our existence, such as our behavior, emotions, and capacity for thought. Thus, we must manage our tension. We have collected 26000 number samples for the suggested system, which measures a person's stress level using three crucial parameters: body temperature, step count, and humidity. The Stress-Lysis.csv dataset contains three physiological states: the neutral, amused, and stressed states. Using machine learning methods such as Linear Regression, K-Nearest Neighbor, Linear SVM, Kernel SVM, Naive Bayes, Random Forest, and Decision Tree, the accuracies for three stress levels (low, normal, and high) were assessed and compared. Using machine learning approaches, three classification issues throughout the study were able to obtain an accuracy of up to 97.65% to 99.88%, respectively. The decision tree algorithm outperformed the others in terms of accuracy; its AUC and training time are 99.88%, 0.99, and 0.00656 seconds, respectively, and it is a key component of cutting-edge methods.

Keywords: Stress level, smart health care, classification, machine learning

**Corresponding author*: ps1452003@gmail.com

Pijush Dutta, Sudip Mandal, Korhan Cengiz, Arindam Sadhu and Gour Gopal Jana (eds.) Design and Forecasting Models for Disease Management (237–252) © 2025 Scrivener Publishing LLC

10.1 Introduction

Stress is among the most prevalent symptoms experienced by humans. Stress may cause a wide range of illnesses, including cancer, lung and respiratory problems, cardiovascular disease, and other ailments [1]. The world's population expansion has increased people's levels of stress. Stress is a disorder that affects everyone these days. People throughout the world have serious, sometimes fatal problems as a result of stress. Stress makes people behave differently than usual, which might interfere with regular routines. We must continuously observe their behavior using machine learning algorithms that are embedded into the system to gather real-time data for an accurate analysis of their behavior within the system. In India, more than 85% of people are suffering from stress [2].

Emotional states like stress and anxiety have a big impact on a person's quality of life. Stress has an impact on one's emotional and physical well-being by resulting in conditions including depression and heart rhythm irregularities, or arrhythmias [3]. A study by the American Institute of Stress [4] found that 80% of workers experience stress at work, and of those, about half say they require assistance in learning stress management techniques, and the remaining half do as well. According to the survey it has been observed that stress, anxiety, and depression may be caused by work-related pressure [5]. Research on both humans and animals indicates that stress might have an impact on the immune system and raise the risk of developing cancer. Medication, if necessary, or tailored therapies can both be used to reduce stress. Psychological and physiological experts traditionally adopted a questionnaire-based analysis to determine stress levels [6]. Because it only depends on people's replies, this strategy is unreliable and fraught with ambiguity. Additionally, many will find it laborious to complete the questionnaire.

From numerous studies, it was identified that stress can create a person's physiological markers. Anything like losing one's job is a psychological issue that can make a situation stressful [7]. Stressful circumstances can be controlled by stress hormones that partially affect physiological issues like stiff muscles, a racing heartbeat, beads of sweat, etc. [4] which causes biosignals. These biosignals quantify an individual's physiological measurements, which aids in the detection of stress. For automated stress detection, a variety of physical sensors were used [8]. The proposed study aims to use physiological data acquired during stressful situations to automatically identify an individual's state of stress. To prevent harmful stress-related disorders, such detection can aid in stress monitoring [9]. To detect the stress level (normal, amused, or stressed) of any person, several AI approaches are examined [10].

To perform this research, a sequence of operations like proper formatting of data structure, including data cleaning, and transforming is the pre-stage before different machine learning approaches are deployed on the publicly accessible Stress-Lysis.csv dataset. Section 10.2 highlights the relevant work, and Section 10.3 discusses the suggested course of action. Section 10.6 summarizes the paper's conclusion and suggests the next directions after Section 10.5 studies the experimental results and conclusions. Section 10.4 examines the experimental setup.

10.2 Related Work

Physicians can use AI algorithms to anticipate stress levels with the use of the concept of a stress detection system. The researchers are putting out their study on the identification of personal stress. There are several learning models that were studied using physiological reactions to predict stress discussed in Table 10.1.

Table 10.1 Review on AI-based stress level detection.

Si no.	Reference	Procedure	Outcomes
1	Philip Schmidt, *et al.* [11]	Proposed five ML Classifier techniques: KNN, LDA, RF, DT, and AdaBoost (AB) for stress state detection.	Maximum accuracy obtained for multiclass problems is 80.34% and for binary class 93.12% for the same feature in both cases.
2	Jacqueline Wijsman, *et al.* [12]	Investigated four ML classifier techniques: Linear (LBN) Quadratic Bayes Normal Classifier (QBN), KNN Classifier, Fisher's Least Square (FLS) Classifier	KNN classifier achieved maximum accuracy is about 80% between stress and non-stress conditions.

(Continued)

Table 10.1 Review on AI-based stress level detection. (*Continued*)

Si no.	Reference	Procedure	Outcomes
3	Virginia Sandulescu, *et al.* [13]	Proposed SVM algorithm for classification of stressed or non-stressed individuals from the collected from Wearable Physiological Sensors,	The maximum accuracy obtained by the classification model is about 82%.
4	Saskia Koldijk *et al.* [14]	We investigated the performance of integrating Unobtrusive Sensors and Machine Learning Algorithms for detecting Stress in Offices.	The accuracy obtained by SVM is about 90%.
5	Md Fahim Rizwan *et al.* [15]	Design of a biosignals-based Based Stress Detection System using Machine Learning Techniques	Maximum accuracy was achieved by using cubic SVM about to 98%
6	G. Giannakakisa *et al.* [16]	Investigated to classification of stress from videos using five algorithms Generalized Likelihood Ratio (GLR), NB, SVM, KNN, and AdaBoost classifier	Maximum accuracy obtained by AdaBoost classifier with 91.68%.

(*Continued*)

Table 10.1 Review on AI-based stress level detection. (*Continued*)

Si no.	Reference	Procedure	Outcomes
7	M. Gjoreski, *et al.* [17]	A binary classification model was proposed for Continuous stress detection from the dataset obtained from a wrist device.	The Random Forest algorithm achieved the maximum accuracy of a binary class problem of about 83%.
8	Elisabeth Andre, *et al.* [18]	An extended linear discriminant analysis (ELDA) model was proposed for physiological changes during music listening.	The accuracy for subject-dependent and subject-independent classification was 95% and 70% respectively.
9	P. Bobade, *et al.* [4]	Multiclass & binary class stress model was proposed on seven classifier models KNN, LDA, RF, DT, AdaBoost (AB), SVM, and Conventional deep learning (DL)	During the study, the accuracy for three classes and binary class accuracy for machine learning is 81.65% and 93.20%, and for deep learning, 84.32% and 95.21% respectively
10	V.R. Archana *et al.* [19]	Three classification model decision trees, Naïve-Bayes, and K-Nearest Neighbor are proposed for classifying stress levels depending upon six attributes	Based on the attributes all the experimental algorithms classify the disease level accurately.

10.3 Architecture Design

The Stress-Lysis system is a stress detection track consisting of a dataset collected from sensors: motion detector, Sweat reduction, and temperature [20]. This is consistent with the hypothesis of our study. The general configuration of the Stress-Track system is shown in detail in Figure 10.1 [21]. By using the Wi-Fi depicted in Figure 10.2 to access sensor data that is directly taken from the human body and saved in the cloud. Stress levels are successfully divided into three different categories using this sophisticated scientific method: low, normal, and high. A cloud storage system accomplishes two things: first, it collects the data in real-time; second, it stores them within predefined time frames. This design offers a thorough comprehension of stress fluctuations over time in addition to enabling the instantaneous retrieval of stress data.

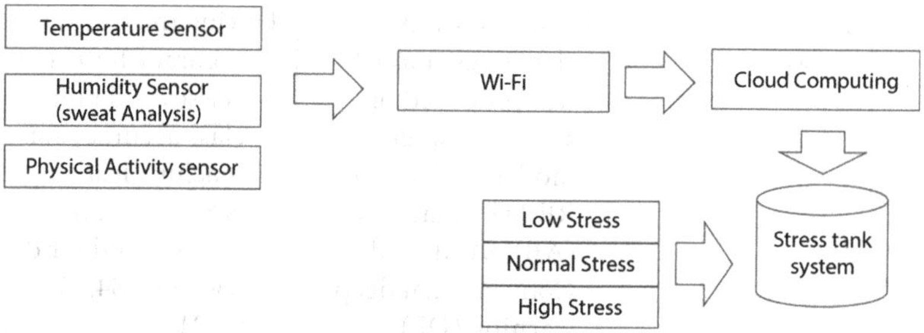

Figure 10.1 Simplified diagram of stress level system.

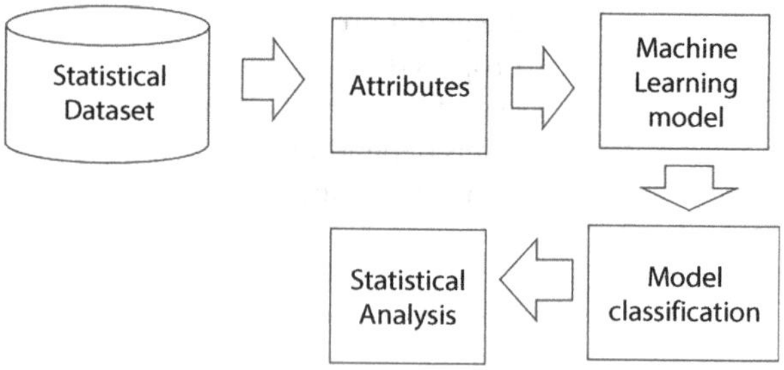

Figure 10.2 System overview.

10.3.1 Body Temperature

Body temperature is an important marker for many diseases, both serious and small. Examining trends in temperature changes makes it feasible to evaluate a person's physical and mental health. The rate at which the body temperature changes over a certain period is known as the temperature rate, and it provides important information. In this case, contact temperature sensors, which are intended to monitor skin temperature were employed for stability and accuracy [20, 21].

10.3.2 Humidity Analysis

Sweat is an essential physiological indication. Sweat is a physical substance that is secreted via the skin's pores in response to heat. When sweat production rises, the current flow between two electrodes also rises, thereby turning the body into a variable resistor. A humidity-detecting sensor is used to measure the sweat secretion level [20, 21].

10.3.3 Step Count Analysis

The rate of force is called acceleration. These forces influence variations of static or dynamic velocity. An accelerometer sensor is used to measure a person's steps along with motions [20, 21].

10.3.4 Dataset

The "Stress-Lysis.csv" dataset [22] consists of electrical signals obtained from four sensors: humidity, temperature, step count sensor, and stress levels sensor. The output of the dataset is prepared corresponding to the different activity of sensors output. It consists of a set of 26000 samples that include information about body temperature, body humidity, steps performed by the user, and the stress levels associated with those steps. Three categories are used to group the stress levels: low, normal, and high. The symbols 0, 1, and 2 stand for low stress, normal stress, and excessive stress, respectively. As Table 10.2 illustrates, the dataset provides a basis for comprehending the correlation between stress levels and other physical activity measures.

Temperature, humidity, and step count are the three properties that we employed in the statistical dataset for the proposed system, as shown in Figure 10.3. Data preparation is done after dividing the data into training

Table 10.2 Range of sensor value.

Sensor	Low stress	Normal stress	High stress
Temperature	79.01-84.00	84.01-95.00	95.01-99.00
Humidity	10.00-15.00	15.01-20.00	20.01-30.00
Steps Count (steps/min)	0- 91	92-129	130-200

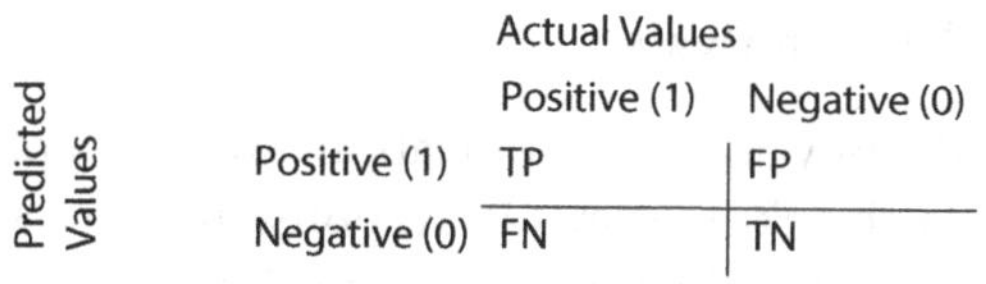

Figure 10.3 Confusion matrix.

and testing datasets based on these techniques. We have encoded the data based on it, and we have separated it into training and testing data, with 80% and 20% of the total data being used for each. The dataset contains 26,000 samples based on sensor output with three different ranges shown in Table 10.2.

10.4 Experiment

- Step 1: Data collection: The dataset is taken from "Stress-Lysis.csv" [22] which contains three input attributes "Humidity–Temperature–Step count" along with one output attribute "Stress levels" which have three different levels, low, normal and high stress.
- Step 2: Data pre-processing: In this stage, the collected data-set is cleaned and checked if there is any missing data available or not.
- Step 3: Feature extraction: unreadable or missing data is replaced by the mean values of the original data.
- Step 4: Classification model: In this stage, six ML classification model techniques are used to classify the model.
- Step 5: During the classification stage datasets are split into 80:20 for the training & testing phase. An algorithm classifies

any model efficiently when several training datasets are kept higher than several testing datasets.

- Step 6: Based on the confusion matrix and training and testing time the best machine learning model is identified for the present research.

10.4.1 Performance Matrices

To evaluate the model performance in this research we used three fundamental metrics: accuracy, F1 measure, and confusion matrix. Accuracy is defined as the proportion of the model's correct predictions. The relationship between actual & predicted samples is summarized in the confusion matrix presentation in Figure 10.4. This study's goal was to classify different stress levels; Table 10.2 lists the predictive measures for this categorization. During the test dataset, we validated how the proposed model validates the stress levels.

The parameters as they are being evaluated are referred to as true positive, false positive, true negative, and false negative. Using Table 10.2, the formulae for accuracy, recall, and precision are obtained as shown in Eqn. (10.1) to Eqn. (10.4).

Accuracy: it is the ratio of correct prediction and overall predictions represented by

$$\frac{TP+TN}{TP+TN+FP+FN} \tag{10.1}$$

Precision: ratio of correct samples to the total number of samples represented by

$$\frac{TP}{TP+FP} \tag{10.2}$$

Recall of Sensitivity: ratio of correct predicted samples to the actual class samples represented by

$$\frac{TP}{TP+FN} \tag{10.3}$$

Table 10.3 Performance of classifier model in three class tasks.

Classifier	AUC	Accuracy	Precision	Recall	F1_score	Testing time (sec.)
LR	0.912	96.23	95.68	94.89	95.27	0.049495
KNN	0.869	96.73	96.23	94.85	92.85	0.008766
L_SVM	0.756	97.1	93.32	91.68	92.65	24.24177
K_SVM	0.852	97.9	91.65	93.32	91.62	1.281648
NB	0.892	98.6	86.32	88.6	91.3	0.00956
RF	0.921	99.21	96.2	89.65	95.6	0.01095
DT	0.995	99.88	97.8	97.6	97.8	0.00656

Specificity: It is the ratio of the actual 0 to the correctly predicted 0.

It can be defined as

$$\frac{TN}{TP+TN} \tag{10.4}$$

F1 score / F measure: Geometric mean of precision and recall when data is imbalanced represented by $\frac{2*(recall*precision)}{recall+precision}$

All the abovementioned metrics are considered in the present research shown in Table 10.3. In this research out of 26000 sample datasets, we consider only 2000 samples to establish a level of significance is 0.05.

10.5 Result Analysis

In the present research, seven classifier algorithms were applied to the "Stress-Lysis" dataset [22]. To find out the best classifier model a comparative study was performed based on metrics obtained from the confusion matrix, computational time, and AUC [23–27]. The effectiveness of the confusion matrix empowers us to gather more information regarding classification outcomes. The confusion matrix helps to determine the value of Precision, F1 score, and Recall of every classifier model shown in Figure 10.4. To identify the best-fitted classification model the overall dataset was distributed into an 80:20 training and testing set [28–34]. Each row represents the actual class while the column represents by predicted value.

A comparative study of the proposed seven classifiers is described in Table 10.3 based on six metrics for evaluating the performance. From the experimental analysis, it was seen that DT achieved maximum accuracy of 99.88% accuracy, furthermore, from the concept of Precision 97.8% performed true positive predictions while 2.2% were false positives. A correctly recognized positive sample is identified by the properties called Recall, from the result it has been observed that DT captured 97.6% of true positives and false positives about 2.4%. F1 Score reflects an overall balance; the highest score of F1 is 97.8% indicating a higher degree of correlation between experimental and actual positive samples. A ROC curve is a metric used to determine model predictiveness plotted with the help of the TP rate and the FP rate. The numerical value of AUC indicates the degree of categorizing the negative and positive classes. The maximum value of AUC is 0.995 achieved by DT indicates this technique categorizes

Predicted Value	Actual Value: 0	Actual Value: 1	Actual Value: 2
0	82	1	0
1	0	165	1
2	1	0	151

Figure 10.4 Confusion matrix for decision tree.

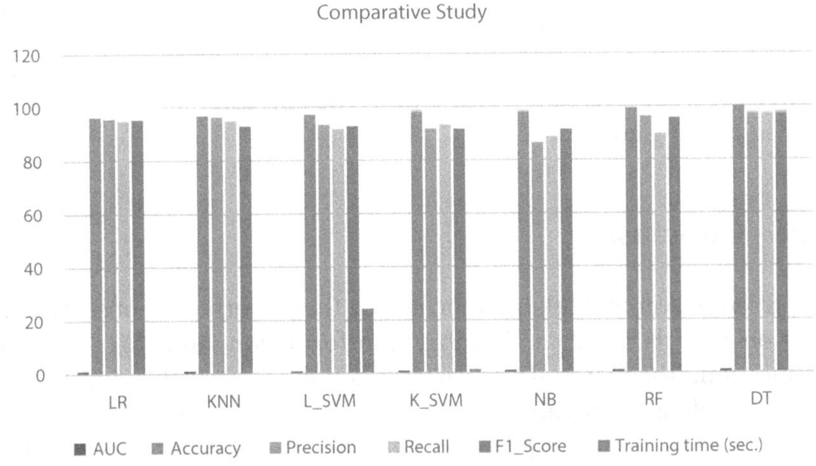

Figure 10.5 Comparative study of seven classifier models.

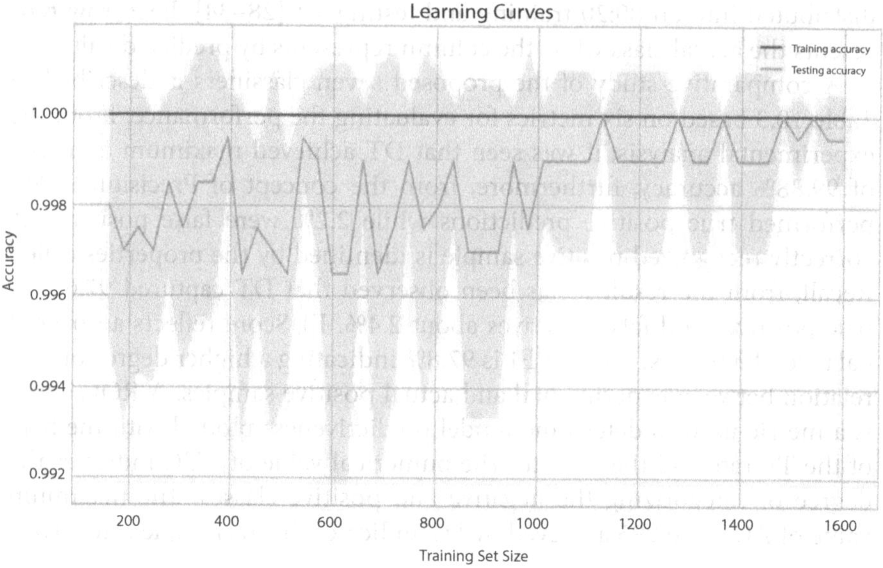

Figure 10.6 Learning curve concerning training data size.

the negative and positive classes more effectively. From the statistical analysis, it is observed that DT is more efficient and LR is the worst classification model for three class-level problems. Graphically this comparative study of seven classifier models is represented in Figure 10.5 on six different metrics.

The learning curve is plotted to visualize how the model's accuracy changes with varying training set sizes shown in Figure 10.6. This helps determine whether the model is overfitting or underfitting and guides the dataset's sufficiency. From the learning curve, we can see that we have a better accuracy over time with a difference of around .001% accuracy between training and test set.

10.6 Conclusion

The present research work has represented the performance of seven classification models for data cleaning, transformation, and three levels of classification. The "Stress-Lysis" dataset is publicly available and used as an example dataset for this research. This dataset contains three different sensing outputs: body temperature, humidity, and several steps, which in turn help to determine the stress level in human beings. In the last stage of the experiment, outcomes are compared.

The proposed system may help in predicting the stress level and taking preventive measures to decrease the stress level. Based on the values, some of the algorithms are Decision Tree, linear SVM, Kernel SVM, K-Nearest Neighbor, Random Forest, Decision Tree, and Naïve Bayes are used to classify the three classes of classification of stress. Six metrics are used to identify the best classify model. From the result analysis, it is concluded that decision trees achieved maximum accuracy while Logistic regression have the least accuracy. In the future scope, a multiple number of wireless body sensors can be used to detect the stress level of human beings. Other than that response of hybrid machine learning in this dataset is an open challenge.

References

1. Dubois-Deruy, E., Peugnet, V., Turkieh, A., Pinet, F., Oxidative stress in cardiovascular diseases. *Antioxidants*, *9*, 9, 864, 2020.
2. Pangtey, R., Basu, S., Meena, G.S., Banerjee, B., Perceived stress and its epidemiological and behavioral correlates in an urban area of Delhi, India: a community-based cross-sectional study. *Indian J. Psychol. Med.*, *42*, 1, 80–86, 2020.
3. Palmer, D. and Change, C., Stress, and Civil Servants' Occupational Health, c. 1967–85, in: *Stress in Post-War Britain, 1945–85*, pp. 95–110, 2016.
4. Bobade, P. and Vani, M., Stress detection with machine learning and deep learning using multimodal physiological data, in: *2020 Second International Conference on Inventive Research in Computing Applications (ICIRCA)*, pp. 51–57, IEEE, 2020.
5. Ayaz, A., A cross-sectional study of the relationships between ethnicity and occupational classes with mental wellbeing in the UK. *Life Res.*, *4*, 1, 4, 2021.
6. Biondi, M. and Zannino, L.G., Psychological stress, neuroimmunomodulation, and susceptibility to infectious diseases in animals and man: a review. *Psychother. Psychosom.*, *66*, 1, 3–26, 1997.
7. Bourne Jr., L.E. and Yaroush, R.A., *Stress and cognition: A cognitive psychological perspective*, NASA/CR, University of Colorado, 2003, (No. IH-045).
8. Ladakis, I. and Chouvarda, I., Overview of biosignal analysis methods for the assessment of stress. *Emerg. Sci. J.*, *5*, 2, 233–244, 2021.
9. Arza, A., Garzón-Rey, J.M., Lázaro, J., Gil, E., Lopez-Anton, R., de la Camara, C., Aguiló, J., Measuring acute stress response through physiological signals: towards a quantitative assessment of stress. *Med. Biol. Eng. Comput.*, *57*, 271–287, 2019.

10. Geetha, R., Krishnamoorthy, N.V., Murugan, K.S., Gnanaprakasam, C., Swarna, M., A Novel Deep Learning based Stress Analysis and Detection Scheme using Characteristic Data, in: *2023 Eighth International Conference on Science Technology Engineering and Mathematics (ICONSTEM)*, pp. 1–8, IEEE, 2023.
11. Schmidt, P., Reiss, A., Duerichen, R., Marberger, C., Van Laerhoven, K., Introducing wesad, a multimodal dataset for wearable stress and affect detection, in: *Proceedings of the 20th ACM international conference on multimodal interaction*, pp. 400–408, 2018.
12. Wijsman, J., Grundlehner, B., Liu, H., Hermens, H., Penders, J., Towards mental stress detection using wearable physiological sensors, in: *2011 Annual International Conference of the IEEE Engineering in Medicine and Biology Society*, pp. 1798–1801, IEEE, 2011, August.
13. Sandulescu, V., Andrews, S., Ellis, D., Bellotto, N., Mozos, O.M., Stress detection using wearable physiological sensors, in: *Artificial Computation in Biology and Medicine: International Work-Conference on the Interplay Between Natural and Artificial Computation, IWINAC 2015, Elche, Spain, June 1-5, 2015, Proceedings, Part I 6*, pp. 526–532, Springer International Publishing, 2015.
14. Koldijk, S., Neerincx, M.A., Kraaij, W., Detecting work stress in offices by combining unobtrusive sensors. *IEEE Trans. Affective Comput.*, *9*, 2, 227–239, 2016.
15. Rizwan, M.F., Farhad, R., Mashuk, F., Islam, F., Imam, M.H., Design of a biosignal based stress detection system using machine learning techniques, in: *2019 international conference on robotics, electrical and signal processing techniques (ICREST)*, pp. 364–368, IEEE, 2019.
16. Giannakakis, G., Pediaditis, M., Manousos, D., Kazantzaki, E., Chiarugi, F., Simos, P.G., Tsiknakis, M., Stress and anxiety detection using facial cues from videos. *Biomed. Signal Process. Control*, *31*, 89–101, 2017.
17. Gjoreski, M., Gjoreski, H., Luštrek, M., Gams, M., Continuous stress detection using a wrist device: in laboratory and real life, in: *Proceedings of the 2016 ACM international joint conference on pervasive and ubiquitous computing: Adjunct*, pp. 1185–1193, 2016.
18. Kim, J. and André, E., Emotion recognition is based on physiological changes in music listening. *IEEE Trans. Pattern Anal. Mach. Intell.*, *30*, 12, 2067–2083, 2008.
19. Archana, V.R. and Devaraju, B.M., Stress detection using machine learning algorithms. *Int. J. Res. Eng. Sci. Manage.*, *3*, 8, 251–256, 2020.
20. Rachakonda, L., Mohanty, S.P., Kougianos, E., Stress-Lysis: An IoMT-enabled device for automatic stress level detection from physical activities, in: *2020 IEEE International Symposium on Smart Electronic Systems (iSES) (Formerly iNiS)*, pp. 204–205, IEEE, 2020.

21. Rachakonda, L., Mohanty, S.P., Kougianos, E., Sundaravadivel, P., Stress-lysis: A DNN-integrated edge device for stress level detection in the IoMT. *IEEE Trans. Consum. Electron.*, *65*, 4, 474–483, 2019.
22. https://www.kaggle.com/datasets/laavanya/stress-level-detection
23. Dutta, P., Paul, S., Kumar, A., Comparative analysis of various supervised machine learning techniques for diagnosis of COVID-19, in: *Electronic devices, circuits, and systems for biomedical applications*, pp. 521–540, Academic Press, 2021.
24. Dutta, P., Paul, S., Majumder, M., *An efficient SMOTE based machine learning classification for prediction & detection of PCOS*, Research Square, USA, 2021.
25. Dutta, P., Paul, S., Obaid, A.J., Pal, S., Mukhopadhyay, K., Feature selection based artificial intelligence techniques for the prediction of COVID like diseases. *J. Phys.: Conf. Ser.*, 1963, 012167, IOP Publishing, 2021.
26. Dutta, P., Paul, S., Cengiz, K., Anand, R., Kumar, A., A predictive method for emotional sentiment analysis by deep learning from EEG of brainwave dataset, in: *Artificial Intelligence for Neurological Disorders*, pp. 25–48, Academic Press, 2023.
27. Dutta, P., Paul, S., Shaw, N., Sen, S., Majumder, M., Heart disease prediction: A comparative study based on a machine-learning approach, in: *Artificial Intelligence and Cybersecurity*, pp. 1–18, CRC Press, 2022.
28. Dutta, P., Pal, S., Kumar, A., Cengiz, K., *Artificial Intelligence for Cognitive Modeling: Theory and Practice*, CRC Press, USA, 2023.
29. Dutta, P., Paul, S., Cengiz, K., Anand, R., Majumder, M., A predictive method for emotional sentiment analysis by machine learning from electroencephalography of brainwave data, in: *Implementation of Smart Healthcare Systems using AI, IoT, and Blockchain*, pp. 109–130, Academic Press, 2023.
30. Dutta, P., Shaw, N., Das, K., Ghosh, L., Early & accurate forecasting of mid term wind energy based on PCA empowered supervised regression model. *Comput. Intell. Mach. Learn.*, *2*, 1, 53–64, 2021.
31. Sen, S., Saha, S., Chaki, S., Saha, P., Dutta, P., Analysis of PCA based AdaBoost machine learning model for predict mid-term weather forecasting. *Comput. Intell. Mach. Learn.*, *2*, 2, 41–52, 2021.
32. Dutta, P., Paul, S., Majumder, M., *Intelligent SMOTE Based Machine learning Classification for Fetal State on Cardiotocography Dataset*, Research Square, USA, 2021.
33. Dutta, P., Paul, S., Jana, G.G., Sadhu, A., Hybrid Genetic Algorithm Random Forest algorithm (HGARF) for improving the missing value Imputation in Hepatitis Medical Dataset, in: *2023 International Symposium on Devices, Circuits, and Systems (ISDCS)*, vol. 1, pp. 01–05, IEEE, 2023.
34. Dutta, P., Paul, S., Sadhu, A., Jana, G.G., Bhattacharjee, P., Performance of Automated Machine Learning Based Neural Network Estimators for the Classification of PCOS, in: *Doctoral Symposium on Human-Centered Computing*, pp. 65–73, Springer Nature Singapore, Singapore, 2023.

11

Diabetes Prediction Using a Hybrid PCA-Based Feature Selection and Computational Machine Learning Algorithm

Sumanta Dey[1]*, Pijush Dutta[1], Gour Gopal Jana[1] and Arindam Sadhu[2]

[1]*Department of Electronics & Communication Engineering, Greater Kolkata College Engineering and Management, Baruipur, Phultala, West Bengal, India*
[2]*Department of Electronics and Communication Engineering, Dr. Sudhir Chandra Sur Institute of Technology and Sports Complex, Kolkata, India*

Abstract
Diabetes Mellitus is normally known as a metabolic issue in which the body can't utilize insulin or store glucose for energy as well as unable to produce insulin. Diabetes patients can suffer various sicknesses that incorporate kidney disappointment, stroke, visual impairment, cardiovascular failures, and lower appendage removal. The earlier classification model could not predict with a higher degree of accuracy. In the proposed work, we have used six different machine learning classifier models that would help to identify the potential chances of getting affected by Diabetes Related Diseases. The overall process is performed in four steps: data pre-processing, features selection, dimensionality reduction method using Principal Component Analysis (PCA), and finally machine learning classify model. From the result analysis, it has been observed that SVM offered a maximum accuracy of about 91% for the selection of specific features. Besides that, present research outcomes show better outcomes than the other state-of-the-art methods.

Keywords: PIMA dataset, classifier, feature selection, dimensionality reduction

**Corresponding author*: sumanta.dey2002@gmail.com

Pijush Dutta, Sudip Mandal, Korhan Cengiz, Arindam Sadhu and Gour Gopal Jana (eds.) Design and Forecasting Models for Disease Management (253–268) © 2025 Scrivener Publishing LLC

11.1 Introduction

According to medical experts, Diabetes patients can be found in two types: type l and type 2 depending upon production and insulin response. Diabetes patients have several main symptoms, including frequent urination, frequent feelings of hunger and thirst, exhaustion, and blurred eyesight. According to WHO, across the world, there are 382 million people affected [1] and every year more than 75% of patients are affected by Type 2 which is a concerned matter. As a researcher, the main motive is to design a model that improves the accuracy level for detecting and protecting the diabetic patient from long-term damage. Earlier several researchers performed a machine learning algorithm to predict diabetes. A decision tree classifier model has been studied for the computation of both numerical and unambiguous data [2] where each node denotes a feature. Naïve Bayes (NB) classifier experimented on the prediction of the diabetes dataset as it generated the probability value of a specific object feature with a historical dataset with the same features [3]. KNN is a classifier model that takes longer time than other classifiers also experimented on diabetes prediction based on the Euclidean distance measurement unit [4]. Due to classifying the structured and random data into preferred forms both RF and SVM are the best classification models. The main mechanism of the RF classifier is it consists of multiple decision trees and the working principle lies in two major concepts: firstly, it can perform a random sampling of data and secondly it creates a large number of subsets depending upon features. SVM is a classification model based on margin maximization principles. It can perform both classification, regression, and outlier detection. In this paper, we experimented on six major classifiers, RF, LR, KNN, XGBoost, SVC, and SVM followed by feature selection and dimensionality reduction applied to the diabetes dataset. This paper is divided into four sections, Section 11.2 of this paper deals with related work and their approach. Section 11.3 outlines the proposed workflow and methodologies. Section 11.4 provides a brief overview of the results and compares various techniques with their metric results and finally Section 11.5 concludes the paper.

11.2 Related Work

Pre-processing is a technique in data analytics that is used to improve the efficiency of any classifier and regressor model. Imputation is one of the common preprocessing stages where the missing value is placed by mean values. Except for imputation, data normalization can also improve the model efficiency. Principal Component Analysis is a common technique during the preprocessing stage used to decrease the non-potential features [5]. Except for dimension reduction, scaling, and data normalization are also performed in preprocessing stage. A decision tree, classifier model has been proposed using the Weka tool for the prediction of Type II PIDD [6]. The proposed model was divided into two parts pre-processing and prediction. A hybrid GA-DT has been proposed to predict the diabetes [7]. GA algorithm is used to find out the potential feature selection and a decision tree is used to classify the model. The results show an improvement of accuracy of about 74.78% and ROC 0.786. A hybrid GA-ANN algorithm was applied to find out the potential attributes from diabetic patients' datasets [8]. The output of this model was also compared with another classifier ANN, and K nearest neighbor. An SVM classifier has been investigated to diagnose PIDD [9]. A Bayes network classifier model has been experimented with to diagnose diabetes patients [10]. Three different classification models Naïve Bayes (NB), SVM, and DT used and implemented for the prediction of diabetes mellitus [11]. It has been noted from the outcome analysis that NB was found efficient among the three with an accuracy of 76.3%. A comparative analysis of three algorithms was used to predict diabetes using three algorithms KNN, GB, and R) classifiers. From the result analysis, it was observed that Gradient Boosting (GB) improved its accuracy depending on the number of feature selections [12]. A feature-selected hybrid three-classifier algorithm DT, RF, and NN was used followed by PCA for the diagnosis of diabetes and the best model was identified based on sensitivity, specificity, and accuracy [13]. The five machine learning classifier algorithms SVM, RF, LR, DT, and KNN were compared [14]. From the result analysis, RF was shown to be more effective, with a maximum accuracy of 75%. A comparative study on four machine learning classifiers Gradient Boost (GB), LR, RF, and Rpart was used for the prediction of diabetes [15]. SVM classifier is used to monitor the health condition of a patient to use as a multi-parameter detector [16], Software-defined radio [17], and identifying most influential parameters for predicting Leukemia Cancer [18]. The fundamental concept and feature selection operation of the Random Forest Classifier are explained [19]. Diabetes Prediction and Classification using RF and SVM algorithms are performed [20]. Comparative study was measured after dimensionality reduction, feature selection was performed and finally, performance was analyzed using accuracy, sensitivity, and specificity.

11.3 Proposed Workflow

In this section, we explain feature selection, dimensional reduction, PCA, and the application to classify the dataset using the Machine Learning (ML) model. Figure 11.1 demonstrates the overall process. The present research was performed based on the PIMA Indian diabetes dataset [21]. The data classification model is analyzed after following the mean values of each parameter are used to fill in the missing values and the selection of the most influencing parameters using Feature selection is performed during the pre-processing step.

11.3.1 Data Pre-Processing

In this stage, the diabetes data set PIDD is collected from the UCI repository [5] which consists 768 samples. In the dataset, there are 9 attributes and 1 output with binary class attribute 0 or 1 where binary 1 indicates

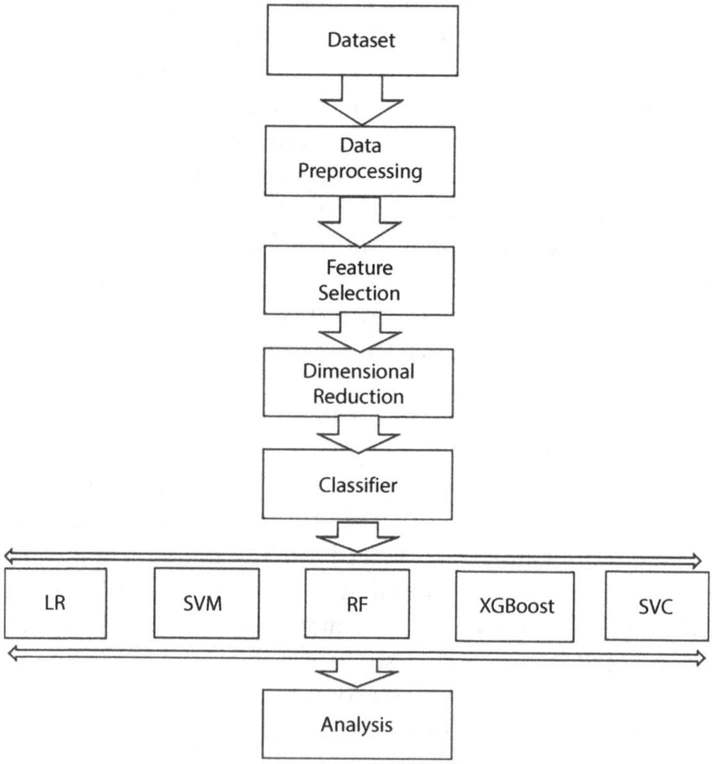

Figure 11.1 Flow diagram for the proposed research.

Table 11.1 Pima Indian diabetes dataset attributes.

Attribute no.	Attribute
1	Number of pregnancies (NP)
2	Glucose concentration (GC)
3	Blood Pressure (BP)
4	Skin thickness (ST)
5	Insulin (mu U/mL) (SI)
6	Body Mass Index (kg/m2) (BMI)
7	Diabetes Pedigree Function (DPF)
8	Age
9	Class 0 and 1 (Diagnosis of PIDD)

patients have positive symptoms and 0 for negative symptoms. A description of the dataset and its features are shown in Table 11.1.

In the data pre-processing stage certain attributes have some missing values which reduce the effectiveness of the dataset [22–24]. The dataset's missing values are replaced with a non-zero mean value. Figure 11.2 represents the boxplot of each attribute of the diabetes dataset.

11.3.2 Feature Selection

It is the most important step in AI where potential attributes can be determined to maximize the performance of any model. This can be achieved by either the filter method or the basis of statistics measures [25]. The feature selection method is independent of the pre-processing step. In this research, we have used a filtering method where redundant & irrelevant features are removed through ranking. Due to less computational time, and unavailability of data overfit filtering methods are very effective [26]. The most famous filtering method is information gain which determines the reduction in entropy while transforming the dataset. In the information gain filtering method, each variable is calculated concerning the target variable [27, 28] present in Equation 11.1:

$$Gain\,(S, a) = Entropy\,(S) - \sum_{v \in values\,(a)} \frac{|S_v|}{|S|} Entropy\,(S_v) \quad (11.1)$$

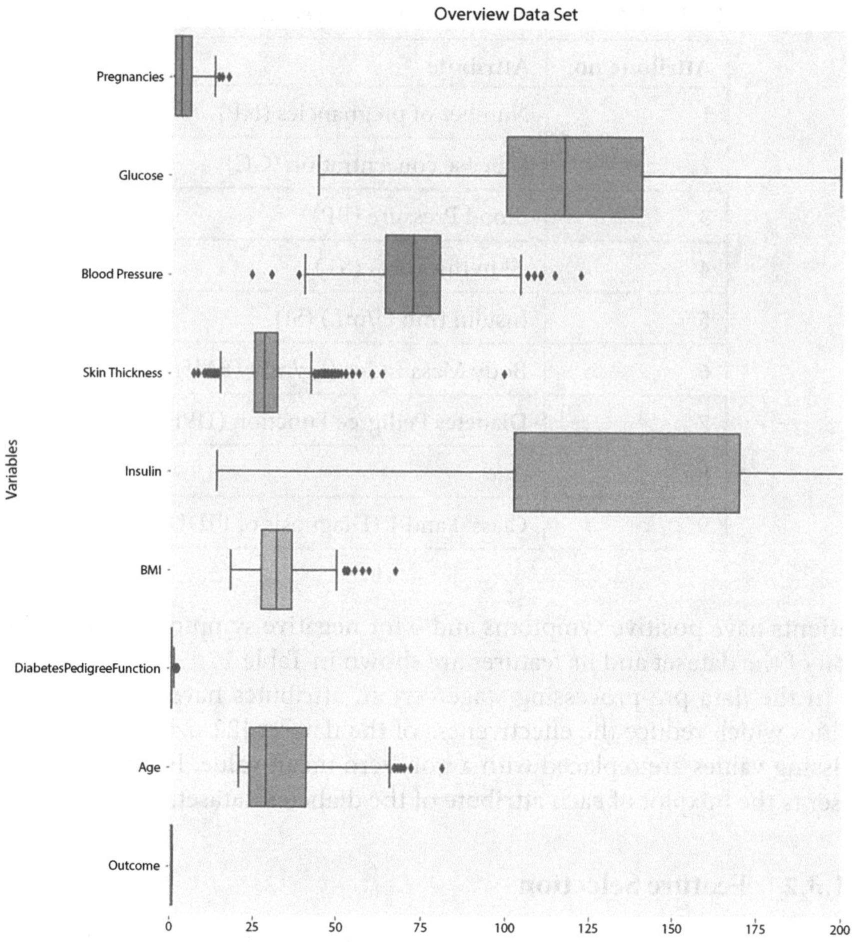

Figure 11.2 Boxplot of different features of PIMA dataset.

Where S is the specific class label, entropy0 (S_v) is the entropy of dataset S; and $\frac{|S_v|}{|S|}$ is the proportion of the values.

11.3.3 Dimensionality Reduction

It is an unsupervised method that is very useful for increasing the model accuracy by reducing nonpotential attributes. In this study, we used PCA to reduce dimensionality while maximizing interpretability and information loss topology [29, 30]. Additionally, these features are quite redundant, adding some noise in the actual dataset and making no sense to have in the training dataset. PCA is known as the covariance method for transforming a matrix of X to another matrix Y using Karhunen Loeve transfer

$$Y = KLT\{X\} \tag{11.2}$$

However, PCA transformation is a linear transformation where principal components decomposed from the feature into an eigenvector.

11.3.4 Classification

It is the last stage of known data classification using machine learning algorithms, and finally, results are analyzed. In the classification stage, a mapping was created between input parameters with output [31]. In this research, we used a binary classification model to predict a person is whether a person is healthy or not. Six algorithms Random Forest, Support Vector Machine, Logistic Regression, XGBoost, Support vector Classifier, and KNN classifications for predicting diabetes. Figure 11.3 represents graphical distribution of frequency versus potential attributes. Figure 11.4 represents different metrics versus outcomes and insulin level, and Figure 11.5 showing the variation of error rate with respect to K in KNN algorithm.

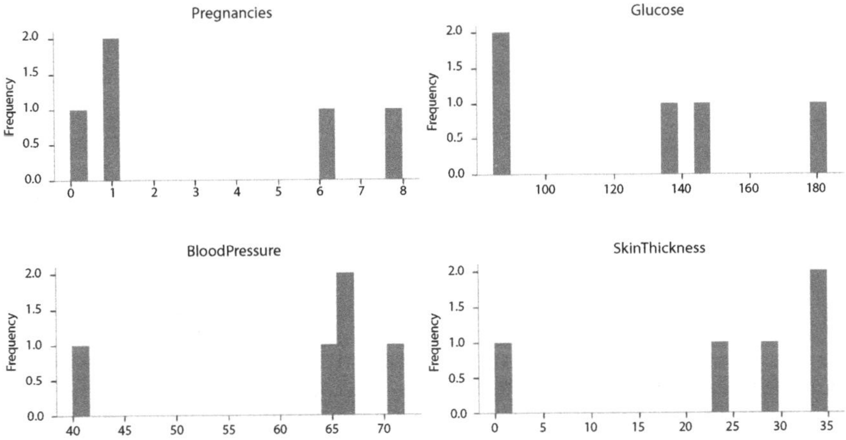

Figure 11.3 Graphical representation for frequency versus attributes.

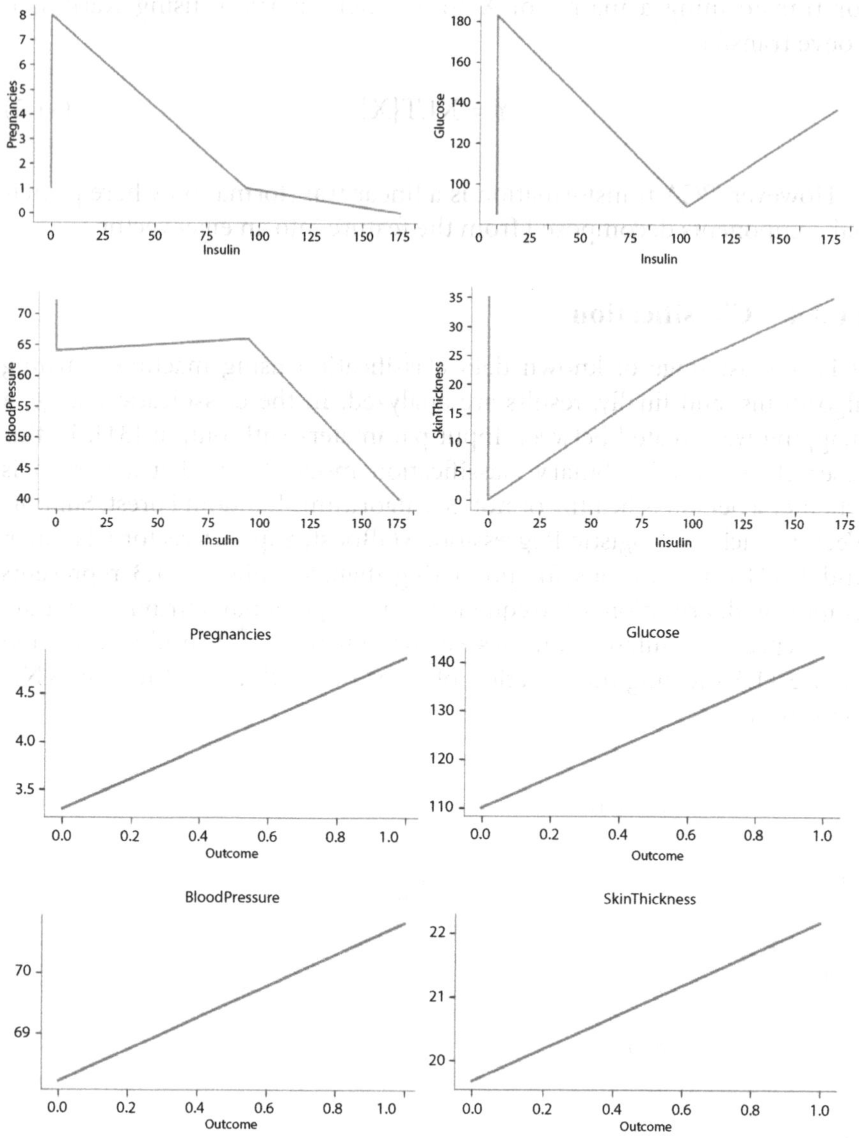

Figure 11.4 Graphical representation for different metrics versus outcome.

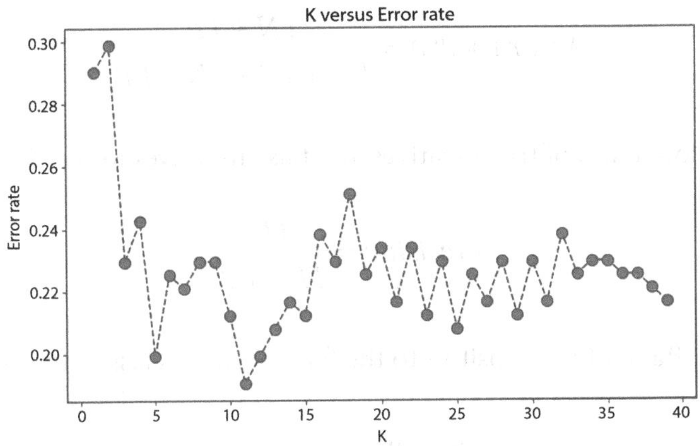

Figure 11.5 Graphical representation of error rate versus value of K in KNN.

11.4 Result Analysis

This section describes and explains the best model out of six machine learning classifiers applied to diabetes prediction. The comparative study was performed based on four basic evaluation criteria obtained from the confusion matrix along ROC and computational time which are briefly explained in the subsection.

11.4.1 Evaluation Criteria

Four fundamental metrics are obtained from the confusion matrix of each machine learning classifier explained in the following section:

True Negative (TN): Identify the number of points that are true negative class.
True Positive (TP): Identify the number of points that are true positive class.
False Negative (FN): Identify the number of points that are incorrect true negative class.
False Positive (FP): Identify the number of points that are incorrect true positive class.
Accuracy: Ratio of the sum of true positives and true negatives class to the total class in the classifier.

$$Accuracy\ (\%) = \frac{TN + TP}{TN + TP + FN + FP} \tag{11.3}$$

Precision: Ratio of true positives and false negatives in the classifier.

$$Precision = \frac{TP}{TP + FP} \tag{11.4}$$

Recall: Ratio of true positive to the false negative class in the classifier.

$$Recall = \frac{TP}{TP + FN} \tag{11.5}$$

F - Measure: It is the geometric mean of Precision and Recall.

$$F - Measure = \frac{2 * Precision * Recall}{Precision + Recall} \tag{11.6}$$

The receiver operating curve (ROC) graph is a graphical measurement for the determination of the performance of a binary classifier depending upon the threshold value.

$$ROC = \frac{1}{2}\left(\frac{TP}{TP + FN} + \frac{TN}{TN + FP}\right) \tag{11.7}$$

We investigated two different classifier models: individual PCA, that is PCA without dimensionality, and PCA-SVM, that is SVM with dimensional reduction. From the study of the results, it has been noted that hybrid PCA-SVM perform outperformed classifier than individual classifier stated in Table 11.2.

Table 11.2 indicates that the highest accuracy of PCA-SVM for five features is around 98.56%, and the ROC value is 0.908. These findings are displayed in Figure 11.6.

In the second phase, we compare the results of our current experiment with other cutting-edge techniques used to forecast and predict diabetes using the same dataset. Figure 11.7 displays some of the most advanced techniques with their accuracy, including the hybrid fuzzy and regression

Table 11.2 Experimental results of PIDD model.

Approach	Total number of feature	Accuracy
SVM	8	82.35
PCA-SVM	5	98.56

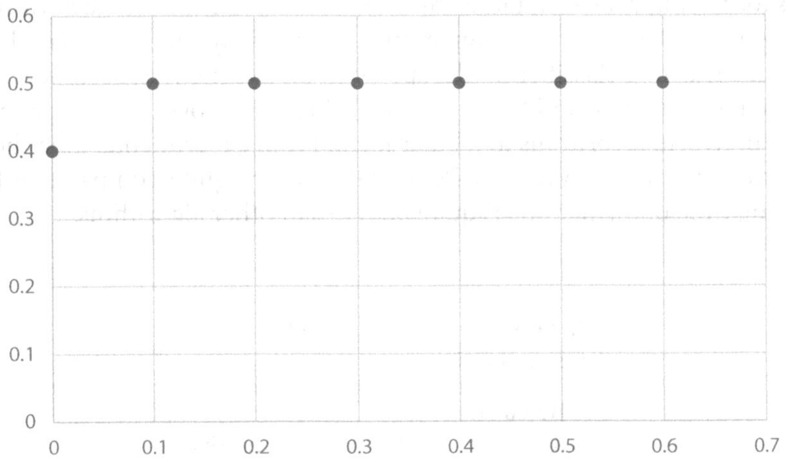

Figure 11.6 ROC curve for PCA-SVM.

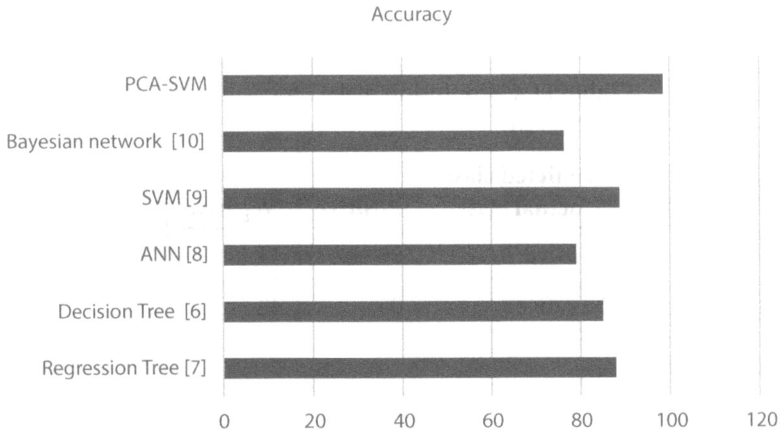

Figure 11.7 A comparison study with the proposed algorithm and state-of-the-art technique.

tree [7], decision tree [6], artificial neural network [8], SVM [9], and Bayesian network [10]. The outcome indicates that, in comparison to alternative methods, our suggested PCA-SVM on the diabetic data set yields a superior accuracy of 98.56%.

For better understanding except accuracy and ROC other three metrics are used to predict the best machine learning model, those are precision, F measure & Recall based on the confusion matrix. The fundamental concept of the confusion matrix for binary classification is shown in Table 11.3.

Table 11.4 shows the confusion matrix for the predicted class & actual class for testing dataset obtained by PCA based SVM classifier algorithm

Figure 11.8 represents comparative study of hybrid PCA based six machine learning classifier model approach on five metrics; accuracy, precision, recall, f-score and training time. Table 11.5 shows the comparative study of hybrid algorithms in tabular form. From experimental it has been observed that PCA-SVM and PCA-XGBoost are quite comparative but training time of former one is quite better than other algorithms.

Table 11.3 Typical example of the confusion matrix.

Predicted class actual class	**Positive**	**Negative**
Positive	TP	FP
Negative	FN	TN

Table 11.4 Confusion matrix for PCA-SVM.

Predicted class actual class	**Positive**	**Negative**
Positive	96	11
Negative	05	42

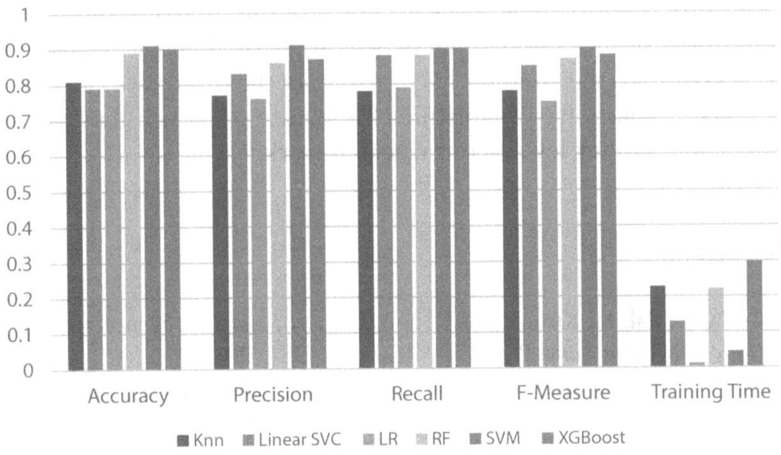

Figure 11.8 Comparison chart of different metrics of algorithm.

Table 11.5 Performance comparison of the proposed model for the testing phase.

	KNN	Linear SVC	LR	RF	SVM	XGBoost
Accuracy	0.81	0.79	0.79	0.89	0.91	0.90
Precision	0.77	0.83	0.76	0.86	0.91	0.87
Recall	0.78	0.88	0.79	0.88	0.90	0.90
F-Measure	0.78	0.85	0.75	0.87	0.90	0.88
Training Time	0.828	0.128	0.012	0.221	0.045	0.298

11.5 Conclusion and Future Work

Due to the presence of duplicate data in the data set the efficiency of the existing algorithm is significantly reduced. There is only way to improve the efficiency of the data classification are feature reduction and dimensionality reduction. In the present research, we proposed a hybrid method to improve the data classification efficiency of diabetes data. The overall process is performed in three steps; in the first phase feature selection is performed by filtering methods followed by dimensional reduction using Kernel-based PCA to reduce redundant data to improve the efficiency of diagnosis of diabetes. In the final stage, a machine learning classifier model has been proposed to achieve the accuracy of the model. From the result analysis, it has been observed that hybrid PCA-SVM performs better than other classifier models used in this research. In the future, we can use PSO, DE, GA, etc. evolutionary algorithms for feature selection steps to further improve the accuracy level of our research work.

References

1. Dutta, P., Pal, S., Kumar, A., Cengiz, K., *Artificial Intelligence for Cognitive Modeling: Theory and Practice*, CRC Press, USA, 2023.
2. Suresh, K., Obulesu, O., Ramudu, B.V., Diabetes prediction using machine learning techniques. *Helix-The Sci. Explorer| Peer Reviewed Bimonthly Int. J.*, *10*, 02, 136–142, 2020.
3. Chen, M., Hao, Y., Hwang, K., Wang, L., Wang, L., Disease prediction by machine learning over big data from healthcare communities. *IEEE Access*, *5*, 8869–8879, 2017.
4. Jakka, A. and Vakula Rani, J., Performance evaluation of machine learning models for diabetes prediction. *Int. J. Innov. Technol. Explor. Eng. (IJITEE)*, *8*, 11, 10–35940, 2019.
5. Wei, S., Zhao, X., Miao, C., A comprehensive exploration to the machine learning techniques for diabetes identification, in: *2018 IEEE 4th World Forum on Internet of Things (WF-IoT)*, pp. 291–295, IEEE, 2018.
6. Al Jarullah, A.A., Decision tree discovery for the diagnosis of type II diabetes, in: *2011 International conference on innovations in information technology*, pp. 303–307, IEEE, 2011.
7. Seera, M. and Lim, C.P., A hybrid intelligent system for medical data classification. *Expert Syst. Appl.*, *41*, 5, 2239–2249, 2014.
8. Pradhan, M. and Sahu, R.K., Predict the onset of diabetes disease using Artificial Neural Network (ANN). *Int. J. Comput. Sci. Emerg. Technol. (E-ISSN 2044-6004)*, *2*, 2, 303–311, 2011.
9. Kumari, V.A. and Chitra, R., Classification of diabetes disease using support vector machine. *Int. J. Eng. Res. Appl.*, *3*, 2, 1797–1801, 2013.
10. Guo, Y., Bai, G., Hu, Y., Using bayes network for prediction of type-2 diabetes, in: *2012 International conference for internet technology and secured transactions*, pp. 471–472, IEEE, 2012.
11. Sisodia, D. and Sisodia, D.S., Prediction of Diabetes using Classification Algorithms. *Procedia Comput. Sci.*, 132, 1578–1585, 2018.
12. Sivaranjani, S., Ananya, S., Aravinth, J., Karthika, R., Diabetes prediction using machine learning algorithms with feature selection and dimensionality reduction, in: *2021 7th International Conference on Advanced Computing and Communication Systems (ICACCS)*, vol. 1, pp. 141–146, IEEE, 2021.

13. Zou, Q., Qu, K., Luo, Y., Yin, D., Ju, Y., Tang, H., Predicting diabetes mellitus with machine learning techniques. *Front. Genet.*, *9*, 515, 2018.
14. Jiby, T.C., A Study on Various Machine Learning Classification Algorithms for Diabetes Prediction. *Int. J. Eng. Res. Technol.*, *10*, 8, 425–427, 2021.
15. Lai, H., Huang, H., Keshavjee, K., Guergachi, A., Gao, X., Predictive models for diabetes mellitus using machine learning techniques. *BMC Endocr. Disord.*, *19*, 1–9, 2019.
16. Bose, S.S.S. and Kumar, C.S., Combining the Multiple Features for Improving the Performance of Multi-Parameter Patient Monitor, in: *2019 5th International Conference on Advanced Computing & Communication Systems (ICACCS)*, pp. 647–651, IEEE, 2019.
17. kumar Abinav, A., Karthika, N., Pratibha, R., Gandhiraj, R., Soman, K.P., SVM based classification of digitally modulated signals for software defined radio, in: *Coimbatore Institute of Technology*, 2010.
18. Kavitha, K.R., Gopinath, A., Gopi, M., Applying improved svm classifier for leukemia cancer classification using FCBF, in: *2017 International Conference on Advances in Computing, Communications and Informatics (ICACCI)*, pp. 61–66, IEEE, 2017.
19. Thambi, S.V., Sreekumar, K.T., Kumar, C.S., Raj, P.R., Random forest algorithm for improving the performance of speech/non-speech detection, in: *2014 First International Conference on Computational Systems and Communications (ICCSC)*, pp. 28–32, IEEE, 2014.
20. Karthika, R. and Parameswaran, L., Study of Gabor wavelet for face recognition invariant to pose and orientation, in: *Proceedings of the International Conference on Soft Computing Systems: ICSCS 2015, Volume 1*, pp. 501–509, Springer India, 2016.
21. Learning, U.M., *Pima Indians diabetes database*, USA, 2016, kaggle.com/uciml/pima-indians-diabetes-database.
22. Dutta, P., Paul, S., Kumar, A., Comparative analysis of various supervised machine learning techniques for diagnosis of COVID-19, in: *Electronic devices, circuits, and systems for biomedical applications*, pp. 521–540, Academic Press, 2021.
23. Dutta, P., Paul, S., Majumder, M., *An efficient SMOTE based machine learning classification for prediction & detection of PCOS*, Research Square, USA, 2021.
24. Dutta, P., Paul, S., Obaid, A.J., Pal, S., Mukhopadhyay, K., Feature selection based artificial intelligence techniques for the prediction of COVID like diseases, *J. Phys.: Conf. Ser.*, 1963, 012167, IOP Publishing, 2021.
25. Dutta, P., Paul, S., Cengiz, K., Anand, R., Kumar, A., A predictive method for emotional sentiment analysis by deep learning from EEG of brainwave dataset, in: *Artificial Intelligence for Neurological Disorders*, pp. 25–48, Academic Press, 2023.
26. Dutta, P., Paul, S., Shaw, N., Sen, S., Majumder, M., Heart disease prediction: A comparative study based on a machine-learning approach, in: *Artificial Intelligence and Cybersecurity*, pp. 1–18, CRC Press, 2022.

27. Dutta, P., Paul, S., Cengiz, K., Anand, R., Majumder, M., A predictive method for emotional sentiment analysis by machine learning from electroencephalography of brainwave data, in: *Implementation of Smart Healthcare Systems using AI, IoT, and Blockchain*, pp. 109–130, Academic Press, 2023.
28. Dutta, P., Shaw, N., Das, K., Ghosh, L., Early & accurate forecasting of mid term wind energy based on PCA empowered supervised regression model. *Comput. Intell. Mach. Learn.*, *2*, 1, 53–64, 2021.
29. Sen, S., Saha, S., Chaki, S., Saha, P., Dutta, P., Analysis of PCA based AdaBoost machine learning model for predict mid-term weather forecasting. *Comput. Intell. Mach. Learn.*, *2*, 2, 41–52, 2021.
30. Dutta, P., Paul, S., Majumder, M., *Intelligent SMOTE Based Machine learning Classification for Fetal State on Cardiotocography Dataset*, Research Square, USA, 2021.
31. Dutta, P., Paul, S., Jana, G.G., Sadhu, A., Hybrid Genetic Algorithm Random Forest algorithm (HGARF) for improving the missing value Imputation in Hepatitis Medical Dataset, in: *2023 International Symposium on Devices, Circuits and Systems (ISDCS)*, vol. 1, pp. 01–05, IEEE, 2023.

12

A Robust IoT-Based Approach to Enhance Cybersecurity and Patient Trust in the Smart Health Care System: Zero-Trust Model

Raghunath Maji, Biswajit Gayen* and Sandeepan Saha

Greater Kolkata College of Engineering and Management, JIS Group of College, Baruipur, Phultala, West Bengal, India

Abstract

The increasing adoption of Smart Healthcare and Internet of Things (IoT) devices has revolutionized the healthcare industries, offering enhanced patient monitoring, diagnosis, and treatment. However, the integration of these devices into the healthcare ecosystem raises serious concerns regarding the security and privacy of sensitive medical records. Cyber-attacks and data breaches on these devices pose significant threats to patient confidentiality and well-being. Therefore, there is a critical need for a robust and automated system that ensures the continuous detection and updating of Smart Health Care IoT devices to safeguard user medical data. If we judge all the old attacks then we will see that most of the cyber-attacks are due to human ignorance so if we use software instead of humans to see all these actions then any system will be more secure. In this paper, we aim to address the challenges associated with securing IoT devices within the healthcare domain by installing the latest security patches and firmware. The key objectives of the system are twofold: to automatically detect potential vulnerabilities in Smart Healthcare IoT devices and to efficiently update the devices with the latest security patches and firmware. The proposed system can significantly contribute to strengthening the overall cyber security framework within the healthcare industry, ensuring patients' trust in Smart Health Care IoT devices, and fostering the continued growth of remote patient monitoring.

**Corresponding author*: bgayenchemistry@gmail.com

Pijush Dutta, Sudip Mandal, Korhan Cengiz, Arindam Sadhu and Gour Gopal Jana (eds.) Design and Forecasting Models for Disease Management (269–282) © 2025 Scrivener Publishing LLC

Keywords: IoT, medical data, data security, smart healthcare, four-dimension model

12.1 Introduction

Now, we are enjoying 5G technology all over the world, this technology makes us faster, smarter, and more intellectual in our society. The health industry also utilized this advanced technology for their smarter development with successful implementation [1]. The adopted technology has a strong impact not only on the medical industry but also its allied industries. It has the responsibility for dealing with huge medical imaging data processing and their transportation, high-definition image and video qualities for better accuracy, and faster decision-makers for expert doctors/medical associates. It has also proved its explicabilities in the field of real-time remote sensing data *via* smart IoT devices [2, 3]. In this aspect, there is no doubt about the successful implementation of smart health care. However, worldwide medical industries have been associated with some serious issues like de-centralized resources, heterogeneous medical equipment, varieties of weathering conditions, updated and modern medical information, and lastly the most important one, a massive number of patients [4, 5]. Now, synchronization of all the above parameters is the biggest challenge for the smart health care and medical industry as this sector deals with the life and death of people. Any minor error in smart healthcare devices, networks, and their implementation may create massive tragedy in both National and International scenarios.

In the new era of smart healthcare, telemedicine, and mobile healthcare are very popular with every person. For sustainable development in this field, different cutting-edge technologies like ultra-high-speed networking systems (5G), IoT, smart sensors, fog computing, and artificial intelligence (AI) have been used to achieve excellent outcomes [6].

Considering the critical need for advanced healthcare solutions, network developers opted for the deployment of the 3rd Generation Partnership Project (3GPP) R16 version of 5G in the year 2020. These R16 or R17 versions of 5G can transmit data at the speed of around 10 Gbps which will help a lot for a patient in critical condition in critical care [7]. According to the Globe Newswire, the smart healthcare sector is anticipated to witness substantial growth in the coming years, with projections indicating a reach of 5.41 trillion US dollars by the year 2032. They also anticipated a compound annual growth rate (CAGR) of 15.7% will be maintained between 2023 and 2032 [www.globenewswire.com]. In the year of 2018, the Hospital of Zhengzhou University first performed ultrasound operations and real-time diagnosis by using smart healthcare supports from remote locations. During the COVID-19 period, many reports have been published on the successful application of smart health care [8]. At present, people can access super-specialty healthcare services sitting in a chair from their balcony. While the focus has been primarily on highlighting the positive aspects of the smart healthcare system, it's important to acknowledge that despite its substantial achievements thus far, there are notable drawbacks in terms of security and privacy considerations. In contrast, traditional healthcare services tend to be more dependable and secure in these aspects [9]. In an online service mode, hospitals find themselves navigating a complex landscape of variables, including diverse users, devices, and a vast data bank. To maintain meticulous oversight over the multitude of inputs and outputs stemming from these variables, and to ensure a secure healthcare experience for consumers, the integration of an airtight security and privacy framework within the smart healthcare system becomes extremely important. Potential security vulnerabilities may greatly endanger both the quality of medical services and the smooth functioning of medical institutions that have adopted the smart healthcare paradigm [10].

As advancements in hospital services aim to elevate patient experiences, these application scenarios concurrently introduce novel security challenges. Given the sensitive nature of medical data and the potential impact on patient safety within certain medical application scenarios, the security facet of 5G medical healthcare emerges as a pivotal factor encompassing national security, network integrity, and information protection [11, 12].

In this book chapter, our objective is to explore a range of significant threats that arise in the application of smart healthcare systems in our modern age. Despite the limitations of our scope, we will endeavor to provide insights into these challenges and offer potential mitigation strategies.

12.2 Security Threats on Smart Healthcare

Smart healthcare services encounter the following significant security threats shown in Figure 12.1.

12.2.1 Medical Data Monitoring and Patient Privacy Information

Smart healthcare is intricately reliant on electronic data directing medical records, medical images, and laboratory information. The manipulation of

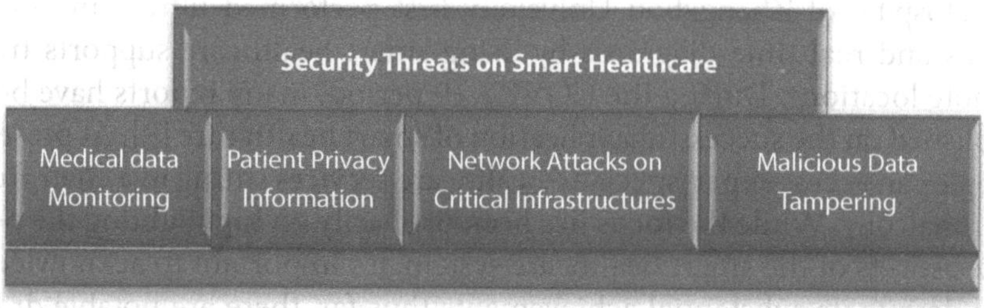

Figure 12.1 Different types of security threats on smart healthcare.

such critical data by malicious hackers stands as a significant peril, posing a grave threat not only to the security of medical information but also to the sanctity of patient confidentiality [12, 13].

12.2.2 Network Attacks on Critical Infrastructures

A secure and ultra-fast network speed stands as a vital prerequisite in the realm of intelligent healthcare, particularly for a multitude of critical medical emergency applications. These include emergency rescues, remote surgeries, and urgent patient diagnoses. In such a pivotal scenario, the occurrence of a network infrastructure breach leading to incapacitation could result in significant repercussions for the well-being of patients [14].

12.2.3 Malicious Data Tampering

The smart healthcare system, along with its interconnected network, possesses the inherent capacity to efficiently assemble and retrieve expansive volumes of medical and health-related information. The collated data and ensuing reports derived from these data play a pivotal role in addressing public health occurrences, including the emergence of epidemics and unknown diseases consistently. Regrettably, the deliberate tampering of these assembled data holds the potential to corrupt vital medical records, thereby undermining surveillance and impeding emergency response mechanisms. The consequences could be dire, potentially catalyzing widespread public health crises [15].

12.3 Smart Healthcare Security and Four-Dimension Model

Despite of serious insecurity, it is possible to provide a true and practical smart healthcare system after securing four dimensions. In the four-dimension model, one can ensure secure smart healthcare by stepwise smart verifications. In this four-dimension model, the four dimensions are considered as a) Subject, b) Object, c) Environment, and d) Behavior [16].

12.3.1 Subject

In a secure smart healthcare system, the subject matter holds utmost importance. This directs identity-based medical practitioners, identity-based medical devices, and the identification of medical records or applications. Ensuring the security of these elements at every stage is pivotal in providing superior healthcare services to individuals. In the proposed model, an increased level of trustworthiness is attributed to specific subjects, particularly Doctors, who possess exclusive authorization to access medical records. This access is granted through verified logins, whereas less trustable subjects, such as IoT-based medical devices, are more limited in their authority [17]. The medical devices themselves do not possess decision-making capabilities within this smart healthcare system. Instead, they act as a medium for data, which is uploaded to a relay system or cloud platform. This data transmission and collection process is carried out under the direction and oversight of authenticated doctors (Figure 12.2).

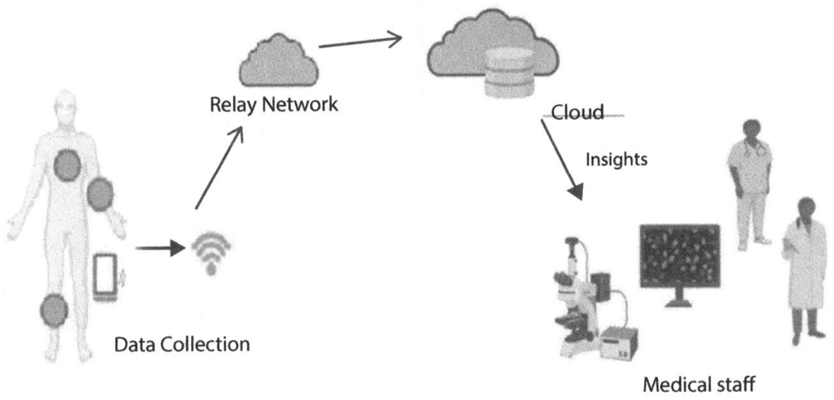

Figure 12.2 A typical IoT-based model healthcare system.

12.3.2 Object

In a smart healthcare system, the concept of objects assumes the second highest importance in ensuring a secure and efficient healthcare environment. Within the framework of the proposed model, these objects consist of a variety of critical elements, including medical data, medical services, and service interfaces (Figure 12.3). Within this ecosystem, these objects are subject to dynamic processes, including uploading, downloading, exchange, and utilization, all controlled by subject matter experts. Primarily, these experts include doctors and medical staff, who play a central role in the operational aspects of these objects [18]. It's crucial to emphasize that the operations involving these objects are subject to severe restrictions, particularly when it comes to subjects such as Doctors and medical staff. This careful regulation and correlation between subjects and objects serve as the solid system's security and safeguard patient privacy during medical services. This includes the handling of sensitive data, the execution of critical service functions, and the interaction with interfaces, all of which can only be accessed by trusted and relevant individuals.

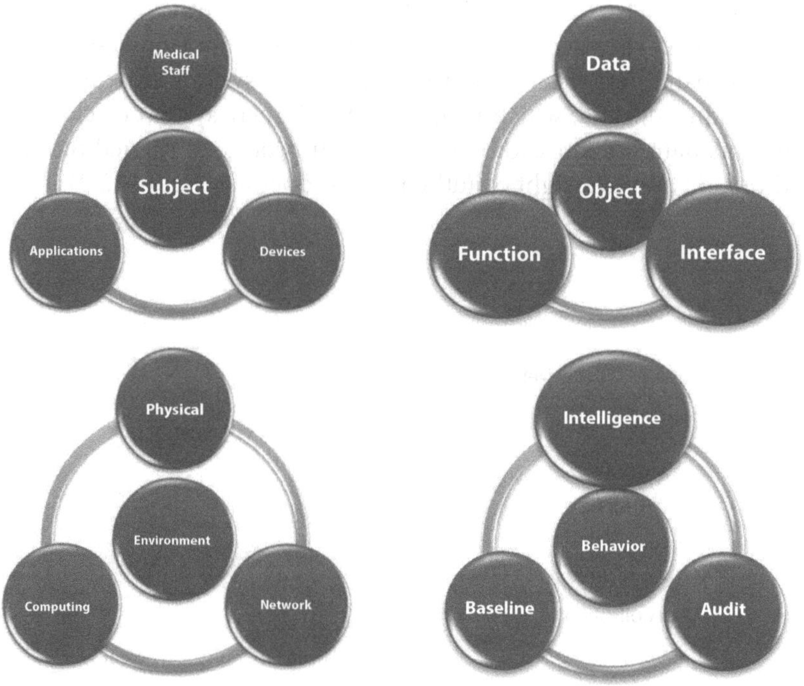

Figure 12.3 Four-dimension smart healthcare system and its different parameters.

12.3.3 Environment

Another crucial aspect of a smart healthcare system is the environment, it includes physical, computing, and network systems. Medical and network pieces of equipment operate within this sophisticated healthcare ecosystem. The security of accessing sensitive medical records is considerably higher within a private treatment area compared to a public setting [19]. The robust infrastructure of smart healthcare environments ensures the confidentiality and integrity of patient data, making it a fundamental pillar of modern healthcare systems (Figure 12.3).

12.3.4 Behavior

The final component of a smart healthcare system is behavior; it includes intelligent inference, behavior baseline, and security audits, among others. Regular updates to the behaviors of the smart healthcare system ensure the trustworthiness of subjects, objects, and environments involved. These real-time adjustments can have a strong impact on the security and privacy of the offering of smart medical services, underscoring the critical importance of maintaining and adapting the system's behavior over time.

12.3.5 Risk Assessment and Security Checking

Risk assessment primarily revolves around two core dimensions: the subject and the environment. It is when we evaluate the trustworthiness of the subject and environmental awareness that the risk assessment mechanism comes into play. Notably, the risks associated with both subjects and environments serve as crucial reference points within the trust assessment process, particularly in ensuring the security of the smart healthcare system.

The risk associated with the implementation of a smart healthcare system is a critical concern, primarily centered on the subject and the environment in which smart medical applications operate. In this context, the controllability of the system plays a pivotal role, and the extent of control hinges on several key factors.

- **Trusty Identification of Networked Medical Equipment:** Ensuring that all networked medical equipment is reliably identified is essential. This involves robust mechanisms for uniquely identifying and authenticating devices to prevent unauthorized access or tampering.

- **Registered Medical Equipment with the System:** Only registered medical equipment should be allowed to interact with the smart healthcare system. This registration process should include stringent validation and verification procedures to guarantee the legitimacy of the equipment.
- **Guaranteed by Trusted Hardware:** The use of trusted hardware is crucial for safeguarding the integrity and security of the smart healthcare system. Trusted hardware components, such as hardware security modules (HSMs), can provide a secure foundation for critical operations.
- **Identity Verification of Medical Personnel:** To maintain control over the system, it's imperative to authenticate and verify the identity of medical personnel accessing or interacting with the system. This ensures that only authorized individuals can make changes or access sensitive information.
- **Authentication for Remote or Local Sources Access:** Whether accessed remotely or locally, the system should enforce strong authentication mechanisms. This prevents unauthorized access from any source, be it within the healthcare facility or from external locations.
- **Assurance of No Interference during Operation:** Ensuring the uninterrupted and interference-free operation of the smart healthcare system is essential for patient safety. Robust safeguards and redundancy measures should be in place to prevent disruptions or tampering.
- **Historical Information/Records for Dynamic Controllability:** Maintaining historical information and records is crucial for achieving dynamic controllability. Analyzing past data can help optimize system performance, identify issues, and offer better healthcare services to patients. It also aids in compliance with regulatory requirements.

The risks associated with smart healthcare systems demand thorough attention to controllability factors, from the identification of networked devices to the verification of personnel and the maintenance of historical data. Only by addressing these aspects comprehensively can we build a secure and reliable smart healthcare ecosystem that prioritizes patient well-being and data security. Concerning the subject of a smart healthcare system, the environmental factors surrounding its operation are equally critical for ensuring its effectiveness and security. These environmental considerations direct a wide range of issues, all aimed at optimizing the functionality and reliability of the system:

- **Operating System Safety and Security:** The foundational layer of any smart healthcare system, the operating system, must be meticulously chosen and configured to be safe and secure. This involves employing security patches, encryption, and access controls to safeguard against potential vulnerabilities.
- **Assessing User Satisfaction with the Operating System:** Continuously evaluating user satisfaction with the operating system is vital. Feedback from healthcare professionals and administrators can facilitate to identification the areas for improvement and enhance the overall user experience.
- **Monitoring for Malware Attacks:** Constant vigilance is required to detect any history of malware attacks or security breaches on the system. This involves deploying intrusion detection systems and antivirus software to prevent and respond to potential threats.
- **Temperature Control for Core Devices and Peripherals:** Maintaining optimal operating temperatures for core medical devices and additional peripherals is crucial to prevent overheating or performance degradation. Effective temperature control mechanisms must be in place.
- **Network Considerations:** Understanding the network environment is essential. This includes knowledge of network location, link type, and capability. Network reliability and bandwidth are fundamental factors in ensuring the seamless operation of the smart healthcare system.
- **Geographical Location and Safe Enclosed Spaces:** In scenarios like ambulance-based healthcare systems, considering the geographical location and the safety of the enclosed space is vital. Ensuring that equipment is secure and functional even in challenging environments is critical for patient care.
- **Special Period Considerations:** The system's performance may vary during special periods, such as public holidays or working hours. Adjusting system operations and resource allocation to meet changing demands during these times is important to ensure uninterrupted service.

By addressing these environmental issues comprehensively, a smart healthcare system can operate optimally, remain resilient to security threats, adapt to varying conditions, and provide reliable healthcare services. This holistic approach to system management ensures that both the subject and the environment are considered and optimized for the benefit of patients and healthcare providers.

A secure smart healthcare system operates under the fundamental principle of "*never trust and always verify*" [16]. This foundational concept emphasizes the importance of rigorous verification and authentication processes to ensure the integrity, confidentiality, and reliability of the system. In the proposed model of a secure healthcare system, several key components and practices are integral to upholding this principle and guaranteeing the safety and efficacy of the system.

- **Data Access from Trusty External Platforms:** To maintain the highest standards of data security and integrity, the system adheres to a stringent approach when accessing data from external platforms. It relies on trusty and authenticated connections to external data sources, be it electronic health record systems, laboratory databases, or other relevant healthcare platforms. This approach minimizes the risk of unauthorized data access and tampering.
- **Accessing Services from User Terminals:** User terminals act as the primary interface for healthcare professionals and authorized personnel to interact with the system. These terminals are equipped with robust security measures, including user authentication, encryption, and access controls. Users must verify their identities before gaining access, ensuring that only authorized personnel can use the system.
- **IoT Device Integration at the Edge:** The system's architecture incorporates Internet of Things (IoT) devices at the edge to monitor and collect vital patient data. These IoT devices are not just connected; they are securely integrated, with data transmissions subject to encryption and rigorous authentication processes. This ensures that data collected from IoT devices is trustworthy.
- **Secured Terminal Access for Safe Operation and Maintenance:** Operation and maintenance of the healthcare system are critical components of its continuous functionality. Secured terminal access protocols are in place to guarantee that only authorized personnel can perform

system maintenance or configuration changes. Multi-factor authentication and audit logs further enhance the system's accountability.
- **Continuous Monitoring and Verification:** The "*never trust and always verify*" principle is not a one-time effort but an ongoing commitment. The system continually monitors user activities, data transactions, and device interactions, verifying the legitimacy of every action. Suspicious or unauthorized activities trigger immediate alerts and responses to maintain the system's security posture.

By adhering to these principles and practices, the proposed model of a secure healthcare system establishes a healthy foundation for trustworthiness, safeguarding patient data, and providing healthcare professionals with a reliable and secure platform for delivering care. This approach not only aligns with the "*never trust and always verify*" principle but also sets a high standard for security in the context of smart healthcare systems.

12.4 Conclusion and Future Prospects

The importance of security awareness and protection systems in modern smart healthcare is steadily on the rise. The convergence of smart healthcare technologies with the associated security challenges is reshaping the landscape of healthcare, accompanied by a new era that calls for an urgent approach to safeguarding sensitive data and critical systems. Within this context, a four-dimensional model, which includes subjects, objects, environments, and behaviors, has surfaced as a promising framework to effectively address the multifaceted security concerns that lie ahead in future healthcare systems. The proposed architectural framework, which seamlessly integrates security awareness and protection systems, has unequivocally proven its effectiveness in implementing healthy security measures. By harnessing the power of network virtualization, promoting data collaboration, implementing precise IoT access controls, and employing integrated network securities, this model has been meticulously accepted in smart healthcare infrastructures. A rigorous process of testing and evaluation has unequivocally validated the functionality and performance of this pioneering approach.

Despite the notable achievements so far, it is essential to acknowledge the persistent challenges that security awareness and protection systems face within the domain of smart healthcare. Technical implementation remains an alarming obstacle, often demanding substantial expertise and resources. Moreover, there is an urgent need to heighten public awareness regarding the critical role of security in healthcare, with individuals and organizations actively engaged in safeguarding their health data.

Looking ahead, one of the most pressing requirements is the maturation of security solutions tailored explicitly to the unique demands of the smart medical industry. Ensuring compatibility with existing security frameworks and standards is yet another important concern that needs immediate attention. As the landscape of smart healthcare continues to evolve, so must the strategies and technologies employed to safeguard its integrity and confidentiality.

The journey toward a more secure and resilient smart healthcare ecosystem is an ongoing process of collaborative efforts from researchers, healthcare practitioners, policymakers, and service holders. A successful terrain of security challenges is imperative to promote a safer future for healthcare innovation in the healthcare industry.

Acknowledgment

The authors express their thanks to the Greater Kolkata College of Engineering and Management for providing scope of work in their Institution. The authors thank Prof. Joydeep Chowdhury from Jadavpur University and Prof. Avijit Banerji from Calcutta University for their valuable suggestions and constant support.

References

1. Ahmad, I., Shahabuddin, S., Kumar, T., Okwuibe, J., Gurtov, A., Ylianttila, M., Security for 5G and beyond. *IEEE Commun. Surv. Tuts.*, 21, 4, 3682–3722, 2019.
2. Akhunzada, A., Islam, S.U., Zeadally, S., Securing cyberspace of future smart cities with 5G technologies. *IEEE Netw.*, 34, 4, 336–342, 2020.
3. Azeez, N.A. and der Vyver, C.V., Security and privacy issues in e-health cloud-based system: A comprehensive content analysis. *Egypt. Informat. J.*, 20, 2, 97–108, 2019.
4. Banchs, A., Gutierrez-Estevez, D.M., Fuentes M. Boldi, M., Provvedi, S., A 5G mobile network architecture to support vertical industries. *IEEE Commun. Mag.*, 57, 12, 38–44, 2019.

5. Buzzi, S., Chih-Lin, I., Klein, T.E., Poor, H.V., Yang, C., Zappone, A., A survey of energy-efficient techniques for 5G networks and challenges ahead. *IEEE J. Sel. Areas Commun.*, 34, 4, 697–709, 2016.
6. Chettri, L. and Bera, R., A comprehensive survey on internet of things (IoT) toward 5G wireless systems. *IEEE Internet Things J.*, 7, 1, 16–32, 2020.
7. Chen, B., Qiao, S., Zhao, J., Liu, D., Shi, X., Lyu, M., Chen, H., Lu, H., A Security Awareness and Protection System for 5G Smart Healthcare Based on Zero-Trust Architecture. *IEEE Internet Things J.*, 8, 13, 10248–10263, 2021.
8. Chen, M., Yang, J., Zhou, J., Hao, Y., Zhang, J., Youn, C.-H., 5G smart diabetes: Toward personalized diabetes diagnosis with healthcare big data clouds. *IEEE Commun. Mag.*, 56, 4, 16–23, 2018.
9. Gupta, R., Tanwar, S., Tyagi, S., Kumar, N., Tactile-Internet-based telesurgery system for healthcare 4.0: An architecture, research challenges, and future directions. *IEEE Netw.*, 33, 6, 22–29, 2019.
10. https://www.globenewswire.com/en/newsrelease/2023/05/01/2658278/0/en/SmartHealthcare-Market-Set-to-Witness-Unprecedented-Growth-of-USD-541-Billion-by-2032-Market-us-Report.html, 2023.
11. Khan, R., Kumar, P., Jayakody, D.N.K., Liyanage, M., A survey on security and privacy of 5G technologies: Potential solutions, recent advancements, and future directions. *IEEE Commun. Surv. Tuts.*, 22, 1, 196–248, 2020.
12. Li, M., Yu, S., Zheng, Y., Ren, K., Lou, W., Scalable and secure sharing of personal health records in cloud computing using attributebased encryption. *IEEE Trans. Parallel Distrib. Syst.*, 24, 1, 131–143, 2023.
13. Loghin, D., *et al.*, The disruptions of 5G on data-driven technologies and applications. *IEEE Trans. Knowl. Data Eng.*, 32, 6, 179–1198, 2020.
14. Pussewalage, H.S.G. and Oleshchuk, V.A., Privacy preserving mechanisms for enforcing security and privacy requirements in e-health solutions. *Int. J. Inf. Manag.*, 36, 6, 1161–1173, 2016.
15. Rupprecht, D., Dabrowski, A., Holz, T., Weippl, E., Pöpper, C., On security research towards future mobile network generations. *IEEE Commun. Surv. Tuts.*, 20, 3, 2518–2542, 2018.
16. Samaniego, M. and Deters, R., Zero-trust hierarchical management in IoT. *IEEE Int. Congr. Internet Things.*, 2018, 88–95, 2018.
17. Simplicio, M.A., Iwaya, L.H., Barros, B.M., Carvalho, T.C.M.B., Näslund, M., SecourHealth: A delay-tolerant security framework for mobile health data collection. *IEEE J. Biomed. Health Informat.*, 19, 2, 761–772, 2015.
18. Vergutz, A., Noubir, G., Nogueira, M., Reliability for smart healthcare: A network slicing perspective. *IEEE Netw.*, 34, 4, 91–97, 2020.
19. Wang, X.A., Ma, J., Xhafa, F., Zhang, M., Luo, X., Cost-effective secure e-health cloud system using identity based cryptographic techniques. *Future Gener. Comput. Syst.*, 67, 242–254, 2017.

13

Safeguarding Digital Health: A Novel Approach to Malicious Device Detection in Smart Healthcare

Raghunath Maji* **and Biswajit Gayen**

Greater Kolkata College of Engineering and Management, JIS Group of College, Baruipur, Phultala, West Bengal, India

Abstract

The proliferation of smart healthcare systems has introduced a new era of patient care, leveraging interconnected medical devices to improve monitoring and treatment. However, this enhanced connectivity also exposes healthcare environments to potential cyber security threats, including malicious devices. Such devices can compromise patient safety, data integrity, and system reliability. To tackle this pressing concern, we present a pioneering approach aimed at detecting and identifying malicious devices within a smart healthcare system by analyzing their behavioral activities. This innovative methodology strives to fortify the system's security measures, ensuring a proactive stance against potential threats and bolstering the overall integrity of the healthcare infrastructure. Our proposed system employs advanced machine learning algorithms, anomaly detection techniques, and behavioral profiling to continuously monitor and analyze the behavior of interconnected devices. By establishing a baseline of normal device behavior, our system can identify deviations indicative of potential malicious intent. Key behavioral parameters, such as data transmission patterns, access frequency, and communication protocols, are scrutinized in real-time to detect suspicious activities.

The main aim of our study offer healthcare providers and system administrators a powerful tool to secure their smart healthcare ecosystems against potential threats. By swiftly detecting and isolating malicious devices, we can safeguard patient safety and data confidentiality, instilling confidence among users and enhancing overall trust in smart healthcare technologies.

**Corresponding author*: writetoraghunath@gmail.com

Pijush Dutta, Sudip Mandal, Korhan Cengiz, Arindam Sadhu and Gour Gopal Jana (eds.) Design and Forecasting Models for Disease Management (283–296) © 2025 Scrivener Publishing LLC

***Keywords*:** Smart healthcare, IoT, machine learning, anomaly detection

13.1 Introduction

In the rapidly evolving landscape of healthcare, the integration of smart healthcare systems has ushered in a transformative era marked by unprecedented connectivity and innovation. These systems leverage interconnected medical devices to revolutionize patient care through enhanced monitoring and treatment strategies. This connectivity, while undeniably advantageous, also introduces a formidable challenge – the heightened exposure of healthcare environments to cybersecurity threats. Among these threats, the emergence of malicious devices poses a particularly menacing risk, capable of jeopardizing patient safety, compromising data integrity, and undermining the reliability of the entire system.

Addressing this critical concern requires innovative approaches that not only harness the potential of cutting-edge technologies but also adapt them to the unique demands of the healthcare setting. In response, this study proposes a novel methodology designed to detect and identify malicious devices within smart healthcare systems by closely analyzing their behavioral activities. By leveraging advanced machine learning algorithms, anomaly detection techniques, and behavioral profiling, this approach aims to create an effective surveillance system capable of continuously monitoring and evaluating the conduct of interconnected devices.

In recent years, substantial progress has been made in the realm of automating patient monitoring, coupled with the facilitation of remote health assessments by medical professionals [1–5]. To ensure the confidentiality of patient information, certain approaches have turned to fog computing, strategically offloading computations to the edge to prevent the transmission of sensitive data to the cloud [6]. Despite these advancements, the escalating concern revolves around the vulnerability of IoT-based healthcare systems to cyber-attacks [7]. Startlingly, statistics reveal that over 90% of healthcare systems have fallen victim to these malicious incursions. This growing threat was underscored in 2017 when White House intelligence disclosed the increasing susceptibility of the healthcare sector to cybercriminal activities [8]. Exacerbating this challenge, the onset of the COVID-19 pandemic marked the advent of an era characterized by an elevated frequency of cyber-attacks, opportunely exploiting the global turbulence [9]. The interconnected and data-rich nature of these systems offers both opportunities and challenges, as the potential benefits of streamlined healthcare delivery are matched by the critical imperative to safeguard patient data and system integrity. In light of this context, comprehensive understandings of the existing vulnerabilities and a proactive approach to mitigating potential threats have become essential.

While conventional Intrusion Detection System (IDS) technology has a well-established presence, its adequacy diminishes significantly in the context of IoT devices and networks [10]. The intrinsic limitations of IoT devices, including constrained processing power and storage capacity, render traditional security solutions insufficient in effectively identifying and countering malicious attacks within the IoT environment [11]. The intricate nature of IoT, characterized by features such as scalability, constrained processing capabilities, limited memory, and power resources, underscores the imperative need for dedicated IoT-oriented IDS solutions.

Amidst these formidable challenges, the imperative quest for the development of Intrusion Detection Systems (IDS) fortified by Internet of Things (IoT) support comes to the forefront. The efficacy of training and testing such systems critically hinges on the availability of a resilient IoT traffic dataset encompassing a comprehensive spectrum of both normal and malicious IoT traffic patterns. Unfortunately, a conspicuous scarcity exists, with only a limited cohort of researchers dedicating efforts to craft bespoke IoT datasets tailored for exhaustive testing and evaluation of IoT-supported IDS.

Predominantly, the evaluation of an application's behavior revolves around two key methodologies: static analysis and dynamic analysis. Static analysis entails scrutinizing the application's characteristics when it is not actively executing, achieved through manual examination by experts or via machine learning algorithms. This approach has led to the formulation of various static analysis techniques, as exemplified by works such as [12–14].

Conversely, dynamic analysis hinges on the appraisal of an application's conduct during its runtime, allowing for an assessment of its behavior while it is actively running. This dynamic approach provides valuable insights into real-time interactions and potential deviations from expected patterns. In the Android context, several dynamic analysis methods have garnered recognition, including notable implementations like [15–17].

In response to the growing concerns and limitations within the realm of Smart Healthcare Systems (SHS), this paper introduces an innovative security framework designed to identify and counteract the presence of malicious devices within the SHS ecosystem. The central premise of our framework is grounded in the recognition that any alterations in a device's functionalities that deviate from the stipulated privacy policy can signify potential threats.

Our framework is meticulously engineered to perpetually monitor the behavioral patterns exhibited by all Internet of Things (IoT) devices operating within the Smart Healthcare System. By closely scrutinizing these behavior patterns, our system is primed to detect any disparities or inconsistencies. Should any deviations be detected, the framework promptly triggers an alert mechanism, which subsequently dispatches an alert message to the relevant authoritative entity.

The major contributions of this paper are summarized as follows:

1. We introduce a novel model designed to construct behavioral templates for IoT devices employed within smart healthcare systems. These templates are generated based on the devices' descriptions and privacy policies.
2. We propose an innovative model capable of actively monitoring and tracking the behavior of individual IoT devices within a smart healthcare environment. This monitoring system continuously observes the activities and interactions of these devices, facilitating the creation of real-time behavioral patterns that depict their ongoing functions and operations.
3. Finally, our model identifying the malicious IoT devices within the smart healthcare system by comparing the established behavioral templates with the observed behavioral patterns.

13.2 Related Work

Rughoobur *et al.* [18] introduced a sophisticated framework meticulously crafted to mitigate the pervasive threat of replay attacks specifically targeting battery-dependent IoT healthcare devices. Their innovative solution revolves around a multi-faceted approach that centers on the analysis of distinctive device identifiers, precise timestamps, and patterns in battery depletion behavior. By amalgamating these factors, the framework not only identifies instances of replay attacks but also implements effective mitigation strategies.

In a distinct yet equally impactful endeavor, Rathore *et al.* [11] devised an approach for combating cyber-attacks within fog-based IoT environments. Leveraging the capabilities of Extreme Learning Machines (ELM), their method exhibits a semi-supervised character, bolstered by the integration of the Fuzzy C-Means technique. Through the process of ELM, the system achieves swift and accurate identification of cyber threats, while the incorporation of Fuzzy C-Means contributes to resolving intricate dataset labeling challenges.

Similarly, Carta *et al.* [19] introduced a pioneering feature engineering technique tailored to enhance the efficiency of conventional Intrusion Detection Systems (IDSs). This approach revolves around the proactive detection of anomalies, thereby elevating the overall performance of the system. By infusing intelligence into the feature engineering process, the proposed methodology demonstrates the potential to significantly improve the accuracy of anomaly detection.

Likewise, Alrashdi *et al.* [20] put forth a robust framework with the explicit goal of detecting malicious activities within fog-based IoT healthcare systems. This framework stands as a testament to their dedication to enhancing security in such critical domains. Through a combination of advanced techniques and strategies, the proposed framework exhibits the capacity to identify and thwart malicious actions effectively, ensuring the integrity and reliability of IoT healthcare environments.

In the study documented in reference [21], a diverse array of computational intelligence techniques was introduced with the explicit aim of addressing the intricate challenge of predicting coronary artery heart disease. This comprehensive exploration not only highlights the complexity of the problem at hand but also underscores the innovative approach taken to harness various computational intelligence methods for enhancing the accuracy and efficacy of coronary artery heart disease prediction models. Notably, the study [22] unveiled the prowess of three distinctive machine learning models: Decision Tree, Random Forest (RF), and XGBoost. These models were adeptly harnessed to forecast cervical cancer probabilities based on behavioral patterns and associated variables. In a similar vein, a comprehensive investigation [23] undertook a comparison of five supervised machine learning techniques: Support Vector Machine (SVM), K-nearest neighbors, Random Forests, Artificial Neural Networks (ANNs), and Logistic Regression. The purpose was to discern their relative effectiveness in addressing classification challenges.

Venturing into a distinct realm, reference [24] delved into the realm of prognostic medicine, employing machine learning algorithms to forecast the 1-year survival rates of patients. This innovative exploration not only anticipates survival outcomes but also delves into the nuanced significance of various features within the predictive process. This dual-pronged approach not only contributes to prognostic accuracy but also aids in extracting insights from the data. Moreover, the development of an expert scheme [25] took center stage in the classification of liver disorders. This scheme, ingeniously employing a combination of Random Forests and Artificial Neural Networks (ANNs), stands as a testament to innovative thinking in the medical domain, showcasing the potential of computational intelligence in healthcare diagnostics.

In [26], a groundbreaking advancement came to fruition with the creation of a centralized, inconspicuous IoT-based system tailored for universal fall detection and rapid response. With its device-type invariant design, this system exhibited exceptional utility in real-time monitoring across a substantial population. Complementing this, a wearable electronic device was ingeniously crafted [27], offering indispensable support to the visually impaired by enabling independent travel. This device, while fostering autonomy, concurrently tracked the real-time location data of these individuals, further enhancing their safety.

In the realm of healthcare data security within IoT-enabled healthcare, a significant stride was taken through the introduction of a novel hybrid encryption scheme [28]. Although this scheme held the promise of bolstering data protection, it was noted that a certain level of computational overhead persisted, warranting further optimization efforts. Meanwhile, the domain of IoT-enabled healthcare systems witnessed robust development in [29] with the formulation of a secure framework catering to SDN-based Edge computing. This framework, intricately designed, stands as a testament to heightened security protocols, ensuring the integrity and confidentiality of data flowing through the IoT ecosystem.

13.3 Our Proposed Framework

Module that creates the template based on actual behavioural activity.

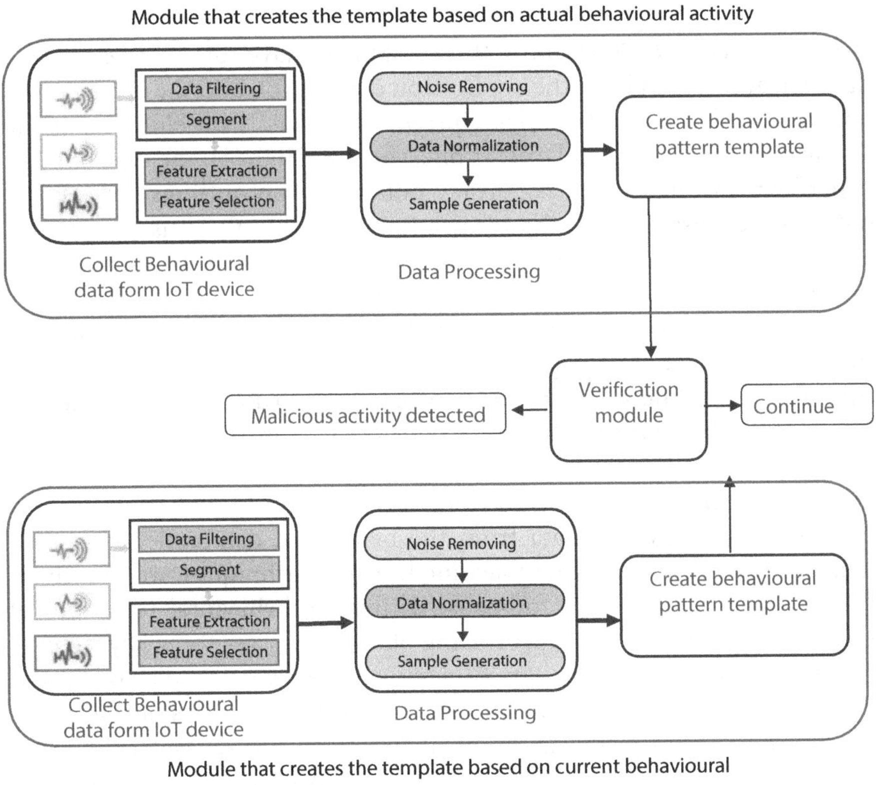

Figure 13.1 Proposed model.

Figure 13.1 represents system architecture of our proposed research.

13.4 Overview of Our Proposed Framework

In this section, we demonstrate the overall architecture of the healthcare framework for the development of digital health safeguarding in smart healthcare and explain our framework methodology. Our proposed framework operates through a structured three-step process, outlined as follows:

- Behavioral Data Collection
- Data Processing

- Creating Behavioral Pattern Templates
- Now, let's delve into each of these steps individually to gain a deeper understanding of their significance and how they contribute to our proposed framework.

***Behavioral Data Collection*:** In this section, we have gathered behavioral data extracted from both the application description and the privacy policy. Within the application description, the focus is on elucidating the application's functions and operations, shedding light on the type of tasks it performs. Conversely, the privacy policy section outlines the specific data and permissions that the application utilizes or abstains from using. It also explicitly states whether the collected data is shared with any third parties for advertising or any other purposes. Within our framework, we incorporate two distinct behavioral data collection modules. The first module is dedicated to acquiring data for training our framework, while the second module is responsible for capturing runtime data to assess whether the behavior of the current IoT device exhibits signs of malicious intent.

***Data Processing*:** Within this section, our framework undertakes the task of collecting behavioral data from the behavioral data collection module. Subsequently, employing machine learning techniques, we selectively extract the data necessary for constructing the behavioral templates of these applications. The data processing phase comprises three key steps.

The first step, known as noise removal, involves standardizing and cleansing the text data by eliminating irrelevant characters and text elements. This process results in a refined and 'clean' text dataset, primed for analysis. We employ deep learning methods to effectively remove noise from the collected data.

Moving on to the second step of data processing, we engage in data normalization. This technique in machine learning transforms data into a consistent format, optimizing its usability for analytics and machine learning algorithms. In our approach, we employ the min-max normalization technique to standardize the data.

Lastly, the third step focuses on sample generation, where we further prepare the data for analysis and model training.

***Creating Behavioral Pattern Template*:** In this crucial section, our model undertakes the task of generating a behavioral pattern template based on the current behavioral data. This template serves as a representation of the application's or IoT device's current behavior. Subsequently, we compare this freshly created template to the previously established behavioral templates derived from the application description and privacy policy data.

The essence of this comparison lies in identifying any disparities between the current and reference templates. If the templates align seamlessly, there is no cause for concern. However, should discrepancies emerge, we raise an alarm, signaling that the application or the IoT device utilizing it may pose a security risk. This step plays a pivotal role in ensuring the safety and integrity of the ecosystem.

13.5 Evaluation Procedure

The evaluation procedure for our framework employs several key metrics to assess its performance. These metrics are crucial for gauging the accuracy and effectiveness of our model in distinguishing between benign and over-privileged or malicious applications. We primarily use four common evaluation metrics: accuracy, precision, recall, and F1 score.

Accuracy: Accuracy provides an overall assessment of the model's performance across all classes. It is calculated as the ratio of correctly identified instances (both True Positives and True Negatives) to the total number of predictions made by the model. In other words, accuracy measures the model's ability to make correct predictions and is a fundamental indicator of its overall effectiveness.

$$\text{Accuracy} = \frac{\text{TP} + \text{TN}}{\text{TP} + \text{FN} + \text{TN} + \text{FP}} * 100$$

Precision: Precision, a pivotal metric in predictive analytics, specifically directs attention to the precision of positive predictions. It is meticulously calculated as the quotient of True Positives divided by the sum of True Positives and False Positives (TP / (TP + FP)). Precision helps us understand how many of the positively predicted cases were actually correct. A high precision score indicates a low rate of false positives.

$$\text{Precision} = \frac{TP}{TP + FP} * 100$$

Recall: Recall, alternatively recognized as sensitivity or the true positive rate, serves as a critical metric evaluating the model's proficiency in recognizing all pertinent instances of a given class. This essential measure is computed by determining the ratio of True Positives to the sum of True Positives and False Negatives (TP / (TP + FN)). Recall helps us understand how well the model captures all actual instances of a particular class, particularly important when dealing with rare or critical cases.

$$\text{Recall} = \frac{TN}{TN + FN} * 100$$

F1 Score: The F1 score is a balance between precision and recall. It takes into account both false positives and false negatives, providing a single metric that combines these aspects. This metric holds particular utility in scenarios characterized by an imbalanced class distribution or when both precision and recall carry significant weight. Its relevance becomes pronounced in contexts where achieving a balanced trade-off between precision and recall is crucial.

$$\text{F1 Score} = 2 * \frac{\text{Precision} * \text{Recall}}{\text{Precision} + \text{Recall}}$$

By utilizing these evaluation metrics, we can comprehensively assess the performance of our framework in identifying the malicious IoT device in smart health care system applications. These metrics collectively provide valuable insights into the model's strengths and weaknesses, enabling us to refine and enhance its capabilities for more accurate predictions and improved security measures.

13.6 Performance Evaluation

In this experiment, we acquired our dataset from Kaggle (https://www.kaggle.com). Our data was subsequently divided into two distinct sections. Seventy percent of the dataset was designated for training our framework, while the remaining 30% was combined with a dataset of malicious applications for testing purposes.

To comprehensively assess the performance of our model, we engaged in a thorough exploration of diverse machine learning algorithms, encompassing the formidable capabilities of Artificial Neural Network (ANN), the decision-making Prowers of Decision Tree (DT), the ensemble robustness of Random Forest (RF), and the proximity-based classification of K-Nearest Neighbors (KNN). Our objective was to assess how well each algorithm performed in identifying benign and malicious activities. The results of our analysis showed that accuracy and F1 score metrics consistently fell within the range of 89% to 94% across different algorithms. Notably, the Decision Tree (DT) algorithm outperformed the others, achieving the highest accuracy and F1 score, both at an impressive 94%. Conversely, the Random Forest (RF) algorithm exhibited the lowest accuracy, with a rate of 89.50%.

Specifically, the accuracy rates for K-Nearest Neighbors (KNN) and Artificial Neural Network (ANN) were 90.5% and 92.5%, respectively. In summary, our framework, aptly named "Safeguarding Digital Health", excelled in accurately identifying benign activities when employing the Decision Tree algorithm, achieving the highest accuracy and F1 score among the algorithms we tested. These results demonstrate the potential effectiveness of our framework in enhancing digital health security.

13.7 Conclusion

In this paper, we have introduced a novel framework named "Safeguarding Digital Health" designed to address the critical challenge of identifying malicious devices within smart healthcare systems and fortifying the security of valuable user data. Our approach involves a meticulously orchestrated three-stage process: Behavioral Data Collection, Data Processing, and Creating Behavioral Pattern Templates. In the pursuit of gauging the efficacy of our framework, a comprehensive evaluation was conducted, employing a diverse array of machine learning algorithms. The selected algorithms encompassed the advanced capabilities of Artificial Neural Network (ANN), the intricate decision-making Prowers of Decision Tree (DT), the ensemble learning strength of Random Forest (RF), and the proximity-based classification methodology of K-Nearest Neighbors (KNN). The overarching goal of our study is to equip healthcare providers and system administrators with a robust toolset to fortify their smart healthcare ecosystems against potential threats and vulnerabilities.

The results of our experiments clearly demonstrate the success of our approach. Our framework outperformed many previous baseline models as well as state-of-the-art approaches in terms of accuracy and performance. By achieving superior results, our "Safeguarding Digital Health" framework showcases its potential to significantly enhance the security and integrity of smart healthcare systems, safeguarding both patient data and the overall functionality of these critical systems.

References

1. Habibzadeh, H., Nussbaum, B.H., Anjomshoa, F., Kantarci, B., Soyata, T.A., survey on cybersecurity, data privacy, and policy issues in cyber-physical system deployments in smart cities. *Sustain. Cities Soc.*, 50, 101660, 2019.
2. Pires, I.M., Hussain, F., Garcia, N.M., Zdravevski, E., Improving Human Activity Monitoring by Imputation of Missing Sensory Data: Experimental Study. *Future Internet*, 12, 155, 2020.
3. Pires, I.M., Hussain, F., Garcia, N.M., Lameski, P., Zdravevski, E., Homogeneous Data Normalization and Deep Learning: A Case Study in Human Activity Classification. *Future Internet*, 12, 194, 2020.
4. Hussain, F., Ehatisham-ul Haq, M., Azam, M.A., Khalid, A., Elderly assistance using wearable sensors by detecting fall and recognizing fall patterns, in: *Proceedings of the 2018 ACM International Joint Conference and 2018 International Symposium on Pervasive and Ubiquitous Computing and Wearable Computers*, Singapore, vol. 8–12, pp. 770–777, 2018.
5. Hussain, F., Umair, M.B., Ehatisham-ul Haq, M., Pires, I.M., Valente, T., Garcia, N.M., Pombo, N., *An Efficient Machine Learning-based Elderly Fall Detection Algorithm*, , arXiv, Cornell University, 2019, arXiv, arXiv:1911.11976.
6. Dimitrievski, A., Zdravevski, E., Lameski, P., Trajkovik, V., Addressing Privacy and Security in Connected Health with Fog Computing, in: *Proceedings of the 5th EAI International Conference on Smart Objects and Technologies for Social Good (GoodTechs '19)*, Valencia, Spain, vol. 25–27, Association for Computing Machinery, New York, NY, USA, pp. 255–260, 2019.
7. Seh, A.H., Zarour, M., Alenezi, M., Sarkar, A.K., Agrawal, A., Kumar, R., Khan, R.A., Healthcare data breaches: Insights and implications. *Healthcare*, 8, 133, 2020.
8. *Worldwide Threat Assessment—The Director National Intelligence's View*, Worldwide Threat Assessment of the US Intelligence Community, US, 2020, Available online: https://www.intelligence.senate.gov/sites/default/files/documents/os-coats-051117.pdf.
9. Lallie, H.S., Shepherd, L.A., Nurse, J.R., Erola, A., Epiphaniou, G., Maple, C., Bellekens, X., Cyber security in the age of covid-19:A timeline and analysis of cyber-crime and cyber-attacks during the pandemic. *Comput. Secur.*, 105, 102248, 2021.
10. Santos, L., Rabadao, C., Gonçalves, R., Intrusion detection systems in Internet of Things: A literature review, in: *Proceedings of the 2018 13th Iberian Conference on Information Systems and Technologies (CISTI)*, Caceres, Spain, vol. 13–16, pp. 1–7, 2018.

11. Rathore, S. and Park, J.H., Semi-supervised learning based distributed attack detection framework for IoT. *Appl. Soft Comput. 2018*, 72, 79–89, 2018.
12. Arp, D., Spreitzenbarth, M., Hubner, M., Gascon, H., Rieck, K., Siemens, C., *DREBIN: effective and explainable detection of android malware in your pocket in NDSS*, Internet society, USA, 2014.
13. La Marra, A., Martinelli, F., Saracino, A., Sheikhalishahi, M., MalProfiler: automatic and effective classification of android malicious apps in behavioral classes, in: *International symposium on foundations and practice of security*, Springer, pp. 3–19, 2016.
14. Bedford, A., Garvin, S., Desharnais, J., Tawbi, N., Ajakan, H., Audet, F., Lebel, B., Andrana: quick and accurate malware detection for android, in: *International symposium on foundations and practice of security*, Springer, pp. 20–35, 2016.
15. Yang, Y., Wei, Z., Xu, Y., He, H., Wang, W., *DroidWard: an effective dynamic analysis method for vetting android applications cluster computing*, pp. 1–11, Springer Nature, USA, 2018.
16. Ferrante, A., Medvet, E., Mercaldo, F., Milosevic, J., Visaggio, C.A., Spotting the malicious moment: characterizing malware behavior using dynamic features, in: *2016 11th international conference on availability, reliability and security (ARES)*, IEEE, pp. 372–381, 2016.
17. Feizollah, A., Anuar, N.B., Salleh, R., Suarez-Tangil, G., Furnell, S., AndroDialysis: analysis of android intent effectiveness in malware detection. *Comput. Secur.*, 65, 121, 2017.
18. Rughoobur, P. and Nagowah, L., A lightweight replay attack detection framework for battery depended IoT devices designed for healthcare, in: *Proceedings of the 2017 International Conference on Infocom Technologies and Unmanned Systems (Trends and Future Directions) (ICTUS)*, Dubai, United Arab Emirates, vol. 18–20, pp. 811–817, 2017.
19. Carta, S., Podda, A.S., Reforgiato Recupero, D.R., Saia, R.A., Local Feature Engineering Strategy to Improve Network Anomaly Detection. *Future Internet 2020*, 12, 177, 2020.
20. Alrashdi, I., Alqazzaz, A., Alharthi, R., Aloufi, E., Zohdy, M.A., Ming, H., FBAD: Fog-based attack detection for IoT healthcare in smart cities, in: *Proceedings of the 2019 IEEE 10th Annual Ubiquitous Computing, Electronics & Mobile Communication Conference (UEMCON)*, New York, NY, USA, vol. 10–12, pp. 0515–0522, 2019.
21. Ayon, S., Islam, M.M., Islam., Hossain, M.R., Coronary artery heart disease prediction: a comparative study of computational intelligence techniques. *IETE J. Res.*, 68, 1–20, 2020. [Online].
22. Akter, L., Prediction of Cervical Cancer from Behavior Risk Using Machine Learning Techniques. *SN Comput. Sci.*, 2, 3, 177, 2021.
23. Islam Md, M., *et al.*, Breast cancer prediction: a comparative study using machine learning techniques. *SN Comput. Sci.*, 1, 5, 290, 2020.

24. Akter, L. and Islam, M.M., Hepatocellular Carcinoma Patient's Survival Prediction Using Oversampling and Machine Learning Techniques, in: *Presented at IEEE 2nd International Conference on Robotics, Electrical and Signal Processing Techniques (ICREST)*, pp. 445–450, 2021.
25. Haque, M.R., Islam, M.M., Iqbal, H., Reza, M.S., Hasan, M.K., Performance evaluation of random forests and artificial neural networks for the classification of liver disorder, in: *Presented at 2018 IEEE International Conference on computer, communication, chemical, material and electronic engineering (IC4ME2)*, pp. 1–5, 2018.
26. Nooruddin, S., Islam, M.M., Falguni, A.S., An IoT-based device-type invariant fall detection system. *Internet Things*, 9, 100130, 2020.
27. Rahman, M., Marufur, *et al.*, Obstacle and fall detection to guide the visually impaired people with real-time monitoring. *SN Comput. Sci.*, 1, 219, 2020.
28. Das, S. and Namasudra, S., A novel hybrid encryption method to Secure Healthcare data in IoT-enabled Healthcare Infrastructure. *Computers Electr. Eng.*, 101, 1–16, 2022.
29. Li, J., Cai, J., Khan, F., Rehman, A.U., Balasubramaniam, V., Jiangfeng, a secured Framework for SDN-Based Edge Computing in IoT-Enabled Healthcare System. *IEEE Access*, 8, 135479–90, 2020.

Index